BIOACTIVE COMPOUNDS AND NUTRACEUTICALS FROM DAIRY, MARINE, AND NONCONVENTIONAL SOURCES

Extraction Technology, Analytical Techniques, and Potential Health Prospects

BIOACTIVE COMPOUNDS AND NUTRACEUTICALS FROM DAIRY, MARINE, AND NONCONVENTIONAL SOURCES

Extraction Technology, Analytical Techniques, and Potential Health Prospects

Edited by

Charanjit S. Riar, PhD

Parmjit S. Panesar, PhD

AAP | APPLE ACADEMIC PRESS

First edition published 2024

Apple Academic Press Inc.
1265 Goldenrod Circle, NE,
Palm Bay, FL 32905 USA

760 Laurentian Drive, Unit 19,
Burlington, ON L7N 0A4, CANADA

CRC Press
2385 NW Executive Center Drive,
Suite 320, Boca Raton FL 33431

4 Park Square, Milton Park,
Abingdon, Oxon, OX14 4RN UK

Library and Archives Canada Cataloguing in Publication

Title: Bioactive compounds and nutraceuticals from dairy, marine, and nonconventional sources: extraction technology, analytical techniques, and potential health prospects / edited by Charanjit Singh Riar, PhD, Parmjit S. Panesar, PhD.
Names: Riar, Charanjit Singh, editor. | Panesar, P. S. (Parmjit Singh), editor.
Description: First edition. | Includes bibliographical references and index.
Identifiers: Canadiana (print) 20240281810 | Canadiana (ebook) 20240281853 | ISBN 9781774914984 (hardcover) | ISBN 9781774914991 (softcover) | ISBN 9781003452768 (ebook)
Subjects: LCSH: Functional foods—Analysis. | LCSH: Bioactive compounds—Analysis.
Classification: LCC QP144.F85 B546 2024 | DDC 613.2—dc23

Library of Congress Cataloging-in-Publication Data

CIP data on file with US Library of Congress

ISBN: 978-1-77491-498-4 (hbk)
ISBN: 978-1-77491-499-1 (pbk)
ISBN: 978-1-00345-276-8 (ebk)

About the Editors

Charanjit S. Riar, PhD
Sant Longowal Institute of Engineering and Technology, Sangrur, Punjab, India

Charanjit Singh Riar, PhD, started his career as a lecturer in the Department of Food Technology, Sant Longowal Institute of Engineering and Technology, Longowal, Sangrur (Punjab), and is presently working as a Professor in the same department. He has 25 years of teaching and research experience. Dr. Riar has published 108 research papers in reputed national and international journals, having 2,642 citations and 'h' index of 31. He has authored/edited four books, presented more than 65 papers in conferences/seminars, and written 20 book chapters. Dr. Riar is a reviewer for national/international publishing houses such as Elsevier, Springer, SAGE, Taylor & Francis, etc. Dr. Riar is on the panel of experts of a number of reputed national universities/institutions in India. Dr. Riar has guided 45 BTech, 20 MTech and has guided/under guidance, nine PhD scholars. His areas of research are starch modification, biodegradable film, bioactive compounds (nutraceuticals) extraction and analysis, seed germination, and characteristics like pasting, rheological, thermal, morphological, molecular, and functional characteristics of food and its products. He completed his graduate and postgraduate degrees in Food Technology at Guru Nanak Dev University, Amritsar (India), and PhD in Food Technology from Punjab Technical University, Jalandhar, India.

Parmjit S. Panesar, PhD
Food Biotechnology Research Laboratory, Department of Food Engineering and Technology, Sant Longowal Institute of Engineering and Technology, Longowal, Sangrur, Punjab, India

Parmjit S. Panesar, PhD, is currently working as a Professor and Dean (Planning and Development) at Sant Longowal Institute of Engineering and

Technology (Deemed to be University: Established by Government of India), Longowal, Punjab. Prof. Panesar has more than 24 years of teaching and research experience and has also served in administrative positions like Dean (Research and Consultancy) and Head of the Department of Food Engineering and Technology, SLIET Longowal. In 2005, he was awarded a BOYSCAST (Better Opportunities for Young Scientists in Chosen Areas of Science and Technology) fellowship by the Department of Science and Technology (DST), Government of India, to carry out advanced research at Chembiotech labs, University of Birmingham Research Park, UK. He was also involved in USA and Australian-sponsored projects during his stay in the UK. In 1999, Prof. Panesar was awarded a Young Scientist Fellowship by the Punjab State Council for Science and Technology, Punjab.

Prof. Panesar has successfully completed seven research projects funded by DBT, CSIR, MHRD, AICTE, New Delhi, and three projects are in progress. He has published more than 160 international/national scientific papers, 50 book reviews in peer-reviewed journals, 40 chapters, and has authored/edited 19 books. He has supervised 14 PhD students and more than 30 MTech students. He is a member of the editorial advisory boards of various national/international journals, including the *International Journal of Biological Macromolecules, Journal of Food Science and Technology*, and *Carbohydrate Polymer Technologies and Applications*. He is serving as a member of the National level Scientific Panel on "Alcoholic Beverages and Water and Non-alcoholic Beverages" constituted by FSSAI and has acted as a member of the international collective of experts of the Foundation for Science and Technology (FCT), Portugal. He has also received advanced training from different institutes/labs, including the International Center for Genetic Engineering and Biotechnology, Italy, and Chembiotech Labs, University of Birmingham Research Park, Birmingham, United Kingdom.

In recognition of his work, Prof. Panesar was elected as 'Fellow 2018' by The Biotech Research Society of India (BRSI) and 'Fellow 2021' by the 'International Society for Energy, Environment, and Sustainability (ISEES).' He has also been selected for the award of the prestigious 'INSA Teachers Award (2020)' by Indian National Science Academy (INSA). Prof. P. S. Panesar was also listed in the most coveted list (2020, 2021) of World Ranking of Top 2% Scientists published by Stanford University, USA. He has served as the National Vice President of the AFSTI during 2019. Prof. Panesar has also served as a member of the Board of Studies (BOS) of various universities, such as the Central University of Punjab, Punjabi

University, Patiala, Punjab Technical University, Jalandhar, HP University, Shimla, Chandigarh University, etc.

He is a life member of the International Forum on Industrial Bioprocesses (IFIBiop), Biotechnology Research Society of India, Association of Microbiologists of India, Association of Food Scientists, Indian Science Congress Association, and Technologists, Indian Society for Technical Education, and Punjab Academy of Sciences. His research is focused on the area of value addition of food industry byproducts, prebiotics, carbohydrate polymers, circular economy, green extraction of bioactives, and nanoencapsulation. He has visited several countries like the UK, USA, Canada, Switzerland, New Zealand, Australia, Germany, France, Singapore, Malaysia, China, Iran, Thailand, etc.

Contents

Contributors

Manab Bandhu Bera
Department of Food Engineering and Technology, Sant Longowal Institute of Engineering and Technology, Longowal, Sangrur, Punjab, India

Farhan Mohiuddin Bhat
CSK Himachal Pradesh Agricultural University, Palampur. Himachal Pradesh, India

Loulouda Bosnea
ELGO-DIMITRA, Institute of Technology of Agricultural Products, Dairy Department, Ethnikis Antistaseos 3, Katsikas GR, Ioannina, Greece

Lucía Ortega Cabello
Biological Systems Department, Metropolitan Autonomous University-Xochimilco Campus, Mexico City, Mexico

Martina Čagalj
University Department of Marine Studies, University of Split, Split, Croatia

Narender K. Chandla
Department of Dairy Engineering, College of Dairy Science & Technology, Guru Angad Dev Veterinary & Animal Sciences University, Ludhiana, Punjab, India

Rouf Ahmad Dar
PG Program of Microbiology, Center of Research for Development, University of Kashmir, Srinagar, Jammu and Kashmir, India; Department of Industrial Microbiology, Sam Higginbottom University of Agriculture, Technology, and Sciences, Prayagraj, Uttar Pradesh, India

Yadwinder Singh Dhaliwal
CSK Himachal Pradesh Agricultural University, Palampur. Himachal Pradesh, India

Arakkaveettil Kabeer Farha
Department of Food Science and Engineering, School of Agriculture and Biology, Shanghai Jiaotong University, Shanghai, China

Roberta Frleta
Center of Excellence for Science and Technology-Integration of Mediterranean Region (STIM), Faculty of Science, University of Split, Split, Croatia

Khalid Gul
Department of Food Process Engineering, National Institute of Technology, Rourkela, Odisha, India

Manpreet Kaur
Department of Biochemistry, Punjab Agricultural University, Ludhiana, Punjab, India

Sunil K. Khatkar
Department of Dairy Engineering, College of Dairy Science & Technology, Guru Angad Dev Veterinary & Animal Sciences University, Ludhiana, Punjab, India

Vasiliki Kontogianni
ELGO-DIMITRA, Institute of Technology of Agricultural Products, Dairy Department, Ethnikis Antistaseos 3, Katsikas GR, Ioannina, Greece

José de Jesús Lira-Ricárdez
Biological Systems Department, Metropolitan Autonomous University-Xochimilco Campus, Mexico City, Mexico

Palak Mahajan
Department of Biotechnology Engineering and Food Technology, University Institute of Engineering (UIE), Chandigarh University, Gharuan, Mohali (Punjab), India

Mawish Mahroof
PG Program of Microbiology, Center of Research for Development, University of Kashmir, Srinagar, Jammu and Kashmir, India

Marios Mataragas
ELGO-DIMITRA, Institute of Technology of Agricultural Products, Dairy Department, Ethnikis Antistaseos 3, Katsikas GR, Ioannina, Greece

Ruqeya Nazir
PG Program of Microbiology, Center of Research for Development, University of Kashmir, Srinagar, Jammu and Kashmir, India

Parmjit S. Panesar
Food Biotechnology Research Laboratory, Department of Food Engineering and Technology, Sant Longowal Institute of Engineering and Technology, Longowal, Sangrur, Punjab, India

Charanjit S. Riar
Sant Longowal Institute of Engineering and Technology, Sangrur, Punjab, India

Vida Šimat
University Department of Marine Studies, University of Split, Split, Croatia

Danijela Skroza
Department of Food Technology and Biotechnology, Faculty of Chemistry and Technology, University of Split, Split, Croatia

Sarana Rose Sommano
Chiang Mai University, Bangkok. Thailand

Sakshi Sukhija
Department of Biotechnology Engineering and FoodTechnology, UIE, Chandigarh University, Mohali, Punjab, India

Antonia Terpou
Department of Agricultural Development, Agri-Food and Natural Resources Management, School of Agri-Cultural Development, Nutrition and Sustainability, National and Kapodistrian University of Athens, Evripos Group GR, Greece

Nitin Wakchaure
Department of Dairy Economics and Business Management, College of Dairy Science & Technology, Guru Angad Dev Veterinary & Animal Sciences University, Ludhiana, Punjab, India

Abbreviations

ACE	angiotensin-converting enzyme
ASTA	American Spice Trade Association
BM	bead milling
CA	caffeic acid
CAM	complementary and alternative medicine
CCP	colloidal calcium phosphate
CE	capillary electrophoresis
CiA	cinnamic acid
CLA	conjugated linoleic acid
CPP	casein phosphopeptides
CVDs	cardiovascular diseases
DAGs	diglycerides
DES	deep eutectic solvents
DSHEA	Dietary Supplement Health and Education Act
EAE	enzyme-assisted extraction
ELF	endolichenic fungi
EO	essential oil
EPS	exopolysaccharides
FA	ferulic acid
FAO	Food and Agriculture Organization
FDA	Food and Drug Administration
FFAs	free fatty acids
FTIR	Fourier transform infrared spectroscopy
GAEs	gallic acid equivalents
GC	gas chromatography
GC–MS	gas chromatography-mass spectrometry
GIT	gastrointestinal tract
GMPs	good manufacturing practices
GRAS	generally regarded as safe
HDL-C	high-density lipoprotein
HMOs	human milk oligosaccharides

HPCCC	high-performance countercurrent chromatography
HPH	high-pressure homogenization
HPLC	high performance liquid chromatography
HPTLC	high performance thin layer chromatography
HSH	high-speed homogenization
IEC	ion-exchange chromatography
IL	ionic liquids
IR	infrared
ISO	International Standard Organization
KBr	potassium bromide
LC	liquid chromatography
LC-MS	liquid chromatography-mass spectrometry
LDL-C	low-density lipoprotein
LF	lactoferrin
LP	lactoperoxidase
LPT	low pasteurization
LQ	liquiritin
LQG	liquiritigenin
LRT	lipid replacement therapy
LSR	liquid to solid ratio
MAE	microwave-assisted extraction
MAGs	mono-glycerides
MALDI-TOF-MS	matrix-assisted laser desorption/ionization-time of flight-mass spectroscopy
MF	microfiltration
MS	mass spectrometry
MW	molecular weight
MWE	microwave extraction
NADES	natural deep eutectic solvents
NBCs	natural bioactive compounds
NDOs	non-digestible oligosaccharides
NMR	nuclear magnetic resonance
PDF	platelet-activating factor
PAN	polyacrylonitrile
PEF	pulsed electric field
PEG	polyethylene glycol
PES	polyether sulfone

PES-PVP	polyether sulfone polyvinyl-pyrrolidone
PLE	pressurized liquid extraction
PMs	primary metabolites
Pre-HPLC	preparative high-performance liquid chromatography
PTFE	polytetrafluoroethylene
PUFA	polyunsaturated fatty acids
PVDF	polyvinylidene fluoride
rhLf	recombinant human lactoferrin
ROS	reactive oxygen species
RSM	response surface method
SAE	shaking-assisted extraction
SAR	structure-activity relationships
SCFAs	short-chain fatty acids
SCWE	subcritical water extraction
SE	solvent extraction
SEC	size exclusion chromatography
SFE	supercritical fluid extraction
SMs	secondary metabolites
SOD	superoxide dismutase
SSR	solvent to solid ratio
SWE	supercritical water extraction
TACs	total anthocyanin compounds
TAGs/TG	triglycerides
TDF	total dietary fiber
TFC	total flavonoid content
TFF	tangential flow filtration
TLC	thin layer chromatography
TPC	total phenolic content
UA	usnic acid
UAE	ultrasound-assisted extraction
UF	ultrafiltration
UPLC-PDA	ultra-performance liquid chromatography-photodiode array
USE	ultrasound extraction
USFDA	U.S. Food and Drug Administration
UV-Vis	ultraviolet-visible spectrometry/UV visible

Preface

Natural products and their bioactive compounds are being increasingly utilized in preventive and therapeutic medication as well as to produce pharmaceutical supplements and, more recently, as food additives to increase the functionality of foods. The demand for functional foods and nutraceuticals is thus on the rise, leaving product development companies racing to improve bioactive compound extraction methods – a key component of functional foods and nutraceutical development.

Natural products from animal and non-conventional sources, either as pure compounds or as standardized extracts, provide unlimited opportunities for new product development and drug development because of the availability of natural biochemical compounds of unmatched diversity. Bioactive peptides, microbial and herbal preparations, and health food development for medicinal usage contain various types of bioactive compounds. Enhanced concern for the quality and safety of food products has increased preference for natural products.

Extraction is the crucial step in the analysis of bioactive compounds because it is necessary to extract the desired chemical components from the naturally available materials for further separation and characterization. This book focuses on the characteristics of functional foods and provides detailed insight into the functionality, extraction technology, and analytical techniques of bioactive compounds from various food and allied sources.

The book comprises eight chapters divided into three sections. The first section provides a detailed discussion of sources, applications, physicochemical, and functional aspects of functional ingredients and nutraceuticals, along with their profiling and chemical characteristics. Subsequent sections comprise the extraction technology and the analytical techniques of qualitative and quantitative analysis and profiling of functional compounds and nutraceuticals and elaborate methodologies, which include the extraction, isolation, and characterization of bioactive and functional ingredients.

The techniques and key adoptions in the extraction, isolation, and characterization of active ingredients from food preparations have been elaborated depending upon the scope of the book and individual requirements of the chapters. Further, the authors enhance and elaborate on the critical aspect of extraction, which is the crucial prerequisite in the analysis of source material since it is necessary to extract the desired chemical components from the bioactive materials for further separation and characterization. Technological aspects of the extraction of bioactive compounds are considered a promising aspect both for researchers and industry, as it gives them first-hand information about the processing and quality control attributes required during handling of numerous sensitive, expensive, and medicinal compounds available in low quantity in natural resources.

Along with conventional methods, numerous new methods have been established, but no single method is regarded as standard for extracting bioactive compounds from plants. From established processes such as steam distillation to emerging techniques like supercritical fluid technology, many methods of extraction are now available and used commercially, such as enzyme, solid-phase micro-extraction, microwave, ultrasound-assisted, pressurized liquid extraction (PLE), supercritical extraction techniques and surfactant-mediated techniques. These along with numerous analytical techniques for qualitative and quantitative estimation of nutraceutical compounds present in various natural food and allied sources by involving techniques such as physicochemical assay, GCMS, HPLC, TLC, immunoassay, and Fourier transform infrared (FTIR) and other available analytical techniques are discussed in this book. The readers will thus get full-scale information about the extraction and analysis of bioactive compounds from natural sources.

This book is divided into sections and chapters, and each section is dedicated to specific sources of bioactive compounds and nutraceuticals such as dairy, marine, and microbial sources, etc. The most important aspect of bioactive compounds, such as their extraction and analysis, is covered and handled in depth with recent trends in extraction and analytical techniques.

—*Editors*

PART I

An Overview About Bioactive Compounds and Nutraceuticals

CHAPTER 1

Bioactive Compounds and Nutraceuticals: Classification, Potential Sources, and Application Status

CHARANJIT S. RIAR and PARMJIT S. PANESAR

Department of Food Engineering and Technology, SLIET, Longowal, Sangrur, Punjab, India

ABSTRACT

Nutraceuticals comprise medicinal foods that can regulate and modulate immunity, have the potential to prevent and cure particular diseases, and maintain overall health. Nutraceuticals include natural food intakes, spices/herbs, and their products, processed products such as fruits, vegetables, cereals, herbs, tubers, dairy foods, marine sources, genetically modified foods, etc. Bioactive compounds are distributed mostly in nature and contribute an important class of compounds in plants/animals. The classification of functional foods containing bioactive ingredients having health prospects is categorized based on food source, nutrient and non-nutrient importance, chemical compound nature, properties, and their action on organs being benefited when consumed in the diet. Crops such as cereals and biofortified food are used as a source for administering nutrients to malnourished people in the approach of a complementary strategy. Dairy or functional food products and health foods that contain ingredients such as probiotic organisms, vitamins, and minerals can be used to improve gut health and are considered a new research area in which by modulating gut microbial composition, prevent disease, and act as antioxidants. The ability of nutraceutical food to control severe diseases like degenerative diseases of various types, type-2 diabetes, etc., must be popularized as having an enormous opportunity in their

Bioactive Compounds and Nutraceuticals from Dairy, Marine, and Nonconventional Sources: Extraction Technology, Analytical Techniques, and Potential Health Prospects.
Charanjit Singh Riar, Parmjit S. Panesar (Eds.)

treatment. Nutraceutical food has the ability to act as an important matter in future therapeutic developments. Phytochemicals have been found to protect plants against spoilage agents such as herbivores, bacteria, fungi, viruses, and are also suitable for use in protecting humans from numerous diseases. Plant foods such as spices and their extracts from cinnamon bark, cranberry, herbs, basil, black currant, cornelian cherry, Açaí palm, thyme, peppermint, cloves, liquorice, etc., are suitable supplies of naturally available antimicrobial agents and disease preventive compounds with appropriate use in food and pharmaceutical industries. This chapter comprises essential content for food scientists and students to get information about bioactive compounds and nutraceuticals, their classification, potential sources, and utilization of bioactive compounds and nutraceuticals in food and non-food applications.

1.1 ABOUT BIOACTIVE COMPOUNDS AND NUTRACEUTICALS

1.1.1 BIOACTIVE COMPOUNDS

The term nutraceutical has been coined by Stephen Defelice which states that "Nutraceuticals is/are any component considered as food, part of food which can provide medical or health promoting benefits, include disease prevention and cure." In simple terms, Nutraceuticals means, nutrient plus pharmaceutical components, food stuffs (containing fortified food or dietary food supplements) that provides health benefits. In medical terminology, a bioactive is a compound that influences [1], can cause a reaction [2], or trigger out a response in the living cell and tissues [3]. The bioactive food components are foods or supplements are responsible for changes in health status [4] in addition to meeting the basic nutritional needs. Here it is vital to know that the bioactive compounds have not be considered nutrients [5], although they are food ingredients and included in their composition. This situation led to arising of many questions, the first being *What are bioactive compounds*? Despite all the relevant and targeted research practices in specific fields, the definition of bioactive compounds remains debatable. A very few references defined "bioactive," and conventionally using a short quote without specifying or elaborating the field of research, thus limiting the very purpose of the term. In a technically and scientific manner, the term "bioactive" is an alternate to "biologically active" compound [6, 7]. A bioactive compound is a substance with some biological activity [8]. It comprises of two words: *bio-* and *active. Bio* is a Greek (Bío) "*bios*" [bio-, -bio], which refers to life and *active* a Latin word meaning "*activus*," or dynamic, full of energy [9, 10], or that has an activity [11]. This action presents all the facts which can reveal the form of life sustainability, function, or any process [12]. For the biological compounds or substances, it is necessary to have biologically active if it can put direct impact on a biological organism who consume them. The positive or negative effect of these components depend on the usage of active substance, the dose, or their bioavailability nature. These compounds could have broader spectrum effects, starting with the proper maintenance of health, healing effects, or can be dangerous. The assimilated dose of bioactive compounds is all time responsive either the effect is positive or adverse. In the view of Biesalski et al. who proposed a definition for the bioactive compounds as the essential compound those natural non-essential compounds which are part of the food system and can influence the human health [13].

As given in the *Encyclopedia of Food and Culture*, "food and bioactive component" refers to the non-essential biomolecules present in foods which can modulate one or several metabolic processes, thus promoting a better health [14]. The beneficial health effects are claimed to have for these extra and non-nutritional components of food. New data on bioactive [15] indicates that these does not normally include the essential food nutrients [16]. Therefore, in response to the various characteristics highlighted, the *bioactive compounds*, they are being exclusively analyzed to depict their beneficial effects on health. The motivation triggering this logical investigation was based on various epidemiologic studies. These studies have authenticated the protecting effects of plant-based diets containing flavonoid compounds or other phenolic compounds for cardiovascular diseases (CVDs) and cancer. The other studies, however, have not shown any similar effects. Various phytochemicals are found in soy foods, whole food grains, flaxseed oil, fruits, and vegetables sources. As phytoestrogens are associated with restricted estrogen agonists and antagonists, therefore, the effects on cancer are very complex. Numerous studies are being carried to understand the beneficial effects of these compounds to prevent CVDs, cancer, and other degenerative disorders. Some bioactive compounds include carotene, lycopene, lignan, resveratrol, tannins, and indoles. These extra-nutritional chemical constituents that specifically are found in small quantity in plant sources. These include fruits, vegetables, nuts, oil seeds, and whole food grains, are explores at large for their health effects [17].

Bioactive compounds also are defined in the manner: the substances that show biological activity, they are living tissues and generates a response in them. Also act in the body and helps in promotion of good health. These living organisms and microorganisms, such as bacteria, fungi, cyanobacteria, bryophytes, lichens, higher plants, and some groups of animals contain a wide range of these substances. Considerably mentioning the term bioactive compound or components (bioactive dietary) particularly having beneficial effects on the humans, such definition considers the bioactive ingredients are the substances stimulating the state of health. Therefore, they have a biological value in addition to their calorie content [18]. In this context, the bioactivity of the compounds has the potential to effect on health of patient in a positive way, which can control livings metabolic activities and shown positive properties such as antioxidant, receptor activities inhibition, enzymes inhibition or induction, and inhibition or induction of gene expression [19].

1.1.2 NUTRACEUTICALS

Nutraceuticals are available from the time of Hippocrates, the father of medicine during period between 460 and 377 BC. Concept of food and medicine originates from China and extended to Japan where now a days nutraceutical science have flourished. The first project of nutraceuticals was set up by Arai in 1984 at University of Tokyo. Dr. Stephen L. De-Felice (1989), the founder chairman of Foundation of Innovation Medicine coined Nutraceutical, which basically combines nutraceuticals and pharmaceuticals. The main motto to emphasize on the nutraceutical food and science is to understand that Let the Food be your Medicine' is a common misquotation attributed to the Hippocrates, the father of Western medicine (as has been considered). The foods naturally contain many bioactive components which are when consumed give health benefits. Indians, Chinese, Egyptians, and Sumerians are the main civilizations who had used food as medicine. Nutraceuticals are functional foods that provide health benefits, by virtue of their physiologically activity in their food components in addition to basic nutritional health benefits. Nutraceuticals are medicinal foods, nutritional supplements, and dietary supplements. The nutraceutical food components having herbal or plant origin can control immunity, increase life expectancy, and/or prevent and cure specific diseases [7]. Nutraceuticals involve a broad range of substances that range from natural or processed foods, herbal products, and fortified foods or genetically modified crops. Nutraceuticals can be categorized as natural or traditional nutraceuticals such as: nutrients, phytochemicals, probiotic microorganisms, herbals, nutraceutical enzymes, etc.) and non-traditional or artificial nutraceuticals such as fortified, recombinant nutraceuticals. Nutraceuticals are considered as products derived from food (plant/animal/marine/microorganisms) sources (natural or processed/modified) that are supposed to provide additional health benefits, over and above basic nutritional value. Based on the jurisdiction, compounds may think to be able to prevent chronic diseases. They can increase life expectancy and delay the aging process. They can expect to improve health or sustain the body structure and functionality [7].

Cosmeceutica, a blend of cosmetics and pharmaceuticals, contains naturally active ingredients. They have the characteristics to enhance medical or drug-like benefits. Are used in lotions, ointment, creams other similar products/ingredients containing plant and marine extracts (peptides, vitamins, and bioactives). These products are included in the health improvement and beauty of skin and hair growth. A wide variety of biologically active components of natural origin are available including vitamins, sun protectors, moisturizer, skin whiteners, free radical scavengers, etc. A nutraceutical or 'bio-ceutical' being a pharmaceutical alternative claim to have functional benefits. In America, "nutraceuticals" are sold in unfettered market, as they occur in the similar category as dietary supplements and food additives as indicated in the FDA regulations, under Federal Food, Drug, and Cosmetic Act. The contemporary nutraceutical markets developed in Japan in the late 1980s. In comparison to the natural herbs and spices which are used as traditional medicine since centuries, the growth of nutraceutical industries is on modern lines [9]. As in traditional Japanese the drug called 'Kampo' is obtained from many medicinal plants. In order to investigate the pharmacological effects of functional foods and Kampo medicine testes have been carried out. The extracts taken from Kampo and functional foods and administered to test animals have found beneficial in reducing pro-inflammatory rates.

1.1.3 FUNCTIONAL FOODS

The definition of functional food can be given as "Food and its related products being taken as a part of the normal diet in order to have their beneficial effects that go beyond the basic nutritional function." Health Canada describes functional foods as those conventional food components that have ingredients added/contained to provide a specific physiological or medical benefit, other than a merely nutritional effect. Functional foods are fortified during processing and are marketed to provide benefits to the public. These include whole, fortified, enriched, or enhanced foods. They have potentially beneficial effect on human health when consumed on regular basis. The functional foods slightly differ from nutraceutical foods. The functional food is the food which is cooked or prepared using some scientifically proven techniques with or without knowledge of how or why it is being used. Thus, functional food provides the body with the required quantity of vitamins, fats, carbohydrates, minerals, etc., for healthy substances. A functional food is called nutraceutical when it aid in the prevention and/or treatment of disease and/or disorder other than anemia. The foods such as fortified dairy products (e.g., milk), and citrus fruits (e.g., orange juice). Occasionally, some complementary nutrients are added, such as vitamin-D to milk. The lactic deficiency is engineered with recombinant human lactoferrin (rhLf) to be able to solve the lactoferrin (LF) deficiency problem to preschool children in the developing and underdeveloped countries [20]. In Japan, the functional foods have to meet the established requirements such as: (i) exist as natural forms, rather than as capsule, tablet, or powder; (ii) as diet used often as daily consumption; and (iii) must regulate a biological process with prospects of improving, controlling or preventing a disease.

1.2 CLASSIFICATION OF BIOACTIVE COMPOUNDS AND NUTRACEUTICALS

1.2.1 BIOACTIVE COMPOUNDS

Classification of functional foods (bioactive compound having health benefits) is broadly based on six categories which include [21]:

- Food source;
- Nutrient and non-nutrient values;
- Chemical nature; and
- Action which the organs have benefits when consuming them in diet.

1. The functional foods classified into three classes based on their origin (source of food):

    - Microbial derived;
    - Plant;
    - Animal derived.

    The functional foods derived from microbial are further classified as: probiotics, prebiotics, symbiotic, and postbiotics. Plant derived functional foods include plant protein, *beta*-glucan, antioxidants, vitamins tocopherols, etc. The animal derived functional foods have omega-3 and omega-6 fatty acids, small peptides, whey, and casein protein in their composition

2. On the basis of nutritional values, the foods rich in lipids, vitamins, carotenes, fatty acids, phenolic components, isoprenoids, and other bioactive components are considered as functional foods.
3. The bioactive compounds classification based on their chemical nature can be further subdivided into following three groups:
   - Terpenes and terpenoids containing about 25,000 compounds;
   - Alkaloids, which contain about 12,000 compounds; and
   - Bioactive compound (PC) which are around 8,000 types [22].

The total classification of bioactive compounds found plant source are given in Figure 1.1.

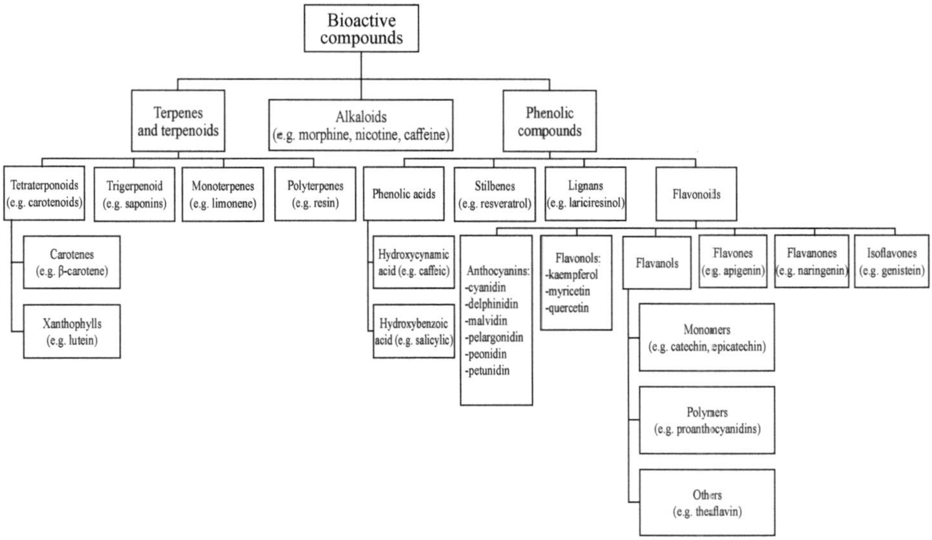

FIGURE 1.1 Plants sources classification of bioactive compounds.
Source: Adapted from Ref. [23]

1. **Bioactive Compounds:** These found distributed widely in nature and are considered as an important group of bioactive compounds plants origin. Polyphenolic compound family comprises about more than 7,500 compounds having variation in their structures. These are commonly present in fruits, vegetables, seed grains, flowers, and plant and their leaves. They are grouped under polyphenolic acids and their derivatives of lignans, tannins, quinones, flavonoids, stilbenes, and other related compounds which are categorized based on ring numbers found in the polyphenolic compound that are the structural elements these rings helps in their linkage. These mono

and polyphenolic compounds biosynthesis from carbohydrates through shikimic acid pathway, phenyl-propanoid, and flavonoid compound pathways [24]. Generally phenolic acids comprise two distinguishing carbon frame works such as hydroxycinnamic and hydroxybenzoic structures as shown in Figure 1.2). The polyphenolic acidic compounds such as ferulic acid (FA), sinapic acid, caffeic acid (CA), chlorogenic, and p-coumaric acids are grouped in this category, while the second one contributes gallic, protocatechuic, syringic, and vanillic acids as bioactives. Polyphenols are also considered as phenolic acids and include gingerol, rosmarinic acid, capsaicin, gossypol, ellagic acid and cynarin [25]. It is indicated that CA, *p*-coumaric acid, vanillic acid, FA, and protocatechuic acids are present in almost all plant sources. Colored fruits, such as blueberries (*Vaccinium corymbosum*), chokeberries (*Aronia melanocarpa*), blackberries (*V. myrtillus*), strawberries (*Fragaria virginiana*), red raspberries (*Rubus idaeus*), elderberries (*S. nigra*) and black currant (*R. nigrum*)} are rich in bioactive compounds. Other polyphenolic compounds are flavonoids comprising derivatives (flavanone, flavanol, flavone) isoflavones and anthocyanins [26, 27]. The most widely present among them are flavonols and their conjugates with 210 in number or more are variable sugar conjugates and their derivatives of kaempferol as polyphenol. In addition to kaempferol other flavonols are myricetin, quercetin, morin, galangin, and isorhamnetin which occur as O-glycosides (Figure 1.3). They are extracted from various fruits (various cherries, berries, and apples), vegetables like tomato and broccoli, beverages line red wine category, spices, and herbs curcumin and caraway. On derivatives including flavones appear in limited number in raw plant materials including pulses (legumes), salads (celery), tea leaves, herbs, and thyme. This group comprises major flavones including apigenin, luteolin, baicalein, chrysin. The derivatives of hydrophobic flavonoids (sinensetin, tangeretin, nobiletin) are also included in this group [24]. Presence of naringenin and hesperetin along with their glycoside flavanones of naringin and hesperidin in citrus fruits (oranges, lemons, lime), medicinal herbs belong to the family: *Rutaceae*, *Rosaceae*, and *Leguminosae* [28]. Another, very important group of flavonoids are flavanols which are simple monomers of (+)-catechin or (-)-epicatechin, as well as associated in hydroxylated (gallo-catechins) and esterified (gallic acid) forms (Figure 1.4). Most important flavonoid derivatives are flavanols (epigallocatechin gallic acid, epigallocatechin, epicatechin gallic acid, catechin, and epicatechin). They are extracted from tea as green tea extracts, apple components, grapes fruits, berry, and cocoa products. Other class of bioactive are characteristic-colored compounds depended on their pH, widely distributed in fruits, flowers, and are grouped in anthocyanin class of phenolic compounds (malvidin, pelargonidin, delphinidin, cyanidin, and peonidin) and their glycosides. In addition to their health benefits, they are widely used as natural food colorings in food and pharmaceuticals [29] (Figure 1.5).

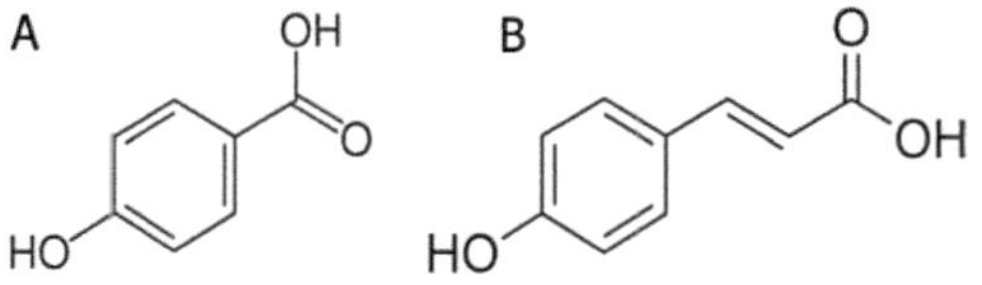

FIGURE 1.2 Structures of polyphenolic acids: (a) 4-hydroxybenzoate; and (b) coumaric acid as bioactive compounds.

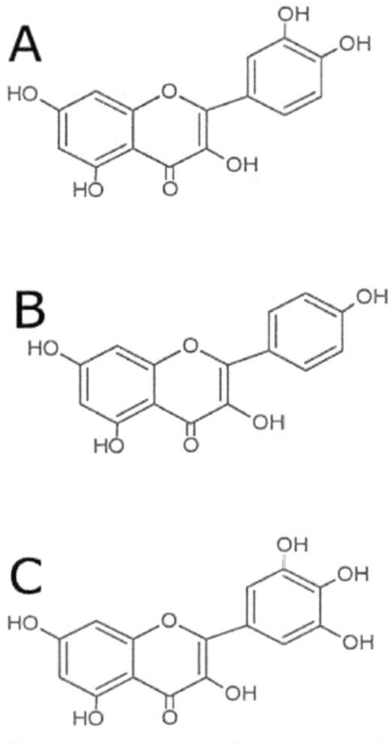

FIGURE 1.3 Structures of polyphenol as flavonoids (flavonol): (a) quercetin; (b) kaempferol; and (c) myricetin, as bioactive compounds.

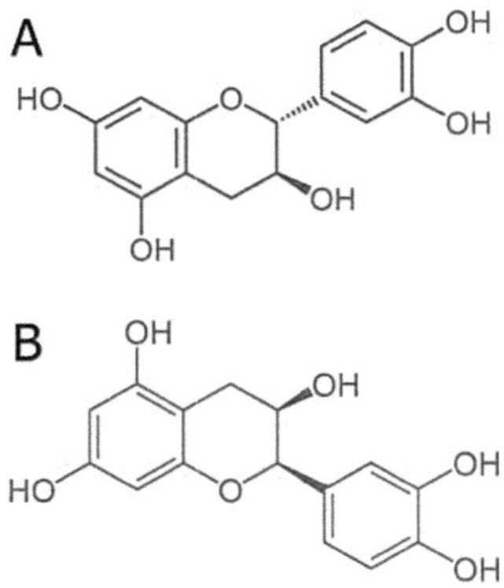

FIGURE 1.4 Phenolic structure of (a) (+)-catechin; and (b) (–)-epicatechin, as bioactive compounds.

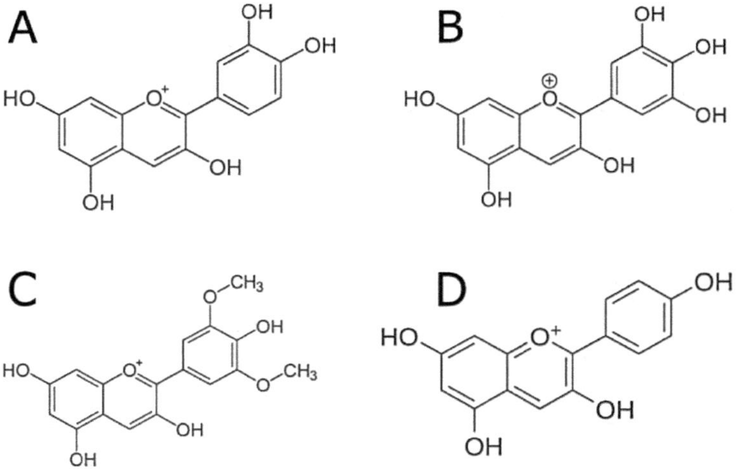

FIGURE 1.5 Phenolic structure of anthocyanin compounds: (a) cyanidin; (b) delphinidin; (c) malvidin; and (d) pelargonidin, as bioactive compounds.

2. **Alkaloids:** These are other group of commonly found phenols found in plants. According to Pelletier who suggested the definition alkaloids in 1983 as "*aromatic compound which contain nitrogen atom in negative oxidation state which is found in limited quantity in living organisms.*" Although of alkaloids are diversified in nature, these compounds are having shown similar in chemical structures: The properties of alkaloids are:

- Nitrogen atom contained in number of cases obtained from many amino acids;
- They exhibit an alkaline pH; and
- They contain no basic forms that of quaternary compounds and their nitrogen oxides [30].

The mostly applied groupings of compounds in alkaloids are:

- True alkaloids (nicotine, atropine, and morphine);
- Protoalkaloids (ephedrine and adrenaline);
- Pseudoalkaloids (theobromine, caffeine, and theacrine).

Contrarily, on biosynthetic pattern basis the alkaloids are categorized as: (i) indole alkaloids (tryptophan derivatives such as, ergometrine); (ii) piperidine (lysine derivatives such as, lobeline); (iii) pyrrolidine (ornithine derivatives such as, hygrine); (iv) amino acid phenylethyl alanine and tyrosine derivatives; and (v) imidazole (derivatives from histidine such as, pilocarpine). It is thought that over 15–21% of plant varities contain alkaloids. Major sources are from plants belonging to the *Solanaceae* families (*Datura stramonium and Nicotiana tabacum* and), *Papaveraceae* (*Papaver somniferum*), *Ranunculaceae* (*Hydrastis canadensis*), *Rubiaceae* (*Pausinystalia johimbe*), *Erythroxylaceae* (*Erythroxylum coca*), and *Campanulaceae* (*Lobelia inflate*). Alkaloids found in plant kingdom act as growth regulators which serve as reserve substances for plant growth [31]. Alkaloid compounds are usually used in pharmacology studies and in therapeutic research. They have a wide range of biological properties and act as anti-depressant, anti-tumor, anti-inflammatory agent, anti-viral agent, anti-hypertensive, anti-microbial, and anti-malarial as well [32].

3. **Terpenes (terpenoids):** These are naturally occurring bioactive compounds class whose existence is most varied. There are around 39,000 different chemical compounds of terpenoids category [33].

They are classified into number of categories based on the carbon atoms numbers as well as based on the presence of isoprene units (IPR):

- Monoterpenes consist of 10 carbon atoms, or two IPR units;
- Sesquiterpenes contain 15 carbon atoms, or three IPR in its composition;
- Diterpenes contain four IPR units or 10 carbon atoms;

- Triterpenes contain six isoprene units or 30 carbon atoms;
- Tetraterpenes containing more than 10 carbon atoms than triterpenes, or two more iso-prenes;
- Polyterpenes composes of large number of IPR units.

The terpenoids are secondary type of metabolites, however, they have special function to play in plants growth, and include gibberellins (diterpenes) which are plant hormones, sterols (triterpene derivatives) which are responsible for plant cell stabilization, while other carotenoids are responsible for protective action against cell photooxidation. The most beneficial among these are compounds are the oxycarotenoids (xanthophylls) carotenoids belonging to tetraterpene family. Carotenoids, named lutein, *beta*-carotene, lycopene, or zeaxanthin, are lipid soluble coloring compounds largely found in fruits and vegetables and impart yellow, orange, and red color to them [24]. Some toxins terpenes are also present in plants and play an important role against mammals, insect creatures. The compound pyrethroids and its monoterpene esters obtained from *Chrysanthemum* spp., have high insecticidal effect. However, the highly valuable from of all terpenoids are the volatile monoterpenes and sesquiterpenes compounds which are well known for their essential oil (EO) value. They are found in herb and spice foods, as well as in some fruits and are responsible for their characteristic aroma. Herbs/spices that are rich in EOs include peppermint (*M. piperita*), rosemary (*R. officinalis*), cinnamon (*C. zeylanicum*), basil (*O. basilicum*), and lemon (*Citrus limon*). These EOs containing compounds are largely known known for their bioactivity and are used in agroindustry, aromatherapist applications and in microbiological studies, etc. It has been reported that terpenes exhibit numerous pharmacological characteristics having anti-inflammatory activity, anti-carcinogenic effect, anti-viral, anti-bacterial nature and as hepatoprotective [34, 35].

4. **Saponins:** The group of compounds that are structurally similar compounds are called saponins such as steroids and triterpenoids. The steroid aglycone and/or triterpene containing compounds which are attached to the chains of sugar molecules of arabinose, galactose, rhamnose, xylose, fucose molecules. They have amphoteric in nature because of presence of hydrophobic aglycone compounds

and chains of hydrophilic sugars [36]. These molecules have unique properties of saponins which include foam formation, emulsifying agent, and detergent. The natural found steroidal saponin sources is *Lilianae* (monocotyledons), while triterpenoid is generally available from *Dicotyledoneae* (dicotyledons) family plans. In the plants, belonging to family of *Rosaceae*, *Caryophyllaceae, Agavaceae*, *Leguminosae*, and *Umbelliferae* they concentrate in their tree bark, roots, and leaves tissues [37]. The main rich sources of saponins are honey locust (*Gleditsia sinensis*), soapbark (*Quillaja saponaria*), Mongolian milkvetch (*Astragalus propinquus*), ginger (*P. ginseng*), soapwort (*Saponaria officinalis*), Yucca (*Yucca schidigera*). They are used as detergents (*Q. saponaria*) in traditional preparations and act as nutrients for animal to reduce their fecal matter off-odors (*Y. schidigera*). The saponins also exhibit molluscicidal, antiviral, antifungal, antibacterial, insecticidal, anthelmintic, activities [38]. They also deliver anticancer activity, the others such as *Androsace umbellate* are reported as cancer cells inhibitors and involved in cell apoptosis induction [39, 40]. Other saponins found in *Yucca plant source* show use to inhibit food-spoilage from yeast (*Debaryomyces hansenii, Candida famata, Hansenula anomala, Zygosaccharomyces rouxii*, etc. [41]. Now a days, saponins are used as a nature friendly, environment friendly additives. In Japan, an additive such as *Y. schidigera* is commonly used prevention of yeasts contaminating in cooked rice, in vegetable pickles, and fish meat preparations [42]. Other studies have shown that with saponin extract treated yeast (*Q. Saponaria) helps in* increasing its cell membrane permeability and speed up yeast salt-free lysates preparation [43].

Since many bioactive compounds found in the plants are either beneficial for human health or constitute an important base material in a variety of associated industry sector. Having their promoting properties, which include antibacterial activity, antioxidative, anti-inflammation property, anticancerogenic agent, antimutagenic activity, antiallergic, anti-fungus activity and other effects, research have been intensified towards materials obtained plants (fruits and vegetables), herbs, and spices which have potential as promising functional ingredient in various food product development as well as in pharmaceutical and in therapeutic applications [7].

1.2.1.1 CLASSIFICATION OF POLYPHENOLS BASED ON THEIR NATURE, MECHANISM OF ACTION AND BENEFITS WHILE CONSUMING IN DAILY DIET OR OTHERWISE [21]

Under the category of bioactive that are classified based on their nature, action mechanism and benefits, constitute different group of products that fall under the category of bioactive compounds as well as nutraceuticals in common. These include:

1. **Dietary Supplements:** The dietary supplements are those that contain inherent nutrients (Table 1.1) derived from food products we consume contains more than one dietary ingredient such as dietary fibers, mineral(s), vitamin(s), amino acid(s), herb or other botanical constituent, metabolites, a dietary substance consumed by humans to supplement the diet by increasing the total daily intake. It also includes concentrates, extracts, or combinations of these ingredients which are available in concentrated liquid form, pills, capsules, and tablets. The dietary supplement health and education act (DSHEA, 1994) defines the dietary supplement as: The dietary supplements are the products other than tobacco compounds that are required in the diet to supplement or contains more than one required dietary ingredients such as vitamins, minerals, herbs or plants ingredients such as amino acids or other required dietary substance for increasing the total dietary intake of them. The others compounds such as metabolite, extracts, concentrated compounds, or different combinations of these ingredients in the form of gel caps, liquids, or powders, soft gels, etc. [44]. It is not as such considered for use as a conventional food, as a whole item of daily meal or diet, not labeled as a "dietary supplement." The dietary supplements containing vitamins or mineral supplements are generally used as pill or capsule form by different companies under different brand name and variants.

   As per the United States DSHEA of 1994 the definition of dietary supplement is: A product taken by mouth that contains a "dietary ingredient" taken as supplement in the diet. The Functional food in health and disease journal cited an article in which it has been observed that mitochondrial activity can be maintained by administering the patients with variable natural supplements which included the lipid replacement therapy (LRT). During the multiple clinical trials on chronic fatigue syndrome, fibromyalgia, and other fatiguing illnesses it has been observed that in a majority of patients having chronic infections, LRT reduced fatigue by 35 to 43% [45]. Since the dietary supplements have not come in the approved list of U.S., Food and Drug Administration (FDA) before being marketed, but for the manufacturers they have to register their manufacturing units with the FDA and follow the latest good manufacturing practices (GMPs). As per regulations, the dietary supplements can be marketed only to support the basic food structure or enhance their functionality and have not claim to be used for the disease treatment or similar conditions and must contain in the label which include: "that these statements have not been evaluated by the FDA of USA. This product is not meant for the diagnose, cure, treatment or prevention of any disease."

i. **Dietary Fibers:** As per natural existence and nature, dietary fibers are classified into two types: (i) water insoluble; and (ii) water soluble; daily recommended intake of total dietary fiber (TDF) is 30–40 gms. They are obtained from number of foods we eat including main natural sources plant foods such as beans, whole grain cereals, wheat, and its products, millets, legumes, etc. The dietary fibers act as roughage are the indigestible part of ingested plant foods. Dietary fiber shows many health benefits which include reduction in type-2 diabetes risk and heart disease ailments [7].

ii. **Chemical Groupings Basis of Categorization of Bioactive Compounds:** Reagents derived from other sources such as chondroitin sulfate, pyruvate, steroid hormone, and their precursors which are used to serve the specific functional purpose, such as fortified conventional foods, weight-loss supplements, sports nutrition, and meal replacements [46], whereas for the nutraceuticals it more emphasizes on expecting outcomes of these useable products, such as deterrence or handling of diseases can also be taken into consideration [44].

TABLE 1.1 Classification based on Products Containing Inherent Nutrients Groups [21]

Sl. No.	Classes of Bioactive(s)	Examples
1.	Inorganic mineral and/or vitamin supplements	Multivitamin and/or mineral supplements with combination of vitamin, minerals as well other ingredients.
2.	Probiotic food	Beneficial bacteria
3.	Prebiotic food	Digestive enzymes/foods
4.	Total dietary fibers	Fibers (insoluble, soluble)
5.	Antioxidants	Natural antioxidant ingredients (phytochemicals)
6.	Phytochemicals	
i.	Essential fatty acids	Omega-3 and 6 fatty acids
ii.	Bioactive compounds	Plant polyphenols
iii.	Isoprenoids	Carotenoids and related supplements
iv.	Dietary lipids	EFAs, sphingolipids,
v.	Dietary proteins	Animal proteins, soya proteins, SCP,
7.	Herbs as functional food ingredients	Some examples of herbs having functional properties are murungai, rauwolfia, basil, tea, ajwain, celery, ginger, lavender, kudzu, radish, sesame garlic, etc.

iii. **Inorganic Mineral Supplement as Bioactive Compounds:** The mineral are the constituents of food which are the part of the bioactive compounds used and consumed and are included in the category such as Ca, Mg, Mn, Br, Co, Zn, P, etc. Mineral supplements containing minerals, and sometimes other ingredients also include vitamins both water and oil soluble. They are sole by many names, such as multiples or simply vitamins. The minerals supplements have unique roles in the body. They are also consumed to treat mineral deficiencies those caused by poor nutrition diet, digestive disorders, pregnancy, illness, medications, and certain conditions. Many people get the required minerals they need by eating routine foods and in certain cases, their doctor prescribes mineral supplement for them [21].

iv. **Probiotics:** These include living microflora (bacteria, or yeasts) with health benefits especially for digestive system. They are also referred to as "helpful" or "good" bacteria since they aid in keeping the healthy human gut thus are helpful to health. Colon is the heavily populated area of the gastrointestinal tract (GIT) and holds an approximate 500 distinct bacterial species as *Lactobacillus rhamnosus, Lactobacilli, Streptococcus, Lactobacillus reuteri, Bifidobacterium breve, Lactobacillus case,* yeast

Saccharomyces cerevisiae, some *Bacillus* species and *E. coli*, etc.

These living microorganisms if administered in required amounts deliberate health aid on the host. Probiotics containing good bacteria help in many ways, get rid off bad bacteria accumulated in excess quantities, thus helping feel better [21, 47]. The signs when probiotics are required by the body include, allergies and asthma, when someone s suffer from one or more mood disorders like depression, anxiety, irritability, and other mood disorders, had food poisoning, taken antibiotics, always getting sick or suffering from skin conditions such as acne and psoriasis [7].

v. **Prebiotic Foods:** Prebiotics are those elements, which reach to colon in undamaged form without getting exhausted by the gastric pH, digestive juices, and their enzymes. These prebiotics specifically stimulate the growth of gut probiotic bacteria, consequently they act as supporters/fertilizers for these bacteria (probiotics). Prebiotics include un-digestible, but a fermentable dietary carbohydrate may specifically accelerate growth of bacterial groups tenant in the colon, such as *Lactobacilli* spp., *Bifidobacteria* spp. deemed to be valuable for the human hold. Inulin, a soluble dietary fiber and tolerant to digestive enzyme and moves to large intestine (colon) effectively undamaged, where it is degraded by fermenting bacteria, *Lactobacilli, and* additional probiotic microflora. Commonly known prebiotics are lactulose, inulin, galacto-oligosaccharides, oligofructose, etc. [7, 21, 47].

vi. **Antioxidants:** Dietary supply of antioxidants and certain embellishment molecules, containing zinc and some vitamins are crucial in sustaining a free radical rummaging approaches for membranes, and enzymes and have biosynthetic capability. Antioxidants (Table 1.2) are discovered in the vegetable seed oils such as soybean, canola, corn, oat, wheat germ, palm, evening prime rose oil, etc. The vitamins like tocopherols (vitamin E) which reduces the vulnerability of LDL oxidation and reduce platelet's role in thrombus formation, also includes thiols (lipoic acid, glutathione), carotenoids (β-carotene, lycopene), vitamin C, enzymes (co-enzyme Q, super oxide dismastase), and minerals (selenium, manganese, copper, and zinc). Hydroxytyrosol is a powerful antioxidant and one of numerous phenolics found in olives and olive oil. Resveratrol a bioactive compound in red wine, nuts, etc., has excellent antithrombotic, antioxidant, and anti-inflammatory properties, and reduces risk of cancer. Lycopene, a carotenoid present mainly in tomatoes and other plant food, is a powerful antioxidant and supposed to guard from prostate and other cancers and hinders tumor cell development in animals. Organosulfur compounds in onions, garlic, isothiocyanates in cruciferous vegetables, and monoterpenes in cherries, herbs, and citrus fruits, act as anticarcinogenic. They also have cardioprotective properties. It is always recommended to have a variety of fruits, vegetables, legumes, oils, whole grains, and nuts a diet in order to reduce the chances of number of diseases. It has been observed that the deficiency of such components can cause diseases like Alzheimer's disease, cancers, cardiovascular diseases (CVDs) and rheumatoid arthritis [7, 21, 47].

Antioxidants as summarized in (Table 1.2) are categorized into three classes which include:

- True antioxidants;
- Reducing agents;
- Antioxidant synergists.

2. **Functional Food:** The constituents of the food that offer health benefits that extend beyond their nutritional value are grouped as functional foods. A process called nutrification is used to enriched or fortified functional food. This technique helps to regain nutrient content back to natural levels in a food probably before the food is finally processed. It also contains some supplementary balancing nutrients, such as vitamin-D added or improved milk. Health Canada characterizes functional food as ordinary food that has elements or materials combined to give it a intended medical or physiological advantage, other than have completely nutritional effect. In Japan, all functional foods essentially meet three traditional necessities viz. (i) food should be present in its naturally state and not in the form of capsule, tablet, or powder; (ii) consumed in the diet as habitually as daily, and third should normalize a biological process in expectations of thwarting or controlling diseases. Some examples include, the foods containing vitamins, minerals, probiotics, or dietary fiber.

TABLE 1.2 Antioxidants Rich Sources Which are Having Therapeutic Evidence

Antioxidant	Rich Sources
	Vitamins
Vitamin-C	Citrus fruits, guava, green chilies, vegetables
Vitamin-E	Wheat germ oil, sunflower, beet greens, collard greens, spinach, safflower, soybean oil, almonds, peanuts, sunflower seeds, peanut butter, red bell pepper, pumpkin.
	Carotenoids
Lycopene	Pink grapefruit, papaya, pink guava, tomatoes, watermelon. Relatively large amounts contained in dried apricots and pureed rosehips.
Beta-carotene	Green leafy vegetables and fruits (e.g., lettuce, spinach, carrots, sweet potatoes, broccoli, tomatoes, cantaloupe, and winter squash), yellow, orange. In general, the more intense the color of the fruit or vegetable, the more the β-carotene.
	Xanthophylls
Betacryptoxanthin	Tangerines, oranges, peaches, papaya, and yellow and orange colored corn. Tropical fruit consumption directly proportionate to blood absorptions of β-cryptoxanthin.
	Flavanoids
Rutin	Rutin (plant pigment) found in specific fruits and vegetables. Japanese pagoda tree, buckwheat, and eucalyptus are rich sources of rutin. Rutin is also obtained from elder flowers, ginkgo, apples, rue, St. John's Wort, lime tree flowers, hawthorn, and other fruits and vegetables.
Luteolin	Fruits and vegetables (e.g., parsley, broccoli, onion leaves, cabbages, celery, apple skins, peppers, carrots, chrysanthemum flowers contain luteolin. Luteolin rich plants have been used as Chinese traditional medicine for cure of inflammatory diseases, hypertension, and cancer.
Quercitin	The primary resources of quercetin include specifically apples, onions, parsley, sage, citrus fruits, tea, red wine. Grapes, dark cherries, olive oil, and dark berries such as blackberries, blueberries, bilberries contain high ranking quercetin and additional flavonoids.
Kaempferol	Green leafy vegetables such as spinach, kale, and herbs (chives, dill, and tarragon are rich sources of kaempferol. The wild leeks leaves of or ramps (100 g fresh weight) hold 50.2 and 32.5 mg of quercetin and kaempferol, respectively.
Liquiritin	Major flavonoids in licorice root (Glycyrrhiza spp.) are liquiritigenin (LQG), liquiritin (LQ) and its aglycone.

Nutrient-rich ingredients like fruits, vegetables, grains, nuts, and seeds, are often considered functional foods as well [7, 47, 48].

3. **Medicinal Food:** The FDA deliberates medical foods to be articulated to be consumed or administered internally under the administration of a physician, and which is envisioned for the definite dietary administration of a disease or conditions for which characteristic nutritional necessities, based on documented scientific principle, are recognized by the medical appraisal. Nutraceuticals and supplements do not meet up these conditions and are not categorized as medical food. Medical food can be consumed orally or use by the people suffering with explicit illness. Medical foods are legalized by the FDA and will be recommended or supervised under medical supervision [7, 21, 47].
4. **Therapeutic Supplements:** It implies to medically significant compounds generated from adapted agricultural crops or animals (using biotechnological approach). This also contain the medically beneficial compound produced from crops (modified agriculturally) or modified animals using biotechnological techniques. To alter a DNA, or to split it into a new DNA, transgene, gene pharming has been used by scientists since long. In this technique, the genetically modified animals (transgenic) are often used to develop human proteins with known medicinal values. The protein encoded by the transgene is generally secreted in the milk, eggs, or blood, and is collected and purified for further use. Animals for example cattle, pigs, goats, sheep, poultry, and rabbits and have been modified to produce different useful proteins and drugs [7, 47, 48].

1.2.2 NUTRACEUTICALS

On the other hand, nutraceuticals are categorized based on availability of food in the market.

1.2.2.1 TRADITIONAL NUTRACEUTICALS

Traditional nutraceuticals are basically natural foods with natural food composition or properties. Food that impart benefits beyond basic nutrition value normally contains several natural components, e.g., omega-3 fatty acids in salmon, lycopene in tomatoes, saponins in soy, vitamins in different foods, and carotenoids in yellow fruits, etc. [47].

Traditional nutraceuticals are grouped based on:

1. Chemical Constituent's Basis:

   i. **Nutrients as Chemical Constitution:** Minerals are recovered from animal and dairy products and some plant, and are functional in anemia, osteoporosis, and developing robust bones, teeth, and muscles, enhance nerve impulses and heart rhythm. Flax seed and fish (salmon) have omega-3 and omega-5 fatty acids which helps to control chronic diseases such as inflammation of tissues, brain function and its maintenance, cholesterol deposition reduction, etc. [48]. Mostly dairy foods, whole grain cereal foods, fruits, vegetables, and animal products such as meat, poultry, contain vitamins. These constituents are sympathetic in improving CVDs, stroke, osteoporosis, diabetes, cataracts, and cancer ailments [49].

   ii. **Herbals as Natural Source:** These are nutraceutical food which provides a great potential to enhance health and thwart chronic diseases owing to presence of herbal components. The examples containing an active component are willow bark (*Salix nigra)*, which as salicin, which is considered as anti-inflammatory, antipyretic, astringent, analgesic, and anti-arthritic [50]. Parsley *(Petroselinum cripsum*) which contain flavonoids (apiol, psoralen) and act as diuretic, carminative, and anti-pyretic. Peppermint (*Mentha piperita*) containing menthol as a functioning component and helps in curing cold and flu [51]. Lavender (*Lavandula angustifolia)* which contain tannin is helpful in reducing conditions like depression, stress, hypertension, cough, cold, and asthma [52]. Cranberries *(Vaccinium erythrocarpum)* containing pro-anthocyanadin are useful in fighting against cancer, ulcers, and urinary tract infections [53, 54]. There are different botanical products obtained from vegetables, fruits, whole grain seed, herbs, whole nuts, and various seeds developed as concentrates and extracts form containing plenty of bioactive compounds like terpenoids, sulfur compounds, pigments, etc.

Some commonly used herbal products and their applications include:

- Aloe vera which are used as anti-inflammatory, emollient, wound healing,
- Oil called evening primrose oil used as dietary supplement of linoleic acid for the therapy of atopic eczema
- Garlic which are used as antibacterial, antifungal, anti-thrombotic, anti-inflammatory
- Ginger which are used as carminative. Also act as anti-emetic, treatment of dizziness
- Ginseng which are used in hepatitis, sleep aid. It also relief from depression and improves mental concentration.
- Green tea acts as antioxidant. It reduces risk of CVDs. Also enhances humoral and cell mediated immunity.
- Ginkgo which are used in dementia and for femory enhancement

iii. **Phytochemicals Pharmacological Conditions:** Agreeing to Best [55] phytochemicals come under the class of nutraceutical compounds. They are categorized on chemical name basis given to them according to their phytochemical functions (Table 1.3). These compounds as carotenoids (isoprenoids) are obtained from various yellow fruits, vegetables, and egg yolk, are anti-carcinogenic, enhance natural killer immune cells and safeguard cornea against ultraviolet light. Legumes (chickpeas and soybeans), grains, palm oil contain non-carotenoids. These compounds are capable in eliminating cholesterol and act as anti-carcinogenic. Polyphenolics flavonoids found in fruits, berries, vegetables, and different legumes are powerful antioxidant substances, phytoestrogens, also inhibit breast and prostate cancer, and control diabetes millets. Non-flavonoid polyphenolics are present in dark grapes, raisins, berries, peanuts, and turmeric roots. They are powerful anti-inflammatory, antioxidants. They act as active anti-clotting vehicles and diminish cholesterol. Many phenolic acids and their compounds are found in blueberries, tomatoes, and bell peppers. They have antioxidant activity and decrease mutagenicity of polycyclic aromatic hydrocarbon compounds. Seeds from spp. of *Barbarea verna* and broccoli contain isothiocyanates (glucosinolates) having anti-tumorigenesis activity [7, 48].

TABLE 1.3 Nutraceutical Importance of Phytochemicals and Their Existence [7, 48, 55]

Phytochemicals	Sources	Characteristics
Tocotrienols and tocopherols	Grains	Suppress the development of varied tumors cell lines via admission of apoptosis and concomitant seizure of cells in the gastrointestinal stage of the cell sequence.
Carotenoids	Fruits and vegetables	Act as antioxidants, safeguards hostile to uterine, prostate, colorectal, lung, and GI tract cancers.
Limonoids (triterpenoid)	Citrus fruit juice and peel.	Hampering phase-I enzymes and stimulating phase-II detoxification of enzymes in liver, deliver safeguard to lung tissues.
Phytosterols	Various plants	Help to reduce inflammation, neoplastic, pyretic, and enhance immune-modulating activity, lessening cholesterol deposition.
Polyphenolic constituents	Various plants, wholegrain	Act as antioxidants, depresses the possibility of CHD, type-2 diabetes, hypertension, etc.
Flavonoid	Grapes, wines	They act against free radicals scavengers, show activity for free radicals mediated cellular communicating against soreness, allergic reaction, platelet accumulation, and hepatotoxins.
Iso-flavonoids	Soybeans	Used in the treatment of cancers and osteoporosis.

iv. **Cofactors:** These are enzymatic reactions the inhibitors and act as absorbents that combine to and eliminate unnecessary element in the intestine and enhance the assimilation and/or permanence of vital nutrients [56]. The existence of Cofactor influences the production and cellular DNA control as well as take part in fatty acid synthesis and energy production in living organism. Cofactors are required during different metabolic pathways, and it is possible for the concentrations of a single type of cofactor to affect the fluxes of number of pathways. Many enzymes are dependent on cofactors for proper functioning. Cofactors can be considered "helper molecules" that assist enzymes in their action in small quantity. Cofactors can be ions or organic molecules (called coenzymes) [7].

2. **Probiotic Microflora:** The scientific curiosity in probiotics research has been enhanced with the study of Metchinkoff who transformed the pathogen microflora of the large intestine into a host-friendly colony of *Bacillus bulgaricus* [57]. 'Probiotics are having life promoting expectancy and are defined as living microorganisms, which when swallowed in sufficient quantities, bestow a health effect on the

consumer ([58]. They are sociable bacteria that stimulate a healthy digestion and absorption of some nutrients in the human beings. They help to eliminate pathogens, such as yeasts, harmful bacteria and viruses which cause disease. Good bacteria develop a mutually advantageous symbiosis with the human GIT and health promoting [59]. Probiotics act as effective antimicrobial agent through modification of the microorganisms, protecting pathogens to adhere to the intestinal epithelium, enhance competition among pathogen for nutrients required for their survival, produce antitoxin effects and reversion of some of the infection consequences by pathogens such as secretory changes and neutrophil migration in the intestinal epithelium. Probiotics helps in the treatment of lactose intolerance by the making certain hydrolyzing enzymes (*ß*-galactosidase) that can breakdown the problematic lactose into its component sugars [60].

3. **Enzymes as Nutraceutical Compounds:** Enzymes are useful for life; without them human body cannot function properly. The persons who suffer from some medical conditions such as hypoglycemia, proper digestion, blood sugar abnormalities, obesity can be eradicated by enzyme additions in their diet plan [7].

1.2.2.2 NON-TRADITIONAL NUTRACEUTICAL FOOD

A group of artificially produced bioactive materials with biotechnology support, food technology is grouped in non-traditional nutraceuticals category. This category contains bioactive components which are engineered to produce products for human-wellness. They are grouped into following class:

1. **Fortified Nutraceutical Food:** These are fortified food from agricultural produce and contain added nutrients and/or ingredients, e.g., orange juice fortified with minerals/vitamins, cereals containing added vitamins or minerals and folic acid, iron fortified flour. Other include milk fortified with vitamin-D [61]. *Bifidobacterium lactis containing* prebiotic drinks and probiotic milk fortified with HN019. These fortifications are helpful in the control of diarrhea, breathing infections, and acute illnesses, in children [22]. Banana enriched with soybean ferritin gene in iron deficiency has also been developed by Kumar et al. [62].

2. **Recombinant Nutraceutical Food:** These consist of formulation of probiotics and the of bioactive components extracts produced by enzyme/fermentation system as well as genetic modified technique. It also includes the foods providing energy, such as bread, alcohol, fermented starch, yogurt, cheese, vinegar, and others produced using modern biotechnology. Examples include cows with LF inadequacy is concocted with rhLf to be able to solve the dilemma of LF deficiency in milch [20].

1.2.2.3 CLASSIFICATION OF NUTRACEUTICALS BASED ON THEIR FUNCTIONALITY

Bioactive compounds belong to one of several families of nutraceuticals classified on the basis of their biosynthetic pattern and functionality (Figure 1.6).

- Pattern of their uses:
  - based on skin cosmeceutical manufactured goods;
  - based on hair cosmeceutical manufactured goods;
  - others.
- Those containing active ingredients and preforming various function such as:
  - retinoids;
  - sunscreens;
  - moisturizers;

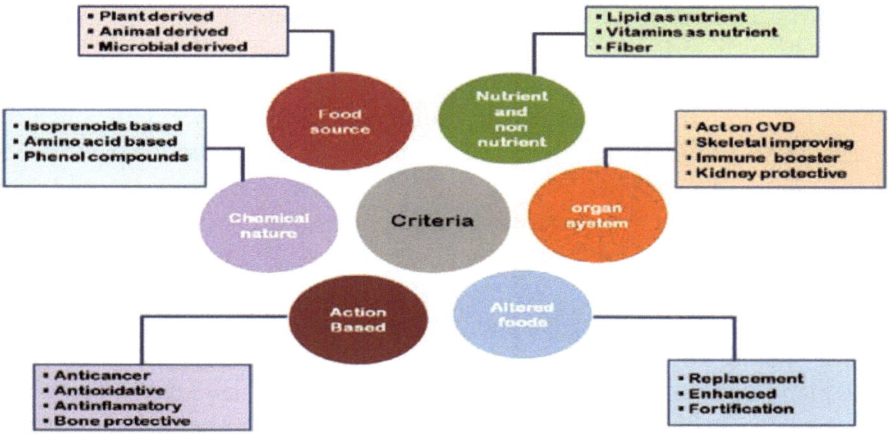

FIGURE 1.6 Classification of nutraceutical foods based on their functionality.

- depigmenting agents;
- exfoliants;
- antioxidants;
- proteins/peptides; and
- growth factors.

Some of other common ways of classifying nutraceuticals is based on their food source, chemical nature, mechanism of action, etc. The natural sources of food as nutraceuticals can be categorized as: probiotics, prebiotics, dietary fiber, polyunsaturated fatty acids (PUFA), polyphenols, spices, and antioxidant vitamins [43].

Nutraceutical compounds can also be largely categorized into following two groups:

- potential nutraceutical food; and
- established nutraceutical food.

A useful nutraceutical food application can be proved only after capable scientific statistics for its health and medical benefits when are achieved and proved with testimony of functionality and effective use [63, 64].

1.3 POTENTIAL SOURCES OF BIOACTIVE COMPOUNDS AND NUTRACEUTICALS

The great biodiversity of plants, animals, and sea foods make them rich sources of bioactive compounds (Table 1.4), and humans can directly acquire the health benefits by ingesting the compound-derived from these food products. Many bioactive compounds have been discovered which vary widely in their chemical structures and functions and are grouped accordingly. Bioactive compounds including flavonoids, are found in all plant sources. They have been investigated comprehensively in cereals, legumes, nuts, olive oil, vegetables, fruits, tea, and red wine. Many bioactive compounds (Table 1.5) have antioxidant properties, and various studies have demonstrated their favorable effects on thrombosis and tumorigenesis and promotion. They can promote good health act interacting with body metabolism [65]. The plants-based nutraceuticals are not mainly incorporated in the terminology of plant bioactive compound but are generated as secondary metabolites (SMs) that are not typically essential for the daily operational functions of the plants such as its growth and development, etc. They are characteristics in competition, defense, attraction, and signaling [66]. Therefore, the bioactive compounds present in the plants are the SMs and are associated with imparting pharmacological and toxicologically beneficial effects in humans and animals [67] (Table 1.6).

TABLE 1.4 Excellent Sources of Nutraceutical Foods and Their Bioactive Components [7, 56]

Nutraceutical Substances	Sources	Nutraceutical Substances	Sources
Psyllium	Psyllium husk	Isothiocynates	Cabbage, broccolini cruciferous vegetables, cauliflower,
Inulin, FOS	Whole grain, onion, garlic	Indoles	Cabbage, broccoli, cauliflower, banana
Lactobacillus	Yogurt, dairy products	cellulose	Plants
Luten	Corn, egg, citrus collards, spinach,	Isoflavines	Soyabean including other legumes
MUFA	Mustard/rape seed, nuts, olive	Ellagic acid	Walnuts, grapes, straw-berry, raspberries,
Catechins	Tea, cocoa, apple, grapes	3-n-butyl phthalide	celery
Resveratrol	Grape, red wine	Quercetin	Onion, grapes, citrus fruit, broccoli, yellow squash
Beta carotene	Carrot, citrus fruits, squashes, pumpkin	Capsaicinoids	Pepper, fruits
Carnosol	Rosemary	EPH and DHA	Fish liver oil
Lycopene	Tomato and its products	Catechin	Teas, berries
Adenosine lignan	Onion, flax, rye	Beta glucan	Flesh and dairy products
Curcumin	Turmeric	Anthocyanins	Red wine

TABLE 1.5 Various Foods and Their Functionalities [7, 46]

Specific Functions	Functional Food Components
Anti-inflammatory	GLA, capsaicin, CLnA. EPA, DHA, quercetin, curcumin
Bone protection	CLA, soy protein, genstein, daidzein, calcioum, inulun, FOS
Anticancer properties	*Lactophillus acidophilis*, capsaicin, gegestein, daidzein, alpha-tocopherol, CLA, Limoene, equal, curcumin, ellagic acid, luetin, carnosol, *L. bulgaricus*
Antioxidants	Glutathione, vitamin-C, beta carotene, polyphenolics, lutolin, catechins, gingerol, chlorogenic acid, tannins, CLA, tocopherol, indole-3 carbonol, alpha tocopherol, ellagic acid, lycopene,
Lipid profile	MUFA, quercetin, omega-3 fatty acids, beta glucan, gamma tocotrtenol, resveratrol, tannins, saponins, guar, pectin

The plants are considered as the main source of bioactive compounds and nutraceutical compounds and their raw materials. Animal food and other living marine organisms (fish, phytoplankton, plankton, marine microorganisms, and microorganisms (bacteria, fungus, algae) also contain these substances [68]; SMs [69], mushroom [70] and in some groups of animals [71]. These sources are many potentially useful components which act as bioactive SMs. The exploration of marine organisms has revealed the presence of numerous proteinaceous nature bioactive compounds in them. These include linear peptides, proteins, cyclic peptides and depsipeptides (mainly found in marine and microbial natural products and are derivatives of peptide, amino acids, and amino acid-like components in which one or more of its amides, -CNHR-, groups are replaced by the corresponding ester, -COR, and have both peptide and ester linkages). The colorful, flavorful fruits and vegetables, containing nutritious components and are part of our diet contain various bioactive compounds. The compounds included in this category are polyphenols, carotenoids, vitamins, phytoestrogens, glucosinolates, and anthocyanins. These fruits and vegetables are being paid increased attention now-a-days because of their potential health benefits.

Metabolites that are produced by living cell in the body are distributed into two main classes: (i) primary metabolites (PMs); and (ii) secondary metabolites (SMs). PMs are the substances needed for growth and development and include proteins, carbohydrates, lipids, and amino acids, SMs are compounds required for plant to help in survival and overcome local challenge combat and increasing overall ability in their interaction with the surroundings. Many bioactive compounds which give plant kingdom their color, characteristic flavor, and aroma are produced during the process of secondary metabolism. Therefore, these act as secondary plant metabolites imparting pharmacological or toxicological effects in human beings, and animals are the plant origin bioactive compound. They are largely used due to their medicinal value and health promoting effects that are found in various herb plants but are also known for their venomous and lethal effects to others [72, 73]. Subsequent table illustrate the different bioactive components produce in various foods.

1.3.1 HERBS AND SPICES AND OIL SEEDS AS SOURCES OF NUTRACEUTICALS AND BIOACTIVE COMPOUNDS

For the maintenance of the human life quality herbs play an important role through the supply of various bioactive ingredients. The bioactive components from herbs are vital class of nutraceuticals in addition supply of minerals, vitamins, and other valuable compounds and have numerous health promoting benefits. The herbs phytochemicals active constitute include polyphenols, terpenoids, flavonoids, and saponin. The people who pursue food supplement for conventional health maintenance mainly uses these herbs based bioactive compounds. In this regard, it and are seen as nutraceutical supplementing herbs chemicals [54] (Table 1.6). The herbs and spices which are frequently taken as functional food or supplements are elaborated in subsequent sections.

1. **Flaxseeds:** These are obtained from plant species *Linum usitatissimum* and belong to family *Linaceae*. The biochemical active constituents of the seed include, delta-linoleate, alfa-linoleate, secoisolariciresinol, lignans, and proteins. The seed constituents are being used in the prevention of mammary, colon, and rectal cancers. It uses also reduces blood pressure in hypertensive patients. The studies have shown that its use has effect on diabetes control and on coronary heart diseases prevention
2. **Ginkgobiloba:** It belongs to family *Ginkgoaceae*. The biochemical active constituents of include ginkgolides type A, B, C, bilobelin, ginkgetin, and iso-ginkgetin, flavanols, It is used in the cure of asthma, impaired memory. The leaves ginkgolides plant can be used to ease the adverse effects of platelets activating factor.
3. **Spirulina:** It is known as *Spindina platensis* or *S. maxima* and belong to family *Oscillatoriaceae*. The bioactive constituents comprise, gamma linoleic acid, oleic acid, glycolipids, and sulfo-lipids. It also contains vitamin-B complex in plentiful amount and beta-carotenes and phycocyanin. Spirulina functions include its immune-stimulant activity, helps in the management of human immunodeficiency virus syndrome and additional viral diseases such as herpes, influenza, cytomegalovirus, and mumps. Also effective in arthritis, atherosclerosis therapy and diabetes, and aging process control.
4. **Bitter Guard (Karela):** It is named as *Momordica charantia* and belong to family: Cucurbitaceae, it is having hypoglycemic effect. The extract of karela has been demonstrated to increases the rate of glycogen synthesis by a level of 4–5-fold in liver.
5. **Turmeric:** It is obtained from the rhizome of curcuma longa and belongs to family *Zingiberaceae*. The bioactive component of

turmeric includes curcumin, desmethoxy curcumin, bis-desmethoxy curcumin. The use of the rhizome comprises due to antimicrobial activity The other outcomes as shown that it has inhibitor activity for integrase enzyme.

6. **Soya Foods:** It is obtained from the plant of glycine max and belong to family: *Leguminoseae. The c*hemical constituents include daidzein, genistein. It is used in the prevention of estrogen-dependent cancers whereas genistein initiate protein tyrosine kinase and DNA Topoisomerase-II inhibitions.
7. **Garlic Extracts:** It is taken from the *Allium sativum* bulbs portion. It herb is from family *Liliaceae.* The chemical constituents include allicin, allin, ajoene. The garlic extracts are applicable for hyperlipidemia cure. It also acts/shows antihypertensive, hypoglycemic, anti-spasmodic activity and helpful in the prevention of colon and lung cancers.
8. **Tomato:** It is obtained from *Lycoper Siconesculentum* plant and belongs to family*: Solanaceae. The c*hemical constituents include Lycopene. It is used in the prevention of digestive tract and prostate cancer types, also diminishes risk of cancers in pancreas, cervix, bladder, and skin.
9. **Fenugreek:** It is obtained from the plant of *Trigonella foenum-graecum*. It belongs to *Leguminoseae* family. The *c*hemical components consist of alkaloids of gentianine and trigonelline, flavonoid compounds, coumarins, proteins (amino acids), saponins (steroids). The applications of spice fenugreek comprise anorexia, gastritis cure. It also has cathartic, expectorant (cough medicine), and emollient properties. It shows hypoglycemic effect and hypo-cholesterol-emic benefits.
10. **Ginseng:** It is obtained from *Panax ginseng* and belongs to family *Araliaceae.* The chemical constituents include B-vitamins (B_1, B_2, B_{12}, pantothenic acid, biotin), polysaccharides (starch), Lipids (sterols, choline), protopanaxadiol, and polyacetylenes.
11. **Peppermint:** The peppermint (*Mentha piperita*) is a natural amalgam of water mint (*M. aquatic*) and spearmint (*M. spicata*) species and is from *Lamiaceae* family. The herb is indigenous to Europe, but now usually grown in whole world. It uses include pharmaceutic and cosmetic material preparations, as well uses as food additive,

primarily for products flavor and aroma enrichment. The conventional uses of peppermint EOs are in traditional medicines. It is also one of the compounds of tea preparations, although the peppermint extracts are predominantly consumed as such and with other dishes [74].

12. **Basil:** The basal *(Ocimum basilicum)* is a medicinal herb which belongs to the family *Lamiaceae*, indigenous to warm tropical climate ranges of Asia (India) Africa, and Mediterranean Europe. Basil herb comprises 150 species, extensively planted globally. It is used for cooking purpose and in traditional medicine preparations used for headaches, coughs treatments, and in kidney malfunction cure. Basil leaves are a prevalent flavoring agent in cousins, a used as chemical additive to traditional medicines, in cosmetic medicines, for fragrance and a food product ingredient [75, 76].
13. **Rosemary:** The rosemary (*Rosmarinus officinalis*) plant which reaches to about one meter high is an perennial flowering shrub having vertical stems, whitish-blue flower with dark green color leaves It belongs to the family of *Lamiaceae*. The major areas of rosemary cultivation are Mediterranean countries of France, Tunisia, Spain. Pine of rosemary shrub are extensively used in traditional medicine and cosmetics due to powerful pleasing smell evocations used as flavoring ingredient in food preparations [77].
14. **Thyme Herb:** The thyme (*Thymus vulgaris*) also known as garden thyme, common thyme or German thyme is from family *Lamiaceae*. This thyme originates from Mediterranean regions of Europe. Its plant is woody evergreen, with pink flowers and scented green colored leaves. The shrubs of thyme make ground covers in gardens, but its important use is as a flavoring agent in preparation of rabbit, and lamb meats in which it develops a characteristic piquant sense of taste. In traditional medicine preparation, thyme herb is used as an concoction or an chemical additive in prevention of skin diseases, carminative, and as a sedative medicine [78].
15. **Nettle Herb:** The nettle (*Urtica dioica*) an herbaceous perennial plant which attain two-meter high is from family *Urticaceae*. The herb originates from countries of continents of Africa, Asia, North America, and Europe, and now grown globally. The plant gets its name from *uro*, signifying 'burn,' or *urere* signifying 'to sting.' Nettle herb has characteristic open sting hairs known as trichomes which

occupy areas of leaves and stem. Trichomes when comes in contact act like needles and infuse active acetylcholine, formate, histamine, serotonin, moroidin, and in human body and cause a burning sensation and develop skin rashes. Nettle herb applications are found in traditional medicines, in food preparations, feed additives and as a source of fibers [79].

16. **Elderberry:** The elderberries (*S. nigra*) a deciduous shrub as reaches up to 6-meter high, belongs to the family *Adoxaceae*. The herb originates from sunlight areas of North America, Asia, Africa, and Europe. It flowers in the structure of white hermaphrodite in every summer and looks like larger corymbs. The plant is known for its very well cherished fruits. Its flower extracts are used as food flavoring agent in beverage industry and in alternative medicine purpose [80].
17. **Cinnamon (*Cinnamomum zeylanicum*):** These plant, an evergreen tropical forest tree is native to tropical climatic regions of India, Burma, and Sri Lanka, and. It belongs to the family *Lauraceae*. The tree reaches up to 10-meters height on an average not exceeding 3 meters. The bark of the tree is used for spice production and is procured two to three occasions in a year during the rainy seasons. The bark after harvest is cut into 3-meter length and around 2.5 cm diameter pieces and followed by short fermentation span with removal of internal and external phloem tissues. The fermented chunks are than dried in the sun light. Through fermentation preparations, the bark achieves its distinguishing yellow-brown color. Cinnamon is used extensively as flavoring in foods and in ethnic medicine around the world [81].
18. **Cloves (*Syzygiu aromaticum*):** It is an aromatic spice is popular in the whole world and belongs to the family *Myrtaceae*. The spice clove gets its name from word '*clavus*' implies nail like. The clove tree belongs to evergreen tropical forests and is native to Indonesia, but is now grown in largely of countries Sri Lanka, India Madagascar, Tanzania, Kenya, and Seychelles. During drying and fumigation, the cloves become dark in color, while the distinguishing aroma flavor exists due to high concentration of compound eugenol. In India and China clove has been used from 2000 years because of presence of EOs. It has been used in China at around 240 BC in the ancient medicine preparations [82].

19. **Licorice Herb (*Glycyrrhiza glabra*):** It is from *Leguminosae* family, and the active component is obtained from roots of Licorice plant. The plan which is a herbaceous perennial plant, grows up to one meter height. The plant is indigenous to southern Europe and parts of Asia, but these days cultivated in many countries of world like India, Iran, Pakistan, Turkey, Afghanistan, Iraq, China, Italy, and England. The herb which is believed to help to augment coherence has been used in Chinese as medicine ingredient. Owing to the medicinal use and sweet taste, licorice finds its use in traditional medicine and as for sweetener and flavor purpose in beverages, candies, tobacco [83].

TABLE 1.6 Plants and Herbs as Sources of Bioactive Compounds and Their Health Prospective [7, 56]

Sl. No.	Common Name	Botanical Name (Family)	Part of the Plant Used	Bioactive Components	Benefits to Human Health
1.	Ginger	*Zingiber officinale* (Zingiberaceae)	Rhizomes	Zingiberene, gingerols	Used in chronic bronchitis and hyperglycemia, have stimulation effect and cure throat pain.
2.	Ginseng	*Panax ginseng* sp. (Araliaceae)	Root	Ginsenosides and panaxosides	It stimulates immune and nervous system.
3.	Garlic	*Allium sativum* sp. (Liliaceae)	Bulbs	Alliin, allicin	Used as anti-inflammatory, antifungal agent antibacterial, and act as antithrombotic, act as antihyperlipidemic and hypotensive.
4.	*Aloe vera*	*Aloe barbadensis* Mill. (Liliaceae)	Gel	Aloins and Aloesin	Benefits to dilates capillaries also act as moisturizer, anti-inflammatory, have wound healing properties.
5.	Turmeric	*Curcuma longa* (Zingiberacae)	Rhizome	Curcumin	Act as anticarcinogenic, have anti-inflammation, antiseptic, and antiarthritic effects.
6.	Onion	*Allium cepa* Linn. (Liliaceae)	Rhizome	Allicin, alliin	Improve hypoglycemic activity, atherosclerosis, and antibiotic.
7.	Maiden hair tree	*Ginkgo biloba* (Ginkgoacea)	Hair of tree leaves	Ginkgolide, bilobalide	Used as antioxidant and to enhance memory, also increased peripheral blood flow, treatment of post-thrombotic syndrome.
8.	Senna	*Cassia angustifolia* (Leguminosae)	Leaves	Sennosides	As a purgative.

TABLE 1.6 *(Continued)*

Sl. No.	Common Name	Botanical Name (Family)	Part of the Plant Used	Bioactive Components	Benefits to Human Health
9.	Echinacea	*Echinacea purpurea* (Asteraceae)	Leaves	Alkylamide, echinacoside	Act as antiviral, anti-inflammatory and immunomodulator.
10.	Liquorice	*Glycyrrhizaglabra* (Leguminosae)	Root	Glycyrrhizin, liquirtin	Act as anti-inflammatory and anti-allergic.
11.	Goldenseal	*Hydrastis canadensis* L. (Ranunculaceae)	Root	Hydrastine berberine, canadine	Use as antimicrobial, astringent, antihemorrhagic agent and used for therapy of mucosal inflammation, dyspepsia, and gastritis
12.	Horehound	*Marrubium vulgare* L. (Lamiaceae)	Whole plant	11-oxomarrubiin and vulgarcoside A	Used as antitussive, expectorant, and choleretic.
13.	White willow	*Salix alba* L. (Salicaceae)	Bark	Salicin	Have analgesic, anti-inflammatory, astringent, antipyretic, property and used in treatment of rheumatic and arthritic.
14.	Saint John's wort	*Hypericum perforatum* (Hypericaceae)	From aerial part	Hypericin and hyperforin	Act as antidepressant for hepatitis-c virus and HIV
15.	Valeriana	*Valeriana officinalis* Linn. (Valerianaceae)	Root	Valerenic acid, valerate	Helps to reduce intestinal cramps, menstrual pain, bronchial spasm, migraine, and tranquilizer.
16.	Bael	*Aeglemarmelos* Corr. (Rutaceae)	Unripe fruits	Marmelosin	Treatment of diarrhea and dysentery
17.	Brahmi	*Centella astatica*	Herb	Asiaticoside, madecassoside	Used as nerve tonic and for anti-anxiety.
18.	Asafoetida	*Ferula asafoetida*	Oleo gum resin	Ferulic acid, umbellic acid	Helps to stimulate, carminative, and also act as expectorant.

1.3.2 FRUITS AS SOURCE OF BIOACTIVE COMPOUNDS AND NUTRACEUTICALS

Various types of bioactive compounds are found in fruit but in relatively small quantity. The health-promoting properties of these secondary metabolites (SMs) (phytochemicals) belong to polyphenolic category, the highly widespread of which are called anthocyanins. These are accountable for purple, black, red, and blue colors in fruit. Other compounds like sterols, stilbenes, glucosinolates, tocopherols, organosulfur, and tannins contained in fruit also have health-benefiting [84]. Various fruits of this category are briefly summarized below:

1. **Cranberry:** The cranberry also known as from America (Latin: *Vaccinium macrocarpon* or *Oxycoccus macrocarpus*) is from *Ericaceae* family is a dwarf evergreen shrub indigenous to north America and nurtured in northern America and Canada all over. The fruit 'cranberry' gets its name '*craneberry*' from popular among European who settles in American continent. The plant has little, pink, or red blossoms looks like head of crane bill. The name *Vaccinium oxycoccos* is prevalent in Europe for cranberry grown in central and northern areas of Europe. European cranberry fruits and leaves are smaller in size, sharp together with acid flavor and invigorating property, while fruits of American cranberry look marginally apple-like. The fruits are either converted to concentrates or consumed fresh, which have various value-added potentials used in various juice preparations [85, 86].
2. **Bilberry:** They are called Eurasian shrubs having small, black berries on plant called bilberry (*V. myrtillus*) and belongs to family *Ericaceae.* The berry gets its name after Danish key word 'bølle' and used for other blueberry (*V. corymbosum*), however there are a little distinction between these two berries. The color of *bilberry* paste is dark purple, yet black, due to the significant quantity of anthocyanins. The blueberry pulp is pale green in color. The shrubs of blueberry have extra evergreen leaves. Blueberries are gathered in clusters while blberries occur in single or pairs on the bushes of the plant. Bilberries were cultivated in the forests of native East, North, Russia, and in Central Europe. Berries are procured in natural environment and are use as flavoring and in functional food preparations. They plant can grow in acidic, nutrient deficient soils, even though commercial cultivation is hard [87].
3. **Black Currant:** It is a type of backyard plant cultivated in Russia in the past 11th century has been described as black currant. In Europe it is grown from 17th century and during World War-II time having extensively nurtured in the United Kingdom. The black currant (*Ribes nigrum*) represents shrub family *Grossulariaceae* which grows up to 2-meters height in numerous parts of the world having temperate climate. Black currant originates from region of central and northern

Europe, northern Asia and is farmed domestically and commercially. The plant is sensitive to waterlogging of soil and droughts but can grow suitable in sandy and forest soils. The plant fruits, bark, roots, leaves, and flowers have been used in traditional medicine because of strongly aromatic property. Syrups obtained from the fruit have been given under two year old children in the past and contributes as a source of vitamin-C [88].

4. **Elderberry:** The elderberry (*Sambucus nigra*) a fruit barring plant is a shrub and grow up to 6 meter tall. It belongs to the family *Adoxaceae*. The plant originates from Asia, Europe, Africa, and North America, were having sunlight-exposed areas [80]. It is believed that elderberry (*S. nigra*) seeds were brought by early settlers from Europe which is famous in North America as *Sambucus canadensis*. The *Sambucus* refers to a type of flute made from twigs of tree and gets its name from '*sambuke*' a Greek name or the Latin word '*sambuca.*' The elderberries fruits are oblong, compact, and green in their early stage and after 6–8 weeks, they progressively develop and transform color from red, purple to black. Fruits on ripening weighing approximately 50–130 mg having diameter of 5 to 7 mm [89].
5. **Cornelian Cherry:** The cherry fruit plant cornelian *(Cornus mas)* belongs to *Cornaceae* family. It is a highly deciduous shrub, or a small tree and grows up to 5 to 8 meter high. The countries of European content (Czech Republic, Belgium, Germany, Slovakia, UK as well as Turkey) are now principal producer of the fruits where cornelian cherry occurs as rare plant in the past. The plant has extraordinary tolerance for various adverse ecological circumstances. There is a saying for the cherry as 'healthy as the Cornelian cherry.' The fruit with sour taste looks cherry-like and long appearance [90].
6. **Açaí:** The Açaí tree (*Euterpe oleracia)* is a palm tree species and belongs to the family *Arecaceae*. The tree is tall, reaching over 25 meters. The countries like America, the flood lands of Amazonian, Brazil, and Columbia are main cultivators of Açaí these days. An increasing demand has been observed now a days for both for palm trees, fruit, and its heart. The fruits with flavor like raspberry which looks like grapes grows about 10 mm in diameter, much smaller in size and are purple black color, produces up to 900 fruits in a single panicle branch. The fruit have low sugar content and are not sweet but have high quality dietary fiber [84].

1.3.3 MICROORGANISMS AS NUTRACEUTICAL FOODS AND THEIR SOURCES

Nutraceuticals are the food components and their products. They have been proven beneficial having positive effect on human health. They are low calories sugars, proteins, and vitamins, etc. Microbes such as bacteria, yeasts, and mold are used for the production food and ingredients such as wine, beer, bakery, vitamins, organic acids, proteins, fermented dairy, and non-dairy products. Microorganisms (Table 1.7), such as probiotic bacterial strains of species Bifidobacterium (adolescentis, animalis, bifidum, breve, and longum) and Lactobacillus (acidophilus, brevis, casei, fermentum, johnsonii, plantarum, delbrueckii, rhamnosus, salivarius, etc.), mold *Aspergillus flavus, Penicillium mold* are most suitable microbial cell factories to produce various nutraceuticals. Medicinal mushrooms having anti-inflammatory and immune-enhancing effect are maitake (*Grifola frondosa*), shiitake (*Lentinus edodes*) and reishi (*Ganoderma lucidum*) [7]. The bacterial nutraceuticals also preserve food through the formation of inhibitory metabolites. These include organic acid (lactic acid, acetic acid, formic acid, propionic acid), ethanol, bacteriocins, etc. These ingredients work in combination with decreased water activity by application of drying or by using salt [91, 92].

TABLE 1.7 Food Sources of Probiotic Microorganisms [7, 59]

Liquid Milk Microorganisms	Fermented Milk Microflora	Related Fermented Products	Human Milk Microflora	Intestinal Tract Organisms	Fruits and Vegetables/ Food Grains Microflora
Propionibacterium freudenreichii	*Bifidobacterium adolescentis*	*Bifidobacterium thermophilum,* and *B. animalis*	*Bifidobacterium infantis, B. longum, B. breve, B. lactis*	*Escherichia coli Nissle strain-1917*	*Lactobacillus brevis* *L. plantarum*
L. lactis and *L. acidophilus*	*Lactobacillus delbrueckii subp bulgaricus*	*Lactobacillus casei, L. curvatus, L. cellobiosus, L. helviticus, L. fermentum, L. farciminis*	*L. salivarius* and *Lactobacillus reuteri* spp.	*Lactobacillus gasseri L. johnsonii*	*Saccharomyces cerevisiae* *S. boulardii*
	Streptococcus thermophilus	*Pediococcus acidilactici* *Enterococcus faecium*			*Leuconstoc mesenteroides*

1.3.4 ENZYMES AS NUTRACEUTICAL FOODS AND THEIR SOURCES

Enzymes are indispensable for health. They are involved in digestion and for a healthy body makeup. They work in association with other body chemicals such as stomach acid and bile. Helps in food break down into absorbing molecules which perform a wide range of body functions. Enzymes having application in nutraceuticals and is available in tablet or capsule which are used as digestive aids by breaking down biomacromolecules to digestible compounds, such as lactases, saccharase. Enzymes are as biocatalysts that regulate metabolic processes necessary to life sustainability. As has been driven by demand in digestive health products preparations, growth of enzyme-based dietary supplements is steady now a days, in line with the larger markets of nutraceuticals like products. There are two types of enzymes mainly categorized as dietary supplements as systemic and digestive form. Systemic enzymes help to promote general body support and to support the body's digestive process, digestive enzymes are used. The digestive enzymes supplements use is more common today (Table 1.8). Foods containing natural digestive enzymes are pineapples, papayas, mangoes, honey, bananas, avocados, kiwifruit, and ginger. Some fermented food containing enzyme are kefir, sauerkraut, kimchi, miso. By adding these foods in our daily diet may help to stimulate digestion and improve gut health. [92, 93].

TABLE 1.8 Nutraceutical Enzymes and Their Sources (Microbes, Plants, and Animals) [7, 55]

Animal Enzymes and Their Source	Microbial Enzymes and Their Source	Plant Enzymes and Their Source
α-amylase obtained from human saliva.	Lactase-β-galactosidase produced from bacterial species.	Glucoamylase obtained from nodule and hold-up cultures of tuber sugar beets and its mature roots.
Trypsin from pancreatic juice.	Amylo-glucosidase from ascomycetes.	α-Galactosidase obtained from brussels sprouts, beans, broccoli, cabbage, asparagus, other vegetables, and whole grains.
Chymotrypsin from All vertebrates.	Glucoamylase from *A. niger, Saccharomycopsis fibuligera*.	*β*-Amylase obtained from Higher plants.
Ox bile from ox.	Hemi-cellulase from microorganisms and mushrooms.	Hemi-cellulase (plant walls).
Pancreo-lipase from Pancreatic juice.	Catalase	Pectinase obtained from cell wall.
Pepsin obtained from animal tracheal secretions.	Cellulase from all living cells.	Bromelain obtained from pineapple.
Lysozyme obtained from egg white, tears, saliva, and from many animal fluids.	Invertase, sucrase from yeast.	Bio-diastase obtained from soybean.

1.4 APPLICATION OF NUTRACEUTICALS

Nutraceuticals symbolize a distinctive mixture of modern-day science, natural vehicles and possibly the best feasible solution for managing lifestyle disorders. In addition to these everyday life disorders, they are also regard as accountable for the predilection of numerous intricate clinical situations [94]. The nutraceuticals and their products are documented not only for their healthiness assistances to diminish the risk of cancer, heart diseases and other associated ailments, but also helps in the prevention or treatment for blood pressure, high cholesterol level, unwarranted bulk, osteoporosis, diabetes, soreness of bones, macular disintegration, cataracts, menopausal indications. With the occurrence of a new set of diseases in modern day work has led to the universally termed as life-style ailments. The main reasons of such types of conditions are inappropriate diet and extraordinary dependency on fast foods, lack of physical movement, non-alignment with biological time, erroneous body position, unnecessary tension, and insufficient sleep. Various marketed nutraceutical alternatives can help to tower block the conversion of lifestyle ailments into fatal ailments. Consumer recognition of nutraceutical grew appreciation from 1980 ahead when scientists-initiated pampering into useful assessment of such products followed by their interpretation in mass media journals, seminars, debate, etc. [95]. The nutraceutical market grows globally around 400 billion USD in 2019 is anticipated to cross 700 billion USD with a *compound annual growth rate* of 8.3% by the year 2027. Fast expansions of nutraceutical products market and recognition along with a beneficial channel of pioneering manufactured goods flourishing the market are all sponsors to the anticipated growth. Although with the entrance of generic manufactured goods in the market bring about a decrease in cost of nutraceutical products but because of high consumer acquiescence, the overall market for such products is states to remains secure [96].

Global countries have regulations regarding ls. In India, regulations for criteria and sale of nutraceutical and their products were implemented in 2010 under FSSAI Act (2006). Under the act, foods are believed as useful food if they have on one or more positive effects in body functioning. They are even now in the form of food but not as a dietary supplement. Occasionally functional foods are fortified or enriched during processing for delivering some valuable effect to the consumers such as dairy products which are fortified as milk, citrus fruits like orange juice, etc., fortified oils, infant food, health/functional food. Having concern involving the safety of functional foods and nutraceuticals, all we know around their existence from thousands of years in the diets of masses of communities of world. They are therefore considered as safe and offer a recognized level of permissibility as matched to modern medicines. Functional foods and nutraceuticals are health wellbeing advantageous since they enhance immune system, defend gastrointestinal tract (GIT) and cardiovascular structure. They also support kidneys, and skeletal structure having anticancer and anti-inflammatory properties. Capsaicin from red chili and pepper, genistein in soy products, omega-3 and omega-6 fatty acids, ellagic acid found in various fruits have anticancer and bone protective properties (Table 1.9). As has been depicted that lipid balance can be enhanced by ingesting beta-glucan from oats, resveratrol of red grapes and blue berries, tannins from chocolates, tea, and berries and saponins accessible in different legumes. Nutraceuticals include nearly all major therapeutic areas including cough, sleeping disorders, anti-arthritic, digestion dilemma and avoidance of a range of cancers, osteoporosis, helps in cholesterol decline, diabetes, hypertension, and beauty treatment related issues. [93].

TABLE 1.9 Bioactive Compound Their Sources and Applications [7, 55]

Bioactive Components	Sources	Applications
Carotenoid (Lycopene)	Tomatoes	Contribute to the prostate maintenance.
Monosaturated fatty acids	Tree nuts	Reduce risk of coronary heart disease.
Flavonols	Onions, apples, tea, broccoli	Helps to neutralize free radical, which may damage cells. Bolster cellular system, and antioxidation defenses.
Beta-carotene	Carrots, various fruits	Helps in the neutralization of free radicals, which may damage cells. Also bolsters cellular antioxidant defenses.
Soy protein	Soybeans and soy-based food.	Reduce risk of coronary heart disease.
Lactobacilli, bifidobacterial	Yogurt, other dairy, and non-dairy applications.	Improve gastrointestinal health, associated with systematic immunity.

Consuming certain medicines or enhancements or applying a variety of chemicals on skin, various detrimental oxidative ions are spawned in the body. Several food products including α-tocopherol, carotenoids from carrot, ascorbic acid as of citrus fruits, glutathione precious foods and gingerol from ginger are shown to be antioxidant property and facilitates the body to get rid of or neutralize these oxidative ions influence [93, 97]. The potential use of traditional nutraceutical and the foods containing them in chronic disease control is given in (Table 1.10).

TABLE 1.10 Traditional Nutraceutical Applications and Chronic Disease Control

Nutraceutical Products	Dosage/ Duration	Beneficial Effects	References
Allenic carotenoid fucoxanthin present brown seaweeds.	2.4 mg per day	It can enhance insulin resistance, declines blood glucose levels as a result of the management of cytokine secretions from white adipose tissues.	[98]
Polyunsaturated fatty acids.	No data	Helps in the prevention of various disorders which have effect on lungs and bronchi.	[99]
Unsaponifiable matter from avocado and soybean oils.	300 mg for 3 years	It could stimulate synthesis of extracellular matrix substance and aggrecan a type-2 collagen there by lowering the build-up of inflammatory pro mediators (IL-8, IL-6) and catabolic (MMP-3) and in osteoarthritis.	[100]
Conjugated linoleic acids.	For 3 months	It can significantly improve the airway hyper responsiveness associated with the reduction in leptin/adiponectin ratio in mild asthma symptoms.	[101]
Siphonaxanthin, a marine carotenoid from green algae.	20 μM for within 6 h of treatment in HL-60 cells.	It is associated with induction of apoptosis in HL-60 cells by decreasing Bcl-2, and increasing of activation of caspase-3 activity.	[102]
Fermented papaya preparation.	6 g per day for 6 months.	Unregulated thioredoxin in liver cirrhosis and TNF-α.	[103]
Monounsaturated fats).	No data	Lowers risk cardiovascular disease and regulate metabolic syndrome problem.	[104]
1,25(OH)2D (calcitriol)	200–600 IU in a 24 h	Regulates the levels of p21 and p27 and increases expression of BRCA-1 and -2 tumor suppressor genes contributing to the repair mechanism of DNA.	[105]
Resveratrol	No relevant data	Helps to control chemosensitizers tumor development by modifying drug carriers, cell proliferative protein, NF-κB and STAT3 signaling pathways members and cell survival proteins.	[106]
Wheat flour (fortified)	About 100–150 μg in a 24 h	Able to lowers occurrence of neural tube defect at time of birth and increase blood cell folate concentrations.	[107]

1.4.1 RECOMBINANT PRODUCTS OF NUTRACEUTICAL APPLICATIONS

In recombinant techniques the products are produced in the laboratory using recombinant gene technology. However, the products manufactured under these techniques are not made from human blood. Recombinant products present a safer option than plasma-derived products because they prevent prospective blood-borne spread of infectious diseases. Recombination in microorganisms take place through three parasexual processes: conjugation, transduction, and transformation. Internal genetic relocations can also occur via translocatable DNA segments (insertion sequences or transposons). Conjugation requires transfer of DNA via cell-to-cell contact. Within the last few decades, however, innovative developments in biotechnology have permitted scientists to precisely modify the DNA of microorganisms, crops, and animals. Conventional methods of modifying plants and animals by careful breeding and cross breeding can take a long time. Genetic engineering permits scientists to move about selected genes from one plant or animal into another or vice versa. In the genetically modified organisms, the practice to create genetically engineered foods is different than selective breeding [92, 93]. Products obtained with recombinant techniques are summarized in Table 1.11.

1.4.2 PLANT SUPPLEMENTS AS NUTRACEUTICAL PRODUCTS

Functional food products are either whole or fortified, enriched, and enhanced foods that deliver healthiness allowances beyond the requirement of essential nutrients and grouped into several classes based on food group they exist, the diseases it inhibits or assuages and physiological consequences. Food additives are categorized as antioxidants, food preservatives, food coloring vehicles, flavoring vehicles, anti-infective agents, excipient, and other analogous properties. Excipients have not much or no medicinal significance but promote mainly the execution of the functional pharmaceutical component and sustain the characteristic, usefulness, security of the formulations comprising solvents, dilutions, suspension, and emulsifiers as well as antioxidants, preservatives, pharmacological agents, coloring mediators, flavoring mediators, carriers, excipients, ointment bases ingredients. Proteins and peptides let the growth of antibodies

TABLE 1.11 Recombinant Microorganisms, Plants, and Animals as Sources for Functional Products

Microorganisms	Targeted Enzymes	Functional Products Produced	References
		A. Recombinant Microorganisms	
Acetobacter xylinum	Beta-glucuronidase	Kombucha beverages	[108]
Escherichia coli K-12	Chymosinogen	Milk-coagulated products	[109]
Fusarium venenatum	Xylanase	Increased solubilization of bran	[110]
Aspergillus oryzae	Esterase, proteinase, glucose oxidase, laccase, pectininase, and lipase	Alcoholic beverages (Koji and Sake)	[111]
Saccharomyces cerevisiae	4-coumaroyl-CoA and stilbene synthase	Compound resveratrol	[112]
Spirulina pacifica spp.	Indoleamine 2,3-dioxygenase	Increasing hemoglobin	[113]
		B. Recombinant Plants	
Plant Sources	**Requirements**	**Recombination Gene**	**References**
Gold kiwifruit	For iron deficiency	Ascorbic acid, zeaxanthin, and carotenoids lutein.	[114]
Potato, cereals, egg	High protein food	Tuber-specific expression of a seed-based protein, *Amaranth albumin*.	[115]
Golden mustard	For vitamin-A deficiency.	Soybean *ferritin* gene.	[116]
Multivitamin from corn products	Multi-vitamins requisite	Ascorbate (rice dehydroascorbate reductase (*dhar*) cDNA), folate (*E. coli fol-E* gene encoding GTP cyclohydrolase (*GCH1*), β-carotene, vitamins, and corn phytoene synthase (*psy1*) cDNA) enzyme.	[117]
Zea maize (corn)	Vitamin-A deficiency	Bacterial genes (*crtB* and *crtI* type)	[118]
Tomato	Folate deficiency	Amino-deoxy-chorismate synthase	[119]
Golden rice	Vitamin A deficiency	Two genes from the daffodil plant and a third from a bacterium	[120]
Iron fortified rice	For iron insufficiency	Soybean *ferritin imparting* gene	[121]
		C. Recombinant Animals	
Fermented soya milk and related products.	For Ca deficiency	*Lactobacillus acidophilus* culture No. 4,962 (American type)	[122]
Ruminants	Lysozyme from human	*rHLZ* phrase vector *pBC2-HLY-NEOR*	[123]
Yogurt	Probiotic microorganisms containing food.	*Lactobacillus acidophilus* LA-5 spp., along with *Bifidobacterium lactis* Bb-12 spp.	[124]
Cows/buffalo	Lactoferrin deficiency	Recombinant human lactoferrin	[20]

and diverse fermentation products, purification developments, whereas the recombination technology shaped probable protein drugs at tolerable cost which can be beneficial in various diseases by applying through oral, transdermal, nasal, pulmonary, ocular, buccal, and rectal routes. Numerous protein pharmaceuticals are accessible for care of rheumatic inflammation, coronary artery blockage, multiple sclerosis, and persistent lymphocytic leukemia. Papain, a plant enzyme, is exceptionally effective in the impediment of atherosclerosis and diabetic heart disease. Natural sedatives, e.g., cocaine, methyl salicylate, capsaicin, piperine (alkaloid), opium have been in use from older generations.

1.4.3 VITAMINS AND VITAMINOIDS AS NUTRACEUTICAL PRODUCTS

Vitamins are categorized by way of solubility in two main groups such as fat-soluble (A, D, E, and K) and water-soluble (B-complex, C, and P) types. Vitamins are organic compounds of low molecular weight (MW), vital for life activity and required in modest quantities for crucial metabolic responses, The deficiency of these vital components gives specific disease indication. The sources of vitamins are plants and microorganisms. Nutraceuticals containing vitamins and other functional ingredients are categorized as phytochemicals, functional foods, medical foods, nutritional supplements, and designer foods and include daily use products as fortified breakfast cereals and bio-yogurts. Vitaminoids belongs to compounds having "vitamin-like" activity such as α-lipoic acid (thioctic acid) contain sulfurous carboxylic acid. It acts as a coenzyme, and works closely with the vitamins B_1, B_2, niacin, and pantothenic acid required for normal cells metabolism. Coenzyme Q_{10}, another vitaminoid referred to as ubiquinone$_{10}$, is a member of the ubiquinones. Ubiquinones widespread in the animal natural world and only Coenzyme Q_{10} is of significance to human body in metabolism. Another vitaminoid, choline (2-hydroxyethyl)-trimethyl ammonium, is a primary amino alcohol and in biological system it is present in the form of phospholipids. Soybeans, egg, and liver are rich of choline or phosphatidylcholine. Choline is associated with the maintenance of normal liver function, normal lipid-metabolism, and normal metabolism of homocysteine [7, 47, 93]. The classes of vitamins, their sources, and their role in the health wellbeing and in the control of various health related activities and their role in maintaining the health balance is summarized in Table 1.12.

TABLE 1.12 Different Vitamins and Their Health Advantages [48, 52]

Name of Vitamins	Sources	Health Advantages
Vitamin (A_1) Vitamin (A_2)	Butter, carrot, spinach, pumpkin, fish liver oil, liver cheese, papaya.	Used as antioxidants for maintaining vision, health skin, and mucous membrane.
Vitamin D	Milk, butter, wheat germ oil, egg yolk.	Suitable for bone and teeth development and helps in dietary calcium absorption.
Vitamin E	Peanut oil, wheat germ oil, cotton seed oil.	Act as antioxidant and helps in blood cell formation, nerve, tissues, and boost immune system and development of muscles, lung.
Vitamin K	Cauliflower, cabbage, tomato	Crucial for blood clotting
Vitamin B_1 (Thiamine)	Cereal and pulses	Required in neurologic work.
Vitamin B_2 (Riboflavin)	Yeast and nuts.	Helps for energy production from food.
Vitamin B_3 (pantothenic acid)	Liver, yeast, and meat.	Energy from food intake and proper brain development.
Folic acid	Green leafy vegetables.	Vital during pregnancy and helps in red blood cell expansion.
Vitamin B_5 (nicotinamide)	Egg, yeast food and milk products.	Necessary for nervous system functioning.
Vitamin B_6 (pyridoxin).	Tomato juice and banana.	Helps in essential protein production and protein energy utilization.
Biotin (Vitamin-H)	Swiss type curd.	For various metabolic functioning.

1.4.4 COMMERCIAL NUTRACEUTICALS

It is thought that new molecules are problematic to unearth and are more costly, riskier, and time consuming and uncertain. Many pharmaceutical companies have now started to manufacture nutraceutical/bioactive compounds and supplements containing bioactives having thought to a very big and flourishing market in years to come. Most of the therapeutic areas are being covered by nutraceuticals such as in digestion help and control of certain cancers, osteoporosis, blood pressure, sleeping disorders, cholesterol control, pain killers, cold, and cough, depression, and diabetes anti-arthritic. The sea foods containing omega-3 and omega-5 fatty acid contribute most promising ingredient in having recognition in human health assistances and act as functional ingredient for disease prevention along with a research base in the past few years [95]. Bio-serae laboratories of France manufactures 'Serenzo certified organic' comprising extract from citrus fruit and have antistress properties, 'Resveravine' comprises resveratrol and improve cardio-vascular safeguard and exhibit anti-aging properties [125]. Shotz Health of UK prepares 'Big Shotz' from ginseng and MEG-3 brand containing prebiotics omega-3 EPA/DHA [110]. A China company 'Guangzhou Lohas Biological Technology Co. Ltd.,' manufactures 'Ginseng Kianpi Pil' make use of Reishi mushroom extract, ginseng extracts, *Rhodiola rosea* extract as dietary supplement [126]. The 'La Casa Agrotech Private Ltd.,' an Indian company produces 'Smrutihills' from Brahmi, Mandukparnee as nerve tonic for intellect and reminiscence. SAB Herbals and Nutraceuticals compiles 'Methoxsalen Xanthotoxin Calcium Sennoside' from phytochemicals used in healing of psoriasis, eczema, and vitiligo [127]. Isha Agro Developers Pvt. Ltd. creates 'Imunohills' from amla, guduchi, gokshura, which promotes cellular and humoral immunity. Bio-Bodyfuelz Ltd. prepares 'PWR Sports' from *Sida cordifolia* extract which assists in boosting stamina and act as muscles rejuvenates and a product 'Fat Burner' from extract of cocoa beans (6% Theobromine). The 'Arctic Sea Super Omega' manufactures by 'Life-style Care' from olive oil, fish oil containing essential omega-3 fatty acids and make 'Forever absorbent C' from oranges and papayas contain bioflavonoids bioactives. Essential'z Energize, Your Health manufactures 'Muscle Juice' from whey protein isolate. Also include whey protein concentrate, calcium caseinate and egg white albumin composite. They helps in the development and nourishment of muscles. Amrutam Life Care Private Ltd. makes 'Obexi' from *Boerhaavia diffusa* which has anti-overweight loss ability [128]. The list of globally available manufactured products is mentioned in Tables 1.13–1.15.

TABLE 1.13 List of Commercial Nutraceuticals [113]

Products	Sources	Categories	Companies
Product as assurance for heart health and energy.	Kiwi, strawberries, tangerine, and melon constituents.	For heart tempo	Assure Food and Beverage
Probiotic wraps and hummus.	Ganeden BC-30 (*Bacillus coagulans*)	Promote digestive health and support immune system functioning.	Cedar's Mediterranean Foods
Vita kids bread	Fortified with MEG-3 (maternally expressed 3) brand omega-3.	Omega-3 EPA/DHA, along with folic acid, vitamin D and C.	Irish Pride Bakeries
Product as splenda essentials.	Whole wheat-based bread.	No calorie sweeteners.	McNeil Nutritionals
Green super food, antioxidant, and greens powder.	Raspberry and rice	Supports brain function, immune system and defend against free radicals.	Amazing Grass Ltd.

TABLE 1.13 *(Continued)*

Products	Sources	Categories	Companies
Phenorex	Bitter orange	Helps in fat burning	Gaspari Nutrition
Collagen and glucosamine complex.	Chicken cartilage	Slowing down of secretion of cartilage joint degrading enzymes.	Jamieson Laboratories
Fish oil+	Salmon	It is a complete brain health food.	Pacific health, Inc.
Good belly probiotic coconut water, super green lives and straight shot.	From coconut water	Probiotics for allergy-prone and vegan consumers.	Next Foods
Product collagen enhance chews	Red grapes wine	Getting age-defying assets	Res Vitále
Forgiven alcohol burner	*Rhodiola rosea* extract	Helps in breaking down of alcohol by-products.	Forgiven Bottling Group
Vectomega™	Salmon	Boosts assimilation of omega-3 FAs	Europharma
BrainShiner	*Ginkgo biloba*, ginseng	Increase memory retention time.	Multifunction supplements
Product fat-free milk containing omega-3 FA and DHA.	Algae oil and vegetarian diets.	Supporting heart, eyes, and brain, at every stage of life.	White Wave Food
Rice protein concentrate.	Brown, unpolished rice.	It is a high-protein food substitute.	Jarrow Formulas
Fuze	Fruits	Good alternative source of antioxidants and vitamins A, C, and E	Whitestone
Daytime restore and nighttime repose	Ginseng (*Ginkgo biloba*)	Helps in rest, full nap	Xigo Health
CogniSure	Proline-rich polypeptides	A beneficial aging care of the brain, in Alzheimers.	Metagenics Inc.
Trident vitality	White tea and mint and	Prevent tooth decay	Company Cadbury of North America Continent
Bite-amins brand	Mixture of vitamin-C, B-complex vitamins, garlic, bromelain, curcumin, *Echinacea*, quercetin, and capsaicin.	Act as anti-mosquito	Vita Pill LLC
Rescue water	Lemon, berry, orange constituents.	Provide necessary nutrients and support liver functioning	AriZona Beverage Company
Milk based products containing calcium.	Fortified milk products.	Reduce risk of osteoporosis in aged population.	Shamrock Farms

TABLE 1.14 Some Commercially Marketed Nutraceuticals [93]

Products	Groups	Contents	Companies
Calcirol D-3®	Calcium supplement	Calcium and vitamins	Cadilla Healthcare Limited, India
Coral calcium	Ca as supplement	Calcium and trace minerals	Nature's answer, Hauppauge, NY, USA
WelLife®	Amino acid supplement	Granulated-L-glutamine	Daesang America Inc., Hackensach, NJ, USA
Omega-woman	Immune supplement	Antioxidants, vitamins, and phytochemicals	Wassen, Surrey, U.K.
Pner-plus	Neuropathic pain supplement	Vitamin and other natural supplement	NeuroHelp, San Antonio, Texas, USA
GRD	Nutritional supplement	Proteins, vitamins, minerals, and carbohydrates	ZydusCadila Ltd. Ahmedabad, India
Proteinex	Protein supplement	Pre-digested proteins, vitamins	Pfizer ltd., Mumbai, India
Daytime rebuilds, and tight time tranquility.	Rest with full of sleep	Ginseng, Ginkgo biloba	Xigo Health
CogniSure	Amino acid supplement	Proline-rich polypeptide complex	Metagenics Inc.
Biovinca	Neurotonic	Vinpocetine	Cyvex nutrition, Irvine, CA, America

Source: Adapted from Refs. [54, 129].

TABLE 1.15 Nutraceutical Products Availability in Market [93, 129]

Brand Name	Bioactive Components	Functions/Potential Benefits
Betatene	Carotenoid compounds	Boosting the immunity
Xangold	Lutein and its esters	Serve for eye health care
Lipoec	α-lipoic acid	Have potential antioxidant activity
Generol	Phytosterol	Coronary heart disease control
Premium probiotics	Probiotic compounds	Prevent GI tract disorder
Soy life	Soyabean phytoestrogen	Improve bone health
Fenulife	Fenugreek galactomannon	Control blood sugar level

1.5 REASONS OF SWIFT OCCURRENCE OF NUTRACEUTICALS AND FUTURE DEVELOPMENTS

It is believed that various standards of nutraceuticals being used on large scale in current periods has been associated with various factors which include:

- Consumers dissatisfaction with drug price;
- Undocumented conventional health care system;
- Untested natural products manifestation in the ailment treatment and inhibition;
- Poor therapeutics alternatives which leads to chronic diseases existence;
- Desired for personalized medicine usage;
- Tendency of larger category of people eliminates aging effects;
- New focus on thwarting the medicine use, public belief that "natural supplements are virtuous," etc.

As a practice, the use of nutraceuticals before long rounds and characteristically during training period may diminish the occurrence of expressions, leads to less loss to inflamed joints during training days, inculcate measures that boost meniscal remedial particularly after surgical reparation. More than 40% of population of developed countries like USA use complementary medicinal remedies like nutraceuticals base supplements and accounts for a considerable portion of population [22].

Increasing awareness levels for health and fitness, emergence of mass media and promoting of many people to lead healthier lifestyles, eat healthy and exercise more are the factors responsible for use of natural food supplements for better lifestyle. The growing nutraceuticals market indicates that consumers are looking for minimally processed food with optimum organoleptic value with extra nutritional benefits. This development in term, is boosting in the expansion of nutraceutical market in the international market. The emergence of functional food and nutraceutical industries can take up the scenery space in the years to come. Its tremendous growth has repercussions for health care, food sector, pharmaceutical sector, and agriculture-based industry. Many scientists think that the enzymes are nutraceuticals components of major concern. Enzymes have been underemployed and are becoming a focus area in the years to come. Fermentation processes which involve microbes for new food products development also signify future possibility. Use of nutraceuticals-based products in sports medicine application is fascinating. Carotenoids and other polyphenols and antioxidants will undoubtedly play a significant role in nutraceutical-based product development. There are plenty of scientific information available which portrays the benefits of consuming natural carotenoids and its disease control properties. The two most powerful natural antioxidant carotenoids are lutein and lycopene. Lycopene offers tremendous nutraceuticals opportunities and Lutein naturally offers protection against macular disintegration an age-related syndrome. This condition occurs around 50–60 years of attainment. This condition makes the aging people a key target. The additional scientific support of lutein lies with its protective benefits against lung, cervical, and breast cancer in women/men. Tocotrienol as a bioactive component is going to be an active ingredient for nutraceutical products because of it can boost human health and have anticancer and cholesterol lowering properties. In the field of sports medicine, athletes are always looking for an edge to avoid injury and quick recovery. Natural treatment attractiveness to athletes now uses nutraceuticals as an alternative regimen for aching joints treatment. The use of nutraceuticals in these areas is future exitance which develops an opportunity for extensive study of their efficacy. The immense concern for health care cost managed care its impact result is an prospective called "*Nutraceutical world.*"

1.6 CONCLUSION

Nutraceuticals and bioactive compounds contained in diets are obtained from nearly all natural food sources and their products mainly herbal plants, but their concentration vary depending upon their source, age, part, and processing conditions and assimilation pattern. The type, quantity, duration of supply of nutraceuticals, age, and sensitivity, immune system responses, all influences on the recovery and health development. By the manipulation of dietary foods and the quantity of active bio-ingredients we can enhance the response and effectiveness and bioactivity. It has been materialized that scheduled uses of diets containing food nutraceuticals in required amount, with practicing stress reduction phenomenon, regularization of exercise, maintenance of healthy body mass index and healthy food habits can minimize health deterioration chances and reduce the adaptation risks of body ailments. The safety of microbial food is considered as the global problem associated with health-related consumer concern. It has been studied that increasing the tolerance of various microflora to the chemicals applied for the technical line disinfection and reduced sensitivity of the personal to the synthetic preservatives, there is need of an alternative source of naturally bioactive antimicrobial compounds. The information gathered in this chapter indicated that various fruit, vegetables, spices, animal food and marine sources are readily used for phytochemicals research, since these sources are having properties that they can act as strong antimicrobial activity agents against food spoilage bacteria, yeasts, and molds and food borne pathogenic microflora. They also have the potential of effective against human pathogenic flora. Along with this they are also associated with health promising benefits and have therapeutic and worth-wile results. From the technological basis various intermediates such as plant extracts, essential oils (EOs), and fruit-based products fresh juice can be used in food products formulations, as preservatives in functional food development. They can be ingredient for flavorings, coloring agents such as anthocyanins, carotenoids present in fruits, herbs, and berries. There natural essential ingredients help to maintain food products stability, can meets the consumers demand and can act as natural food additives. Nutraceuticals have been proven in providing effective health benefits mainly helps in prevention and treatment of chronic human diseases. Although the development of functional food and nutraceutical involves its safety, quality, toxicity, supplementation, and clinical trials for their long-term adverse effects on humans. It is implied that commercial nutraceuticals and their supplements need to pass through strict regulatory controls for their authenticity and beneficial impact on human health.

KEYWORDS

- **bioactive compounds**
- **essential oils**
- **nutraceuticals**
- **plant extracts**
- **sensitivity**
- **supplementation**
- **toxicity**

REFERENCES

1. *The American Heritage Medical Dictionary*, (2007). p. 47. Houghton Mifflin Company.
2. *Mosby's Dictionary of Medicine, Nursing and Health Professions*, (2013). (9th edn., p. 83). Mosby Creator.
3. Miller-Keane, & Marie, T. O., (2005). *Miller-Keane Encyclopedia and Dictionary of Medicine, Nursing and Allied Health* (7th edn., p. 62). Saunders.
4. Studdert, V. P., et al., (2011). *Saunders Comprehensive Veterinary Dictionary* (4th edn., p. 79). Elsevier Health Sciences UK.

5. Kris-Etherton, P. M., et al., (2004). Bioactive compounds in nutrition and health-research methodologies for establishing biological function: The antioxidant and anti-inflammatory effects of flavonoids on atherosclerosis. *Annual Review of Nutrition, 24*, 511–538.
6. Cammack, R., et al., (2006). *Oxford Dictionary of Biochemistry and Molecular Biology* (2nd edn., pp. 74, 75). Oxford University Press.
7. https://www.google.com/webhp (accessed on 30 September 2023).
8. *Dictionary of Food Science and Technology*, (2009). (2nd edn., pp. 47, 48). International Food Information Service (IFIS Editor).
9. Bernard, G., & Dromard, A., (2011). Book of etymology and medical terminology: Lexicon etymology (in French). *Livretd'etymologieet de Terminologiemedicale: Lexiqued'etymologie*. (pp. 1–4).
10. *Online Etymology Dictionary*, (2013). (website): http://www.etymonline.com (accessed on 14 September 2023).
11. Alain, R., (1994). *Le Robert Micro-Poche* (2nd edn., pp. 14, 126). In French.
12. Anonymous (1994). *Larousse de Poche* (p. 8). (In French), Les Editions Francaises Inc.
13. Biesalski, H. K., et al., (2009). Bioactive compounds: Definition and assessment of activity. *Nutrition, 25*(11, 12), 1202–1205.
14. Solomon, H. K., & William, W. W. (2003). *Bioactive Food Components, Encyclopedia of Food & Culture* (2nd edn., pp. 1-201). Acceptance to Food Politics, B Letter, Charles Scribner's Sons.
15. EuroFIR-BASIS – a combined composition and biological activity database for bioactive compounds in plant-based foods – ScienceDirect.
16. Nahler, G., (2013). *Dictionary of Pharmaceutical Medicine* (3rd edn., pp. 19–28). B letter, Springer-Verlag Wien.
17. Kris-Etherton, P. M., et al., (2002). Bioactive compounds in foods: Their role in the prevention of cardiovascular disease and cancer. *American Journal of Medicine, 113*(S9B), 71S–88S.
18. Bioactive compounds, (2012). *Food Bioactives Centre, University of Gothenburg* (website): http://www.chalmers.se/chem/fbac-en (accessed on 13 September 2023).
19. Juana, M., Carbonell-Capella, Francisco, J. B., María, J. E., & Ana, F., (2013). Quality parameters, bioactive compounds, and their correlation with antioxidant capacity of commercial fruit-based baby foods. *Food Science and Technology International, 20*(7).
20. Hyvonen, P., Suojala, L., Orro, T., Haaranen, J., Simola, O., Rontved, C., & Pyorala, S. P., (2006). Transgenic cows that produce recombinant human lactoferrin in milk are not protected from experimental *Escherichia coli* intramammary infection. *Infect. Immun, 74*, 6206–6212.
21. Rohit, K. V., Manisha, P., & Shubhini, A. S., (2010). Nutraceuticals: New era of medicine and health. *Asian Journal of Clinical Nutrition., 3*(1), 11–15. ISSN 0974-2441.
22. Sazawal, S., Dhingra, U., Hiremath, G., Sarkar, A., Dhingra, P., Dutta, A., Verma, P., et al., (2010). Prebiotic, and probiotic fortified milk in prevention of morbidities among children: Community-based, randomized, double-blind, controlled trial. *PLoS One, 5*, e12164.
23. Hubert, A., & Dorota, K., (2017). Food preservatives from plants. In: *Food Additives*, 70090.
24. Charles, D. J., (2013). *Antioxidant Properties of Spices, Herbs and Other Sources.* Springer, New York, USA.

25. Robbins, R. J., (2003). Phenolic acids in foods: An overview of analytical methodology. *J. of Agricultural and Food Chem.*, *51*(10), 2866–2887.
26. Jakobek, L., Šeruga, M., Medvidović-Kosanović, M., & Novak, I., (2007). Antioxidant activity and polyphenols of *Aronia* in comparison to other berry species. *Agriculturae Conspectus Scientificus*, *72*(4), 301–306.
27. Nijveldt, R. J., Van, N. E., Van, H. D. E., Boelens, P. G., Van, N. K., & Van, L. P. A., (2001). Flavonoids: A review of probable mechanisms of action and potential applications. *American J. of Clinical Nutrition*, *74*(4), 418–425.
28. Ninomiya, M., & Koketsu, M., (2013). Minor flavonoids (chalcones, flavanones, dihydrochalcones, and aurones). In: Ramawat, K. G., & Mérillon, J. M., (eds.), *Natural Products* (p. 1867). Springer, New York, USA.
29. Castañeda-Ovando, A., Pacheco-Hernández, M., Páez-Hernández, M. E., Rodríguez, J. A., Galán-Vidal, C. A., (2009). Chemical studies of anthocyanins: A review. *Food Chem.*, *113*, 859–871.
30. Ullah, N., Abbas, A., & Hider, F., (2016). Comparative study of alkaloids in selected medical plants of Mansehra district. *International J. of Applied Pharmaceutical and Biological Research*, *1*(3), 59–65.
31. Khatoon, S., (2017). A novel histological approach for identification of alkaloid bearing plants. *International J. of Botany*, *13*, 28–36.
32. De Sousa, F. H., Leite, J. A., Barbosa-Filho, J. M., De Athayde-Filho, P. F., De Oliveira, C. M. C., Moura, M. D., Ferreira, A. L., et al., (2008). Gastric and duodenal antiulcer activity of alkaloids: A review. *Molecules*, *13*(12), 3198–3223.
33. Lu, X., Tang, K., & Li, P., (2017). Plant metabolic engineering strategies for the production of pharmaceutical terpenoids. *Frontiers in Plant Science*, *7*, 1647.
34. Croteau, R., Kutchan, T. M., & Lewis, N. G., (2000). Natural products (secondary metabolites). In: Buchanan, B., Gruissem, W., & Jones, R., (eds.), *Biochemistry and Molecular Biology of Plants* (pp. 1250–1318). American Society of Plant Physiologists, New Jersey, USA.
35. Tiaz, L., & Zeiger, E., (2006). Secondary metabolites and plant defense. In: *Plant Physiology* (pp. 283–308). Sinauer Associates, Sunderland, Massachusetts, USA.
36. Moses, T., Papadopoulou, K. K., & Osbourn, A., (2014). Metabolic and functional diversity of saponins, biosynthetic intermediates and semi-synthetic derivatives. *Critical Reviews in Biochemistry and Molecular Biology*, *49*(6), 439–462.
37. Kregiel, D., Berlowska, J., Witonska, I., Antolak, H., Proestos, C., Babic, M., Babic, L., & Zhang, B., (2017). Saponin-based, biological-active surfactants from plants. In: Najja, R., (ed.), *Surfactants and Detergents*. In Tech.
38. Güçlü-Ustündağ, O., & Mazza, G., (2007). Saponins: Properties, applications, and processing. *Critical Reviews in Food Science and Nutrition*, *47*(3), 231–258.
39. Du, J. R., Long, F. Y., & Chen, C., (2014). Research progress on natural triterpenoid saponins in the chemoprevention and chemotherapy of cancer. In: Bathaie, S. Z., & Tamanoi, F., (eds.), *Natural Products and Cancer Signaling: Isoprenoids, Polyphenols and Flavonoids. The Enzymes* (Vol. 36, pp. 95–130). Elsevier Inc. Cambridge, Massachusetts, USA.
40. Bissinger, R., Modicano, P., Alzoubi, K., Honisch, S., Faggio, C., Abed, M., & Lang, F., (2014). Effect of saponin on erythrocytes. *International J. of Hematology*, *100*(1), 51–59.

41. Miyakoshi, M., Tamura, Y., Masuda, H., Mizutani, K., Tanaka, O., Ikeda, T., Ohtani, K., et al., (2000). Antiyeast steroidal saponins from *Yucca schidigera* (*Mohave yucca*), a new anti-food-deteriorating agent. *Journal of Natural Products*, *63*(3), 332-338.
42. Sucharzewska, D., Stochmal, A., & Oleszek, W., (2003). The effect of *Yucca schidigera* extract on the physical structure and on the oxidative stability of sugar-candy foam products. *Lebensmittel Wissenschaft and Technologie.*, *36*, 347–351.
43. Berlowska, J., Dudkiewicz, M., Kregiel, D., Czyzowska, A., & Witonska, I., (2015). Cell lysis induced by membrane-damaging detergent saponins from *Quillaja saponaria*. *Enzyme and Microbial Tech.*, *75, 76*, 44–48.
44. EFSA (European Food Safety Authority), (2008). Scientific opinion of then panel on dietetic products, nutrition, and allergies on a request from the EC on food-based dietary guidelines. *The EFSA J.*, 1–44.
45. German, J. B., & Walzem, R. L., (2000). The health benefits of wine. *Annual Review of Nutrition*, *20*(7), 561–593.
46. Eamonn, M. M. Q., (2010). Review on prebiotics and probiotics; modifying and mining the microbiota. *Pharmacological Research*, *61*(3), 213–218.
47. Onyeka, K. N., & Kingsley, I. U., (2020). Nutraceuticals: History, classification, and market demand. In: Egbuna, C., & Dable-Tupas, G., (eds.), *Functional Foods and Nutraceuticals* (pp. 13–22).
48. www.medicinenet.com (accessed on 13 September 2023).
49. www.cdc.gov (accessed on 13 September 2023).
50. Ehrlich, S. D., (2008). *Willow Bark, Private Practice Specializing in Complementary and Alternative Medicine*. Phoenix, AZ. Review, Veri Med Healthcare Network.
51. Ehrlich, S. D., (2009). *Peppermint (Mentha Piperita), Private Practice Specializing in Complementary and Alternative Medicine*. Phoenix, AZ. Review, Veri Med Healthcare Network.
52. www.essentialoils.co.za (accessed on 13 September 2023).
53. https://discover.hubpages.com/food/ (accessed on 30 September 2023).
54. Chauhan, B., Kumar, G., Kalam, N., & Ansari, S. H., (2013). Current concepts and prospects of herbal nutraceutical: A review. *J. Adv. Pharm. Technol. Res.*, *4*(1), 4–8.
55. www.benbest.com (accessed on 13 September 2023).
56. Zhao, J., (2007). *Nutraceuticals, Nutritional Therapy, Phytonutrients, and Phytotherapy for Improvement of Human Health: A Perspective on Plant Biotechnology Application*. Bentham Science Publishers.
57. Hord, N. G., (2008). Eukaryotic microbiotic crosstalk: Potential mechanisms for health benefits of prebiotics and probiotics. *Annu. Rev. Nutr*, *28*, 215–231.
58. Michail, S., Sylvester, F., Fuchs, G., & Issenma, R., (2006). Clinical efficacy of probiotics: Review of the evidence with focus on children, clinical practice guideline. *J. Pediatr. Gastroenterol. Nutr.*, *43*(4).
59. Holzapfel, W. H., Haberer, P., Geisen, R., Bjorkroth, J., & Schillinger, U., (2001). Taxonomy and important features of probiotic microorganisms in food and nutrition. *Am. J. Clin. Nutr*, *73*, 365S–373S.
60. Oak, S. J., & Jha, R., (2019). The effects of probiotics in lactose intolerance: A systematic review. *Crit. Rev. in Food Sci. Nutr.*, *59*(11), 1675–1683.
61. Casey, C. F., Slawson, D. C., & Neal, L. R., (2010). Vitamin D supplementation in infants, children, and adolescents. *Am. Fam*, *81*, 745–748.

62. Kumar, G. B., Srinivas, L., & Ganapathi, T. R., (2010). Iron fortification of banana by the expression of soybean ferritin. *Biol. Trace Elem. Res.*
63. Grundy, S. M., et al., (2005). Diagnosis and management of the metabolic syndrome: An american heart association/national heart, lung, and blood institute scientific statement. *Circulation, 112*, 2735–2752.
64. Hanif, R., et al., (2006). Use of vegetables as nutritional food: Role in human health. *J. Agric. Boil, Sci., 1*, 18.
65. NCI Dictionary of Cancer Terms, National Cancer Institute, Letter B, (website): https://www.cancer.gov/publications/dictionaries (accessed on 30 September 2023).
66. Dudareva, N., & Pichersky, E., (2000). Biochemical and molecular genetic aspects of floral scent. *Plant Physiology, 122*(3), 627–633.
67. Aksel, B., (2010). A brief review on bioactive compounds in plants. In: *Bioactive Compounds in Plants – Benefits and Risks for Man and Animals* (pp. 11–17). Oslo: The Norwegian Academy of Science and Letters.
68. Karlovsky, P., (2008). *Secondary Metabolites in Soil Ecology, Part II Bacterial Metabolites* (pp. 93–127). Springer.
69. Jaiganesh, R., & Sampath-Kumar, N. S., (2012). Marine bacterial sources of bioactive compounds. *Advances in Food and Nutrition Research, 65*, 389.
70. Elga, Š., et al., (2013). Sea buckthorn vegetative parts – a good source of bioactive compounds. *Proceedings of the Latvian Academy of Sciences Section B Natural, Exact, and Applied Sciences, 2*(67), 101–108.
71. Suarez-Jimenez, et al., (2012). Bioactive peptides and depsipeptides with anticancer potential: Sources from marine animals. *Marine Drugs, 10*(5), 963–986.
72. Meltzer, H. M., (2010). Bioactive compounds through food, nutraceuticals, or pills? In: Berbhoft, A., (ed.), *A Brief Review on Bioactive Compounds in Plants* (pp. 205–222). Oslo, the Norwegian Academy of Science and Letters.
73. Azmir, J., Zaidul, I. S. M., Rahman, M. M., Sharif, K. M., Mohamed, A., Sahena, F., Jahurul, M. H. A., et al., (2013). Techniques for extraction of bioactive compounds from plant materials: A review. *J. of Food Engineering, 117*, 426–436.
74. Lv, J., Huang, H., Yu, L., Whent, M., Niu, Y., Shi, H., Wang, T. T. Y., et al., (2012). Phenolic composition and nutraceutical properties of organic and conventional cinnamon and peppermint. *Food Chem., 132*(3), 1442–1450.
75. Dambolena, J. S., Zunino, M. P., López, A. G., Rubinstein, H. R., Zygadlo, J. A., Mwangi, J. W., Thoithi, G. N., et al., (2010). Essential oils composition of *Ocimum basilicum* L. and *Ocimum gratissimum* L. from Kenya and their inhibitory effects on growth and fumonisin production by *Fusarium verticillioide. Innovative Food Sci. Emerging Tech., 11*, 410–414.
76. Kwee, E. M., & Niemeyer, E. D., (2011). Variations in phenolic composition and antioxidant properties among 15 basil (*Ocimum basilicum* L.) cultivars. *Food Chem., 128*, 1044–1050.
77. Szumny, A., Figiel, A., Gutiérrez-Ortíz, A., & Carbonell-Barrachina, A. A., (2010). Composition of rosemary essential oil (*Rosmarinus officinalis*) as affected by drying method. *J. Food Engg., 97*, 253–260.
78. Fachini-Queiroz, F. C., Kummer, R., Estevão-Silva, C. F., De Barros, C. M. D., Cunha, J. M., Grespan, R., Bersani-Amado, C. A., & Cuman, R. K. N., (2012). Effects of thymol and carvacrol, constituents of *Thymus vulgaris* L. essential oil, on the inflammatory response. *Evidence-Based Complementary and Alternative Medicine*, 657026.

79. Upton, R., & Dayu, R. H., (2013). Stinging nettles leaf (*Urtica dioica* L.): Extraordinary vegetable medicine. *J. Herbal Medicine*, *3*(1), 9–38.
80. Veberic, R., Jakopic, J., Stampar, F., & Schmitzer, V., (2009). European elderberry (*Sambucus nigra* L.) rich in sugars, organic acids, anthocyanins, and selected polyphenols. *Food Chem.*, *114*, 511–515.
81. Unlu, M., Ergene, E., Unlu, G. V., Zeytinoglu, H. S., & Vural, N., (2010). Composition, antimicrobial activity, and in vitro cytotoxicity of essential oil from *Cinnamomum zeylanicum* Blume (*Lauraceae*). *Food Chem Toxic.*, *48*(11), 3274–3280.
82. Bhowmik, D., Kumar, K. P. S., Yadav, A., Srivastava, S., Paswan, S., & Dutta, A. S., (2012). Recent trends in Indian traditional herbs *Syzygium aromaticum* and its health benefits. *J. Pharmacognosy and Phytochemistry, 1*, 13–22.
83. Fu, Y., Chen, J., Li, Y. J., Zheng, Y. F., & Li, P., (2013). Antioxidant, and anti-inflammatory activities of six flavonoids separated from licorice. *Food Chem.*, *141*(2), 1063–1071.
84. Costa, A. G. V., Garcia-Diaz, D. F., Jimenez, P., & Silva, P. I., (2013). Bioactive compounds, and health benefits of exotic tropical red–black berries. *J. Functional Foods, 5*, 539–549.
85. Vattem, D. A., Ghaedian, R., & Shetty, K., (2005). Enhancing health benefits of berries through phenolic antioxidant enrichment: Focus on cranberry. *Asia Pacific J. Clinical Nutrition, 14*(2), 120–130.
86. Antolak, H., Kregiel, D., & Czyzowska, A., (2015). Adhesion of *Asaia bogorensis* to glass and polystyrene in the presence of cranberry juice. *J. Food Protection, 78*(6), 1186–1190.
87. Moze, S., Polak, T., Gasperlin, L., Koron, D., Vanzo, A., Poklar, U. N., & Abram, V., (2011). Phenolics in Slovenian bilberries (*Vaccinium myrtillus* L.) and blueberries (*Vaccinium corymbosum* L.). *J. Agri. Food Chem.*, *59*(13), 6998–7004.
88. Donno, D., Beccaro, G. L., Mellano, M. G., Cerutti, A. K., Marconi, V., & Bounous, G., (2013). Botanicals in *Ribes nigrum* bud-preparations: An analytical fingerprinting to evaluate the bioactive contribution to total phytocomplex. *Pharmaceutical Bio.*, *51*(10), 1282–1292.
89. Charlebois, D., Byers, P. L., Finn, C. E., & Thomas, A. L., (2010). Elderberry: Botany, horticulture, potential. In: Janick, J., (ed.), *Horticultural Reviews* (pp. 213–280). Wiley-Blackwell, New Jersey, USA.
90. Rop, O., Mlcek, J., Kramarova, D., & Jurikova, T., (2010). Selected cultivars of cornelian cherry (*Cornus mas* L.) as a new food source for human nutrition. *African J. Biotech.*, *8*, 1205–1210.
91. Ruchi, S., & Kumari, M., (2012). Bacterial nutraceuticals – a beneficial health supplement. *Cibtech. J. Biotechn.*, *1*(2, 3), 13–17.
92. Sajal, K. J., Partha, R., & Subhasis, C., (2021). Nutraceuticals with pharmaceutics: Its importance and their applications. *International J. Drug Development and Research, 13*(S3: 002).
93. Jagtar, S., & Shweta, S., (2012). Classification, regulatory acts and applications of nutraceuticals *International J. Pharmacy and Biological Sci.*, *2*(1), 177–187.
94. Sharma, M., & Majumdar, P. K., (2009). Occupational lifestyle diseases: An emerging issue. *Indian J. Occup. Environ. Med., 13*, 109.
95. Pandey, M., Verma, R. K., & Saraf, S. A., (2010). Nutraceuticals: New era of medicine and health. *A Review: Asian J. Pharmaceutical and Clinical Res., 3*(1), 11–15.

96. Ahmad, M. F., Ashraf, S. A., Ahmad, F. A., Ansari, J. A., & Siddiquee, M. R., (2011). Nutraceuticals, market, and its regulation. *Am. J. Food Technol.*, *6*, 342–347.
97. https://www.fao.org/common-pages/search/en/?q=probiotics (accessed on 30 September 2023).
98. Miyashita, K., Nishikawa, S., Beppu, F., Tsukui, T., Abe, M., & Hosokawa, M., (2011). The allenic carotenoid fucoxanthin, a novel marine nutraceutical from brown seaweeds. *J. Sci. Food Agric.*, *91*, 1166–1174.
99. Fasano, E., Serini, S., Piccioni, E., Innocenti, I., & Calviello, G., (2010). Chemoprevention of lung pathologies by dietary n-3 polyunsaturated fatty acids. *Curr. Med. Chem.*, *17*, 3358–3376.
100. Henrotin, Y., Lambert, C., Couchourel, D., Ripoll, C., & Chiotelli, E., (2011). Nutraceuticals: Do they represent a new era in the management of osteoarthritis? – A narrative review from the lessons taken with five products. *Os. Car.*, *19*, 1–21.
101. MacRedmond, R., Singhera, G., Attridge, S., Bahzad, M., Fava, C., Lai, Y., Hallstrand, T. S., & Dorscheid, D. R., (2010). Conjugated linoleic acid improves airway hyper-reactivity in overweight mild asthmatics. *Clin. Exp. Allergy*, *40*, 1071–1078.
102. Ganesan, P., Noda, K., Manabe, Y., Ohkubo, T., Tanaka, Y., Maoka, T., Sugawara, T., & Hirata, T., (2011). Siphonaxanthin, a marine carotenoid from green algae, effectively induces apoptosis in human leukemia (HL-60) cells. *Biochim. Biophys. Acta, 1810*, 497–503.
103. Marotta, F., Chui, D. H., Jain, S., Polimeni, A., Koike, K., Zhou, L., Lorenzetti, A., et al., (2011). Effect of a fermented nutraceutical on thioredoxin level and TNF-alpha signaling in cirrhotic patients. *J. Biol. Regul. Homeost. Agents*, *25*, 37–45.
104. Kastorini, C. M., Milionis, H. J., Esposito, K., Giugliano, D., Goudevenos, J. A., & Panagiotakos, D. B., (2011). The effect of Mediterranean diet on metabolic syndrome and its components: A meta-analysis of 50 studies and 534,906 individuals. *J. Am. Coll. Cardiol*, *57*, 1299–1313.
105. Bulathsinghala, P., Syrigos, K. N., & Saif, M. W., (2011). Role of vitamin D in the prevention of pancreatic cancer. *J. Nutr. Metab, 2010*, ID: 721365, doi: 10.1155/2010/721365.
106. Gupta, S. C., Kannappan, R., Reuter, S., Kim, J. H., & Aggarwal, B. B., (2011). Chemosensitization of tumors by resveratrol. *Ann. NY Acad. Sci, 1215*, 150–160.
107. Berry, R. J., Bailey, L., Mulinare, J., & Bower, C., (2010). Folic acid working group, fortification of flour with folic acid. *Food Nutr. Bull*, *31*, S22–S35.
108. Malbasa, R. V., Loncar, E. S., Vitas, J. S., & Canadanovic-Brunet, J. M., (2011). Influence of starter cultures on the antioxidant activity of kombucha beverage. *Food Chem*, *127*, 1727–1731.
109. El-Sohaimy, S. A., Elsayed, E., & El-Saadani, H. M. A., (2010). Cloning and in vitro-transcription of chymosin gene in *E. coli*. *The Open Nutraceuticals Journal*, *3*, 63–68.
110. Sibbesen, O., & Sorensen, J. F., (2010). *Polypeptides with Xylanase Activity*. WIPO Patent Application.
111. Ghorai, S., Banik, S. P., Verma, D., Chowdhury, S., Mukherjee, S., & Khowala, S., (2009). Fungal biotechnology in food and feed processing. *Food Res. Int.*
112. White, E., (2009). Current status of metabolically engineered resveratrol, MMG 445. *Basic Biotech* (p. 5).
113. http://www.nutraceuticalsworld.com (accessed on 13 September 2023).

114. Beck, K., Conlon, C. A., Kruger, R., Coad, J., & Stonehouse, W., (2011). Gold kiwifruit consumed with an iron-fortified breakfast cereal meal improves iron status in women with low iron stores: A 16-week randomized controlled trial. *Br. J. Nutr*, *105*, 101–109.
115. Chakraborty, S., Chakraborty, N., Agrawal, L., Ghosh, S., Narula, K., Shekhar, S., Naik, P. S., et al., (2010). Next-generation protein-rich potato expressing the seed protein gene AmA1 is a result of proteome rebalancing in transgenic tuber. *Proc. Nat. Acad. Sci.*, *107*, 17533–17538.
116. Chow, J., Klein, E. Y., & Laxminarayan, R., (2010). Cost-effectiveness of "golden mustard" for treating vitamin A deficiency in India. *PLoS One*, *10*, 5.
117. Naqvi, S., Zhu, C., Farre, G., Ramessar, K., Bassie, L., Breitenbach, J. D., Conesa, P., et al., (2009). Transgenic multivitamin corn through biofortification of endosperm with three vitamins representing three distinct metabolic pathways. *Proc. Nat. Acad. Sci. U S A*, *106*, 7762–7767.
118. Aluru, M., Xu, Y., Guo, R., Wang, Z., Li, S., White, W., Wang, K., & Rodermel, S., (2008). Generation of transgenic maize with enhanced provitamin A content. *J. Exp. Bot.*, *59*, 3551–3562.
119. De La Garza, D. R. I., Gregory, J. F., & Hanson, A. D., (2007). Folate biofortification of tomato fruit. *Proc. Nat. Acad. Sci.*, *104*, 4218–4222.
122. Cheung, A. L., Wilcox, G., Walker, K. Z., Shah, N. P., Strauss, B., Ashton, J. F., & Stojanovska, L., (2011). Fermentation of calcium-fortified soya milk does not appear to enhance acute calcium absorption in osteopenic post-menopausal women. *Br. J. Nutr.*, *105*, 282–286.
123. Yang, B., Wang, J., Tang, B., Liu, Y., Guo, C., Yang, P., Yu, T., et al., (2011). Characterization of bioactive recombinant human lysozyme expressed in milk of cloned transgenic cattle. *PLoS One*, *6*, e17593.
124. Allgeyer, L. C., Miller, M. J., & Lee, S. Y., (2010). Sensory and microbiological quality of yogurt drinks with prebiotics and probiotics. *J. Dairy Sci.*, *93*, 4471–4479.
120. https://www.rockefellerfoundation.org/blog/can-policy-change-diet-for-better-health-yes-it-can-now/ (accessed on 30 September 2023).
121. www.biotech.nature.com (accessed on 13 September 2023).
125. https://www.nutraingredients.com/Health/Healthy-ageing (accessed on 30 September 2023).
126. https://www.tradeindia.com/search.html?keyword=ginseng (accessed on 30 September 2023).
127. https://www.hotfrog.in/company/1100160320344064 (accessed on 30 September 2023).
128. https://trade.indiamart.com/offer/drugs-medicines/ (accessed on 30 September 2023).
129. https://creativecommons.org/licenses/by/4.0 (accessed on 13 September 2023).

CHAPTER 2

Extraction Technology of Bioactive Compounds and Nutraceuticals: Factors Influencing its Extraction and Bioactivity

PALAK MAHAJAN,[1] PARMJIT S. PANESAR,[2] and MANAB BANDHU BERA[2]

[1]*Department of Biotechnology Engineering and Food Technology, University Institute of Engineering (UIE), Chandigarh University, Gharuan, Mohali (Punjab), India*

[2]*Department of Food Engineering and Technology, Sant Longowal Institute of Engineering and Technology, Longowal, Sangrur, Punjab, India*

ABSTRACT

Bioactive compounds are extra-nutritional substances that are present in foods in minute quantities and are recognized for their health-promoting effects, such as immune-boosting, anti-aging, anticancer, and healing properties. These advantages have increased their demand in industries such as the food, pharmaceutical, and chemical sectors. Utilizing efficient techniques not only helps to enhance the worth of the food waste but also helps in reducing the cost of the formulated products which would otherwise be beyond the reach of the common man. The primary requirement for their extraction requires the disruption of the cell matrix, in which these compounds are present. Various extraction methods are solid-liquid extraction and liquid-liquid extraction, subcritical water extraction (SCWE), supercritical fluid extraction (SFE), microwave, and ultrasound-assisted extraction, and enzyme-assisted extraction. These methods facilitate the

Bioactive Compounds and Nutraceuticals from Dairy, Marine, and Nonconventional Sources: Extraction Technology, Analytical Techniques, and Potential Health Prospects.
Charanjit Singh Riar, Parmjit S. Panesar (Eds.)

disintegration of the cell matrix and the penetration of the solvent, hence increasing the yield of these compounds. These methods, in single or in combinations, can be used to boost the production of bioactive substances for use as nutraceuticals or functional food ingredients. This chapter thus discusses the various extraction methodologies, their related underlying mechanisms, and the factors that affect their extraction yield. In addition to these, various techniques for the characterization of these compounds are also discussed.

2.1 INTRODUCTION

Over the last few years, the curiosity of food researchers about bioactive compounds has been increased. This has eventually led to the increased interest in the new research area related to the food resources such as spices, herbs, condiments, fruits, and vegetables, and many more food items used in daily life. This has caused beneficial and disease-oriented research findings related to alkaloids, tannins, flavonoids, and phenolic compounds. Almost every plant and its parts contain these compounds, either in small or large amounts. Usually, bioactive compounds are referred to as secondary metabolites (SMs), which are not essential for normal survival but are required for various functions such as defense, competition, and signaling. The word "bioactive" is derived from the two words, "bio" meaning "life" and "active" means "energetic" [1]. These biologically active compounds can have a positive or negative effect on human health depending upon their intake concentration, which also then depend upon the type of the components, their complexes with other biological substances, exposure routes, or the bioavailability of these compounds. The numerous health benefits of the bioactive compounds such as immunity booster, antiaging, anticancerous, healing properties, antipyretic, and so on, however, after a certain dose, they may impose adverse health effects. These compounds affect the metabolic processes for better human health such as induction and inhibition of the gene expression, inhibition or induction of enzymes, antioxidant activity, and inhibition of receptor activities [2].

Nutraceuticals are medicinal food materials that have a significant effect on increasing and maintaining human health, improving immunity, maintain wellbeing, and preventing specific diseases. Nowadays, there is increasing demand for the development of functional and nutraceutical food products in the food industry. This new specific class of food products possesses huge consideration and acceptance in the food-based market because of the increased consumer interest in healthy foods. Both food and pharmaceutical domains possess a common interest in the development of natural bioactive compounds (NBCs) that can be further employed in the application of functional foods, nutraceuticals, and drugs. Naturally occurring bioactive compounds are mostly in their natural matrices without further alterations. The main sources for the extraction of bioactive compounds are plant tissues, microorganisms, algae, and microalgae, and food waste. Natural compounds from plant sources can be categorized into primary and SMs. The former includes sugars, fatty acids, amino acids, nucleic acids, and all other constituents which are ubiquitous for the growth and development of all plants. The latter performs various assorted functions throughout the lifecycle of the plant. The SMs can be divided into five groups depending upon their origin of biosynthesis. These are alkaloids, phenylpropanoids, isoprenoids, flavonoids, and polyketides. Some of the useful plant SMs are ellagic acid, carotenoids, terpenoids, farnesene, vinblastine, and many others and are commonly used for the production and development of pharmaceuticals and nutraceuticals [3]. In the last 25 years, about 60–70% of the newly approved drugs for the treatment of cancer and infectious diseases were derived from naturally bioactive compounds [4].

Microorganisms use the biological system for producing important molecules [5]. About 23,000 naturally occurring bioactive compounds of microbial origin have been identified and have different functions such as protecting against environmental stress, lethal photooxidation, and acting as cofactors in the enzymatic reactions [6]. There are several fungal varieties such as Streptomyces, Penicillium, and Aspergillus, that are exploited to produce bioactive compounds such as enzymes, organic acids, and antibiotics [7]. These compounds also act as flavor enhancers, texturizers, thickeners, surfactants, nutritional supplements, and preservatives. The basic reason for using microbes to produce bioactive compounds is their simplicity for obtaining high yields and the opportunity for environmental and genetic manipulations [8]. Microalgae are the main constituents of many ecosystems such as marine, freshwater environments, and desert sands. About 30,000 distinct microalgae species might inhabit the earth, and over 15,000 novel compounds have been extracted from them [9]. Algae are the natural source for the extraction of new compounds and possess antimicrobial, antioxidant, and antiviral activities [10]. This organism resides in extreme conditions and produces a large variety of bioactive SMs which take part in the natural defense mechanism [9].

The cheapest source for bioactive extraction is food waste, especially the waste generated from fruits and vegetables. Table 2.1 shows the various bioactive compounds extracted from the food wastes. Fruits and vegetables contain polyphenols and carotenoids that have antioxidant activity and reduce the risk of various cancer types. Bioactive phytochemicals such as carotenes, tocopherol, terpenes, sterols, and polyphenols extracted from the tomato waste have antioxidant activities. Animal-based waste comprises waste from the seafood, poultry, and dairy industry. The recovered biomolecules and their derived by-products can be utilized to produce functional foods in food manufacturing and/or pharmaceutical preparations. Besides these applications, these natural antioxidants can be used as an additive in food formulations to extend their shelf life. These compounds are also effective in the prevention of the degenerative process via their incorporation in functional foods and nutraceuticals.

TABLE 2.1 Bioactive Compounds Present in Various Food Waste

Sources	Residue	Bioactive Compounds	References
Avocado	Seeds and peel	Cyanidin 3-glucoside, epicatechin, gallic acid, homogentisic acid, chlorogenic acid.	[11]
Wheat	Germs and bran	Antioxidants, phenolic acids	[12]
Guava	Seeds and skin	Galangin, gallic acid, cyaniding 3-glucoside, kaempferol	[11]
Mango	Kernel	Ellagic acid, condensed tannins, gallates, gallic acid	[13, 14]
Pomegranate	Pericarp and peel	Gallic acid, cyanidine-3,5-diglucoside, cyanidin-3-diglucoside	[15]
Cucumber	Peel	Phellandrene, chlorophyll, caryophyllene	[16]
Rice	Bran	Bran oil, γ-oryzanol	[17]
Apple	Pomace and peel	Catechin, quercetin, procyanidins, epicatechin, phloretin glycoside	[18]
Grapes	Skin and seed	Caffeic acid, vanillic acid, resveratrol, coumaric acid, quercetin, gallic acid, ferulic acid	[19]
Tchi	Seeds, pericarp	Gallic acid, epicatechin-3-gallate, cyanidine-3-glucoside, malvidin-3-glucoside	[20]
Carrot	Peel	Beta-carotene, phenols	[21]
Barley	Bran	β-glucan	[22]

2.2 TYPES OF BIOACTIVE COMPOUNDS

There are various types of bioactive compounds based on (i) their constituents such as carotenoids, tocopherols, organo-sulfur compounds, and phytosterols; (ii) phobities such as hydrophilic and lipophilic; (iii) presence in plant parts vary with the species; and (iv) their concentration range in the plants [23]. Phenolic compounds are the SMs derived either from the pentose phosphate pathway or phenylpropanoid pathways in plants [24]. Half of these compounds are flavonoids, are responsible for the large spectrum of defense mechanisms against harmful ultraviolet rays, and act as antioxidant, anticarcinogenic, and antimutagenic agents [25]. The most important phenolic compounds can be divided into several classes such as phenolic acids, tannins, flavonoids, lignin, and stilbenes. The structure of the phenolic compounds has a key role in determining their radical scavenging activity and is referred to as structure-activity relationships (SAR). Their antioxidant activity depends upon the number and position of the hydroxyl groups concerning the carboxyl functional groups. For instance, the monohydroxy benzoic acids with -OH moiety at the para and ortho position show no antioxidant activity, whereas the meta hydroxybenzoic acid has antioxidant properties [26].

Flavonoids are the most ubiquitously found phenol in nature. Besides providing color and taste to food products, flavonoids prevent fat oxidation and maintain the quality and quantity of the vitamins and useful enzymes in the food [27]. Its structure contains two aromatic rings, A and B, joined via three carbon heterocyclic rings, C [28]. The structural changes in these rings due to changes in their oxidation states further categorize the flavonoids into flavonols, flavones, isoflavones, flavonols, and anthocyanins (Figure 2.1). Most common flavonoids include quercetin (flavonol present in onion, apple, and broccoli); catechin (green tea); cyanidin glycoside (anthocyanin in berries); naringenin (flavanone in grapefruit).

Based on its structure and properties, anthoxanthin is divided into two groups: flavones and flavonols. Flavones perform a variety of functions that makes the plants compatible for the terrestrial adaptation such as pathogen resistance, protection against ultraviolet rays, oxidative stress, and herbivory and allelopathy [27, 29, 30]. Also, these flavones possess biological impacts on humans and animals by providing important nutritional value. On the other hand, flavonols are the result of the modification of flavones and are referred to as 3-hydroxyflavones. A large number of flavonols are present in the form of O-glycosides and rarely in the form

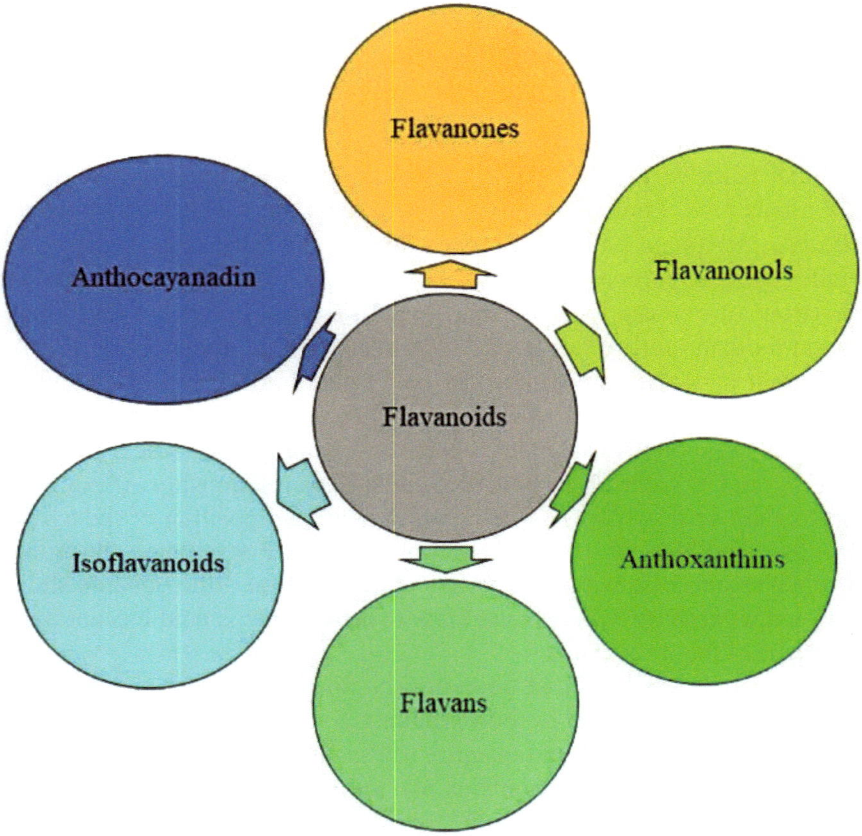

FIGURE 2.1 Classification of flavonoids.

of C-glycosides. Anthocyanidins are water-soluble and impart purple, blue, red, or orange color to the plants and their parts. They are polymethoxy and polyhydroxyglycosides that are derived from 2-phenylbenzopyrylium [31]. These compounds act as phytoalexins, antioxidants, and antibacterial agents and are referred useful in providing resistance to the plants against biotic stress.

Flavanols and their polymerized products are known to be the functional ingredients in beverages, herbal remedies, supplements, whole, and processed food products. Their presence affects food quality and their related parameters such as sourness, bitterness, astringency, aroma, sweetness, and color formation [32]. These compounds result in the astringency of the food products and beverages by interacting strongly with the proteins [33]. The metal chelating property of these compounds makes them significant in limiting the growth of the invasive microbes by resulting in the depletion of the essential minerals, whereas the presence of the epicatechins in the plants' tissues possesses antifungal properties. Tannins, high molecular weight (MW) compounds, can form complexes with proteins, alkaloids, and proteins. Tannins are further categorized into condensed and hydrolyzable tannins. The condensed tannins, also referred to as proanthocyanidins, are the polymers of the polyhydroxy flavanol monomers. These compounds are commonly found in chickpea, lentils, cocoa, tea, coffee, chocolate, grapes, peach, walnuts, blackberry, strawberries, apple, pomegranate, raspberries, etc. The structure of the stilbenes contains a 1,2-diphenylethylene with the hydroxyl group present at the aromatic ring. The most important well-known compound of this class is trans-resveratrol, which has a trihydroxy stilbene skeleton. The most important sources of these compounds are soy, wine, grapes, peanuts, and its products [34]. Lignin is the most important metabolite that is produced in plants by the phenylalanine/tyrosine metabolism. It is the second most prolific biopolymer, contributing about 30% of organic carbon in the biosphere [35]. Lignin enhances the cell wall rigidity of the plants, mineral transport, hydrophobicity, response against various abiotic stress, and resistance towards pests and pathogens. Lignin and its derivatives have been found to possess prevention against various human diseases as it works as antidiabetic, antiviral, anticoagulant, antiobesity, and antiemphysema agents [36–39].

2.3 EXTRACTION METHODS OF THE BIOACTIVE COMPOUNDS

The key steps involved in the analysis of bioactive constituents involve their extraction from the raw material and are strongly influenced by the strength and weakness of the extraction techniques employed. According to the reports of WHO, about 20,000 medicinal plants are present in 91 countries with mega biodiversity. The steps involved in the utilization of these compounds from the raw materials involve their extraction, isolation, and characterizations, toxicological estimation, and their clinical evaluation. A summary of common approaches applied in the extraction and characterization of the bioactive compounds is shown in Figure 2.2.

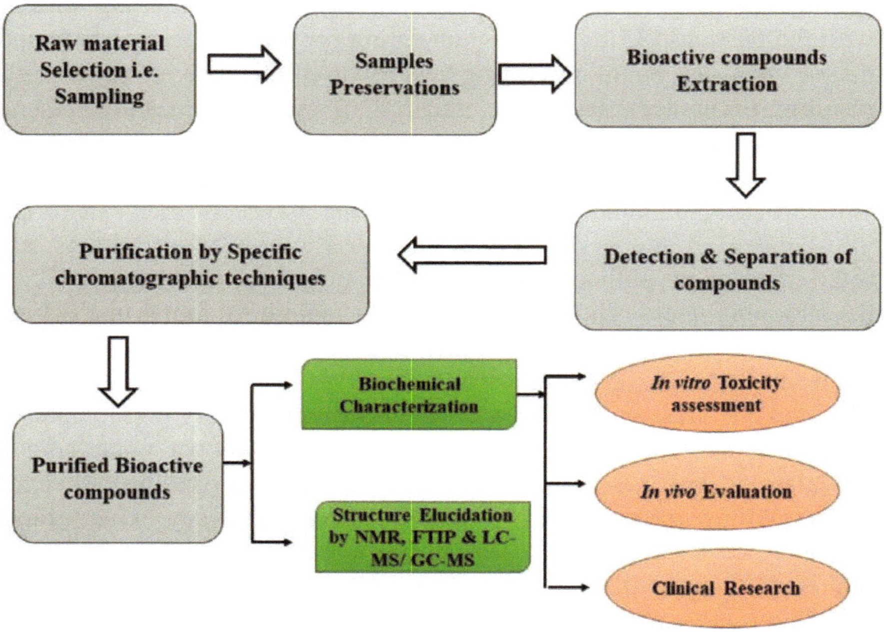

FIGURE 2.2 General steps involved in the extraction and characterization of the bioactive compounds.

2.3.1 EXTRACTION METHODS

Extraction is the most important stage in the analysis of these valuable plant compounds. The common steps involved are prewashing, drying, homogenizing or grinding, increasing the sample contact with the solvent. The choice and type of solvent used for the extraction of the compounds depend upon the unambiguous nature of the bioactive compounds being extracted [40]. For the isolation or extraction of the hydrophilic compounds, methanol, ethanol, ethyl acetate is used while for the hydrophobic ones, dichloromethane or its mixture with the methanol are used. Also, sometimes, the extraction of chlorophyll is done using hexane [41]. Besides, the extraction solvents, multiple methods such as Soxhlet extraction, heating under reflux, and sonification are used commonly for the extraction of plant compounds. These plant extracts can also be prepared by the maceration or percolation of the fresh or dried crushed plant materials in the water or organic solvents. Bioactive compounds present in the raw agricultural material or their waste can be recovered employing various extraction techniques. Based on the research literature survey, the extraction techniques employ solvent extraction (SE), ultrasound extraction (USE), microwave extraction (MWE), use of enzymes, SCWE, and SFE (Figure 2.3). These extraction techniques are discussed in the following section, taking reference from recent research studies.

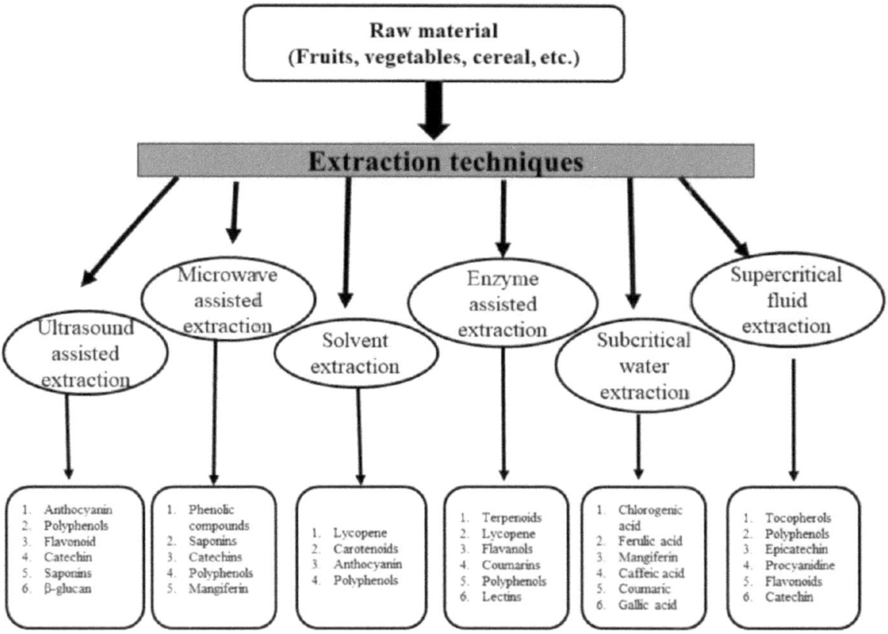

FIGURE 2.3 Different extraction techniques for extracting the bioactive compounds from the raw materials.

2.3.1.1 SOLVENT EXTRACTION (SE) TECHNIQUE

In this type of extraction, the suitable size of the raw materials is exposed to the different or suitable organic solvents. These solvents then take up or solubilize the like compounds of interest, and also some of the flavoring and coloring agents (anthocyanins) that have anti-inflammatory and anticancerous properties. Usually, the sample under study is centrifuged and filtered to remove the solid residues and the obtained extract can be used as an additive, food supplement, and/or in the preparation of functional foods. This technique is much more beneficial than any other technique because of its low cost and ease of operation. However, the main drawback lies in the use of the harmful solvents that further require additional heating for the evaporation and/or condensation for their recovery, this also leads to extended time besides toxicity. Also, there are chances of degradation of these compounds at high temperatures, and these drawbacks can be improved by employing other techniques, such as MWE, USE, or SFE, to yield better yield. Extraction of lycopene and β-carotene from the tomato pomace and seeds was carried out using ethanol. The use of 50% acetone yields the highest content of the polyphenols from the powders of *Parkia speciosa* pod than the other solvents such as ethanol, methanol, hexane, and ethyl acetate [42]. Out of the various solvent used for the extraction of the bioactive compounds, ethanol was found most efficient as it yielded the highest extraction rates and hexane showed the lowest yield. Further, the yield of the extracted compounds increased with the increase in the extraction time.

2.3.1.1.1 Factors Affecting the SE

The Soxhlet extraction method is the simplest extraction technique; thus, its extraction is mainly based on the type of solvent used. Different solvents yield different extracts and their composition [43].

2.3.1.1.2 Type of Solvent

The most widely used solvent for the Soxhlet extraction is hexane for the extraction of the targeted bioactive compound and nutraceutical. The use of other solvents such as ethanol, water, hydrocarbons, and isopropanol has increased health and environmental safety concerns. However, these other solvents result mostly in less recovery because of their decreased molecular affinity among the solute and solvent. Sometimes a co-solvent is required to enhance the polarity of the liquid phase. For example, a mixture of isopropanol and hexane has been found to enhance the extraction yield and kinetics [44].

2.3.1.1.3 Characteristics of Matrix

The matrix characteristics, such as particle size, can strongly influence the extraction methodology because its related internal diffusion may be the limiting step for the extraction process. For extracting total fat from the oleaginous seeds, 12 hrs. extraction resulted in 99% extraction efficiency with the particle size of 2.0 mm. However, the same efficiency can be achieved within 2 hr if the particle size was reduced to 0.4 mm [45].

2.3.1.1.4 Operating Conditions

The solvent used is usually recovered via evaporation during the Soxhlet extraction. The temperature used for the extraction and evaporation possesses a noteworthy effect on the final product quality. The d-limonene extracted rice bran oil was found to have a high slightly darker color as compared to the hexane-based extracted oil. This might be probably due to the higher evaporation and extraction temperature used in the case of d-limonene SE. However, the very high boiling temperature, if used for the recovery of solvent, can be reduced by using a vacuum or via membrane separation for the solvent recovery [46].

2.3.1.2 SUPERCRITICAL FLUID EXTRACTION (SFE)

The supercritical extraction method, an environment-friendly method, was described as an important substitute to conventional extraction for the extraction of biologically active compounds from natural sources like plants, food raw materials and their by-products, and (micro)algae [47]. SFE using carbon dioxide (SC-CO_2) is an attractive alternative to the use of organic solvents because of the non-explosive, inexpensive, and non-toxic characteristics of this technique. It also has the capability to solubilize the lipophilic compounds and thus can be effortlessly detached from the ultimate products [47]. Some conditions optimized for the SFE of compounds from the plant sources are shown in the Table 2.2.

For this technique, the raw material is placed in the extraction vessel supplied with pressure and temperature regulators to sustain the essential condition. The vessel is filled by the fluid via a pump. Fluids and the dissolved compounds are then transported to the separators, and the products are then collected in the lower part of the separator through a tap. Finally, the fluid is then regenerated, cycled, or released into the environment. The choice and selection of the supercritical fluids is an important factor for the proper performance of the process, and a variety of solvents can be used in this process. SC-CO_2 modified with ethanol gave a better yield of the naringin, a flavonoid, extracted from the citrus waste, as compared to the pure SC-CO_2 [48]. Polyphenols and procyanidins were extracted from the grade seeds using SFE using methanol as a modifier. Methanol-modified CO_2 was found to release more than 79% catechin and epicatechin [49]. Apart from these compounds, antioxidant-rich oil was extracted from the wheat bran using SC-CO_2 and Soxhlet extraction. Oil extracted from the SC-CO_2 showed higher resistance towards oxidative rancidity and radical scavenging activity [50].

TABLE 2.2 Extraction of Bioactive Compounds Using Supercritical Fluid Extraction

Sources	Bioactive Compounds	Co-Solvent	Temperature (°C)	Pressure (Bar)	References
Apricot pomace	Carotenoids	Dimethoxy propane	39.85–59.85	304–507	[51]
Citrus peel	Naringin	Ethanol	58.6	95	[48]
Guava seeds	Phenolic compound	Ethyl acetate	40–60	100–300	[52]
Tomato skin	Carotenoids	Ethanol	75	350	[53]
Green tea leaves	Catechins	Ethanol	60	310	[54]
Spearmint leaves	Flavonoids	Ethanol	40–60	100–300	[55]
Grape peel	Antioxidants, phenolic, anthocyanins	Ethanol	37–46	137–167	[56]
Blueberry residue	Anthocyanins	–	40	150–300	[57]
Tomato waste	Trans-lycopene	–	40–80	200–300	[58]

2.3.1.2.1 *Factors Affecting SFE*

Apart from the type of solvent used, the ability of this process to remove the solute from the solid matrix hinge upon numerous factors such as its location in the matrix, porosity of the extractor bed, solubility of the solute in the supercritical fluid, and its interaction with the solid matrix [59].

1. **Location of the Solute in the Matrix:** Solutes can be located over the cell surface whereas most of them are in the intracellular spacings. A pre-treatment method such as milling and comminution is often required to ease the extraction process. For example, the volatile oils are mostly located inside the plant tissues in a special secretory structure, this makes their contact difficult. In these cases, the extraction efficiency can be improved by lowering the mass transfer resistance because of the cell structures [59].

2. **Porosity and Particle Size of the Solid Material:** These factors influence the mass transfer rate of the method. To increase the extraction rate, the matrix needs to be ground well, as this will help in enhancing the surface area. However, very small particles should be exempted as they can block the extractor bed and cause resistance towards the internal mass transfer. This thus lowers the extraction rate. Bed porosity also influences the mass transfer of the extraction process. Generally, an increase in the porosity of the matrix promotes efficient and rapid extraction [59].
3. **Solvent Flow Rate:** A high solvent flow rate generally enhances the extraction capacity. However, the use of an extremely high solvent flow rate typically depresses the yield because of insufficient interaction time between the solvent and solute [60]. Also, a high solvent flow rate enhances the capital and operational costs, which is an important fact to be considered from an industrial viewpoint.
4. **Pressure and Temperature:** The selection of these operating conditions depends on the type of the target compounds. For example, the EOs are mainly extracted at pressure 10–25 MPa and temperature 303–333 K. An increase in temperature and pressure is essential for the extraction of heavy compounds. Temperature and pressure affect the solvation tendency of the solutes depending upon the solvent density. Generally, at constant pressure, a temperature increase lowers the solvent density and its solvation power, while at a constant temperature, the rise in the pressure upsurges the solvation power. It states that the solubility of the solutes is directly proportional to the solvent density in the SFE [59].

2.3.1.3 SUBCRITICAL WATER EXTRACTION (SCWE)

This extraction technique is also regarded as a growing alternative technology used for the extraction of phenolic compounds from different food sources. Besides using carbon dioxide, water is also used as the extraction medium. Subcritical water is called water at a temperature between 100 and 374°C and high pressure that can maintain its liquid state (especially below the 22 MPa critical pressure). SCWE offers various advantages over the other conventional extraction methods such as low solvent cost, short extraction time, environmentally friendly, and higher quality of extraction [61]. These advantages make this technique a most promising engineering approach for the extraction of various biologically active compounds [62].

Phenolic compounds were extracted from the mango peels using SCWE and it was concluded that extraction was higher in the case of SCWE than by using conventional Soxhlet extraction [63]. The technique was also used for extracting polyphenols from *Terminalia chebula Retz.* Fruits and the contents of the extracted ellagic and gallic acids were found to increase with the increase in the temperature of the subcritical water to 180°C [64]. Eight phenolic compounds were extracted from the potato peels via SCWE. These phenolic compounds were recovered most (81.83 mg/100 g sample) at 180°C and at the 30 min extraction time. At 180°C, the main phenols extracted from the potato peels were gallic (29.56 mg/100 g) and chlorogenic acid (14.59 mg/100 g) [65]. The study also concluded that subcritical water at 6 MPa, 160–180°C, and 60 min can be a good option to extract phenols from the potato peel rather than by using conventional organic solvents. Coffee waste residues were used for the extraction of the total phenolic compounds using the SCWE, with the highest yield (26.64 mgGAE/g coffee powder) at 22.5 MPa and 200°C [66]. The green technology SCWE was used for extracting phenolic compounds from the petals of *Crocus sativus*. The optimized conditions (159°C, 54 min., and 36 ml/g water: solid) were used for the extraction of total flavonol content and total phenolics [67].

2.3.1.3.1 Factors Affecting SCWE

To quantify the bioactive compounds and to characterize them, it is very important to select an appropriate and effective method. Various factors can affect the extraction as matrix properties, solvent: matrix, type of solvent used, processing conditions such as pressure, time, and temperature.

1. **Temperature:** It is an important factor that influences the extraction of bioactive compounds, extraction efficiency, and selectivity. It not only affects the physicochemical properties of the water used but also leads to the degradation of the thermally labile analytes [68]. During the process, the temperature is usually kept higher than the boiling point (b.p.) of the fluid used for the study. The subcritical water at high temperatures possesses low viscosity, high diffusivity, and low surface tension. High temperature alters the nature of subcritical water and changes from polar to non-polar with the increase in temperature. This promotes the dissolution of fewer polar compounds in the water [69]. However, the use of high temperature causes the degradation

of the bioactive compounds and increases the chances for oxidation and hydrolysis to occur. For instance, during the extraction of terpene from the oregano and basil leaves, the terpene content decreased between 200 and 250°C [70]. However, the use of a high temperature of more than 300°C can cause solubility of some of the non-polar compounds which can thereby improve their efficiency. So, it is important to wisely select the suitable temperature depending upon the bioactive compound to be extracted.

2. **Pressure:** By changing and adjusting the pressure levels, the influence of changing the water phases can also be achieved. Mostly, medium pressure such as 15 bar (200°C) and 85 bar (300°C) is required for maintaining the liquid phase of water. Unlike temperature, change in pressure does not show any significant effect on the recovery of the bioactive compounds using subcritical water [71]. Also, some studies have shown that the change in pressure or using different pressure didn't improve their recovery during the SWE process [72].
3. **Particle Size:** This is also an important factor that possesses an impact on the extraction process [73]. Generally, the extraction yield is higher, and the time requirement is low for the small particle size, while for large particle sizes, the extraction efficiency is low, and the time required is also extended [74]. This is because the smaller particle size increases the surface contact area between the solvent or extractant and the sample matrix, which causes a lowering of the extraction time and an increment in the extraction efficiency.
4. **Solvent Flow Rate:** The kinetics of the extraction of compounds depends on the extent of their solubility in the solvent. An increase in the flow rate can improve the extraction rate of the compounds [75]. A high flow rate not only improves the extraction efficiency but also lessens the residence time of the compounds in the high-temperature water. Also, a high rate of extractant may cause excessive dilution of extract, which requires an additional step, i.e., concentration at the end of the extraction process.

2.3.1.4 ENZYME ASSISTED EXTRACTION

Enzymes are used widely for extracting bioactive compounds from food waste. Plant tissues are the main source of antioxidant extraction.

Polysaccharides such as pectin, cellulose, and hemicellulose are present in the plant cell, and these components inhibit the release and loss of intracellular substances. Enzymes such as pectinase and cellulase help in the degradation of the cell wall and depolymerize the cell wall polysaccharides thus facilitating the release of the attached compounds [76]. Enzyme-assisted extraction (EAE) is regarded as the most eco-friendly method for extracting bioactive components and oils [77]. Enzyme assisted extraction was used for extracting lycopene from the tomato peels and it was concluded that the use of a mixture of enzymes possessing pectinolytic and cellulolytic activities can greatly improve lycopene recovery [78]. Lycopene-rich extracts from the tomato waste were prepared at 50°C using enzyme cellulase (20 U/g) and pectinase (30 U/g) [79]. Enzyme-based extraction has been regarded as a green technology that can be a promising substitute for conventional SEs. The methodology is based upon the enzymes' ability to catalyze the reaction in the aqueous solutions under mild processing conditions [80].

2.3.1.5 MICROWAVE-ASSISTED EXTRACTION (MAE)

This extraction technique combines both microwaves and conventional SE methods. The method possesses various advantages such as shorter time duration, less solvent requirement, high extraction rates, and yields, and low cost as compared to the other traditional methods. One of the main advantages of this method over the widely used Soxhlet extraction and ultrasound-based extraction is that it can be used for extracting plant metabolites with shorter extraction times [81]. Saponins were extracted from the chickpea using this methodology and it was reported that this method is superior to the Soxhlet extraction in terms of the quantity of the solvent used and time and energy expended [82]. Mangiferin was extracted from the *Curcuma amada* using MAE and ethanol as a solvent. The quantity of mangiferin was found to increase till 500 W, however, further increase in the microwave power decreased the content. An optimal mangiferin yield, 41 μg/ml, was found at 500 W microwave power and an extraction time of 15.32 s [83]. MAE was optimized using the response surface method (RSM) for the extraction of basil polyphenols. Optimum extraction conditions were 442 W microwave power, 50%ethanol, and 15 min extraction time, and the extracted basil liquid extract contained 0.85 g catechin Equiv./100 g total flavonoids and 4.29 g GAE/100 g total polyphenols [84]. The studies concluded that the MAE has numerous benefits over the other methodology such as high extraction efficiency, less extraction time, low labor cost, and high extraction selectivity.

All of these make this method a favorable option highly recommended for the extraction of biologically active components [85].

2.3.1.5.1 Factors Affecting MAE

Several researchers have been conducted for optimizing the factors to achieve an optimal yield of the bioactive compounds. Some of the important factors are solvent-to-feed ratio, type of solvent, the effect of stirring, extraction temperature, irradiation time, and microwave energy density. These factors and operating parameters are essential in determining the efficiency of the extraction process [86]. Thus, their understanding is of paramount importance.

1. **Solvent Type and Solvent to Feed Ratio:** The selection of adequate solvent will result in an efficient extraction process. The solubility of the compounds, their mass transfer kinetics, and solvent penetration are the inevitable parameters [87]. The selection of the extraction solvent depends upon the solvent's capacity to absorb the microwave energy [88]. Solvents such as water, ethanol, and methanol are considered excellent microwave-absorbing solvents that can have sufficient polarity required for heating via microwave power. Also, the addition of water to the polar solvent even in small quantities can result in higher water diffusion into the cell matrix, which can lead to effective heating and thereby facilitate the transfer of the bioactive compounds into the solvent used at higher mass transfer rates [87, 88]. In addition, the volume of the solvent used is also an important factor, as a large solvent volume needs more energy and time to undergo condensation during the purification process.
2. **Stirring Effect:** The stirring can enhance the mass transfer rate in the solvent phase because upon stirring the equilibrium between the vapor and aqueous phase can be achieved more quickly. Further, the application of the stirrer in the process can accelerate the extraction efficiency and rate by increasing the dissolution and desorption of the bioactive compounds [89]. Thorough stirring can minimize the drawback, which would otherwise result in using a solvent-to-solid ratio (SSR). This could also minimize the mass transfer barrier [90].
3. **Time of Irradiation:** It is an important factor that affects the MAE. Less time consumption is one of the important advantages of the

MAE. The usual time requirement ranges from a few minutes to half an hour, based upon the type of the matrix, however, the irradiation time is usually affected by the dielectric property of the solvent [90]. Solvents may heat up quickly upon longer exposure, which can cause degradation of the thermolabile compounds present in the extracts [88]. An increase in the irradiation time improves the recovery yield, but it can decline upon the prolonged irradiation time [91].

4. **Microwave Power and Temperature:** Both these factors affect the extraction yield. High microwave power increases the temperature, causing an increase in the extraction yield. An increase in the temperature results in an increase in the solvent power due to a reduction in surface tension and viscosity. This increases the solvent interaction to solubilize the solutes and improve the matrix penetration [90]. Microwave power is associated with the sample quantity and required extraction time. The power offers localized heating in the solid matrix and acts as a driving force to destroy the plant matrix such that the solute can diffuse and dissolve in the solvent. An increment in the microwave power lowers the shorter extraction time and improves the extraction yield [90, 92]. While high microwave power can lower the extraction yield causing degradation of the temperature-sensitive compounds present in the food matrix [93]. Thus, it becomes important to select an appropriate microwave power to lower the extraction time needed to acquire a required set temperature and avoid the bumping phenomenon [90].

5. **Microwave Energy Density:** The three heating operational modes used in evaluating the performance of the microwave-assisted extraction (MAE) are constant temperature heating mode, constant power heating mode, and intermittent heating mode [94]. Microwave power alone does not provide a satisfactory description of how the energy is absorbed in the extraction medium. Terigar et al. [92] concluded that the constant power heating mode is regarded as the standard practice for the extraction of temperature-sensitive active compounds. The study related to the interrelationship of the microwave energy density and the extraction yield suggests that the microwave energy density is an important parameter that affects the extraction efficiency of the bioactive compounds in the case of MAE.

2.3.1.6 ULTRASOUND-ASSISTED EXTRACTION (UAE)

Extraction based on ultrasounds has been considered the simplest and most effective technique in contrast to the traditional/conventional extraction methods. The technique persuades diffusion of the solvents in the cellular materials, which thereby improves the mass transfer. It disrupts the cell wall and thus facilitates the release of biologically active compounds. The basic mechanism involved in the UAE is acoustic cavitation. The collapse of the cavitation bubbles and the sound waves induces fragmentation, pore formation, increased absorption, and swelling. The bubbles, upon collapsing, generate shockwaves and accelerate the collision among the particles. This causes fragmentation and thereby increased solubilization because of a decrease in the particle size and increase in surface area. The sound waves cause damage to specific locations and are termed erosion. The phenomenon increases the contact of the matrix with the solvent and thus increases the extraction efficiency and yield. The extraction yield of these compounds is strongly influenced by the ultrasound frequency, based on the nature of the plant material that has to be extracted [12]. Anthocyanins and phenolic compounds were extracted using ultrasounds from the grape peels [56]. Stilbenes were extracted from the grape canes using this ultrasound-assisted technique at the 75°C and 60% ethanol concentration for about 10 min [95]. The study also concluded that the by-products of the grape cane are the potential source for the extraction of the bioactive compounds. Water-soluble polysaccharides were extracted from the dried and millet by-products of the *Agaricus bisporus*. B-glucans were extracted from the mushroom by-products and the yield was found highest (4.7%) at the conditions as 15 min extraction time and 80%ethanol concentration [96].

2.3.1.6.1 Factors Affecting UAE

Ultrasound waves lead to the degradation or disruption of the matrix tissues through the physical forces developed through acoustic cavitation. This helps in the release of the extractable bioactive compounds into the solvent in less time through the enhancement of mass transport. The most important variables or parameters associated with the UAE are frequency, temperature, solvent type, power, and liquid: solid ratio [97].

1. **Ultrasonic Power:** It is expressed as amplitude percentage in the range of 0–100% where 100% amplitude refers to the rated power of

that equipment [98] and power density (W/ml) is referred to as power dissipated per unit extraction medium volume. The ultrasonic power applied for the bioactive compounds varies with the compound to be extracted and the type of the sild matrix opted for the extraction and the range varies from 20–700 W [99, 100]. Yield and efficiency of the bioactive compounds increase with the rise in the power and decline after reaching a peak [101–104]. This is due to the increase in the violent cavitation bubble collapse with the rise in power. The size of the bubble is directly related to the power of the ultrasonic, and the size increases its impact on the implosion also intensifies. This causes fragmentation, the formation of pores, and the mixing of the tissues, which thereby causes an increase in the extraction yield [104]. Hydrodynamic forces also increase with the increase in power, leading to tissue disruption [99]. However, the high intensity of the ultrasound may degrade the bioactive compounds, thereby decreasing their yield [97]. The yield of the bioactive compound increased with the increase in the power when the processing time was low, whereas, in the case of long duration, the yield lowered with the increase in power [105].

2. **Frequency of Ultrasonic:** UAE frequency lies between 20 and 120 kHz for the extraction of bioactive compounds. Most of the studies focus on the constant frequency, while only a few focus on the effect of the varying frequency on the extraction quality and yield [97]. Low-frequency, high-intensity UAE generates the strong mechanical and shear force required in the extraction process. On the other hand, high-frequency, low power leads to the formation of a large number of free radicals. A constant frequency of 20 kHz was applied for the pectin extraction from orange peel [106], grapefruit peel [101], pomegranate peel [107], mango peel [108], and for extraction of tartaric and malic acid from the grape seeds, a 24 kHz frequency was applied [98]. The selection of low frequency might be due to the formation of a smaller number of cavitation bubbles with a large diameter [109]. The frequency of the ultrasonics is indirectly related to the duration of the rarefaction phase. Thus, at the high frequency, the bubble gets less time to grow which impedes their implosion effect. High frequency results in a large number of bubbles, which causes resistance to the mass transfer.

3. **Type of Solvent:** Various solvents such as acidified water, alcohol, ethanol, and ethanol are used for extracting the bioactive compounds using the ultrasonication method. Ethanol possesses the highest affinity for phenolic compounds and is thus the first choice of researchers [110]. The other reasons for its acceptance are its affordability, origination from renewable sources such as sugarcane, and its consideration as GRAS (generally regarded as safe) [111]. An increase in the ethanol concentration leads to an increase in the extraction yield. However, a very high concentration of ethanol may harm the yield. This might be due to the increase in the solubility and diffusivity of the phenolic compounds because of the lowering of the dielectric constant of the solvent with the rise in ethanol concentration. Around 100% ethanol concentration leads to dehydration of the plant tissues in addition to the protein denaturation causing lower yield. In contrast, at a fixed temperature and time for the extraction, the increase in the ethanol concentration led to an increase in the extraction yield of the total phenols from the grape seeds and acquires maximum value at the maximum ethanol concentration [112].
4. **Liquid to Solid Ratio (LSR) or Solvent to Solid Ratio (SSR):** The LSR is the volume of solvent per the sample quantity taken for the extraction. An increase in the LSR causes an increase in the yield of the bioactive compounds upon reaching a peak. In the case of low LSR, the viscosity of the solution is high, which causes difficulty in the cavitation effect; however, with the initial increase in the LSR, concentration and extract viscosity decrease, causing greater cavitation. Further increase in the LSR causes more fragmentation and pore formation, leading to higher yield [97].
5. **Temperature:** An initial increase in the temperature causes an increase in the yield of the UAE; however, a further increase in the temperature causes a lowering of the yield, similar to the case of a power increase. An increase in the temperature causes an increase in the desorption property and solubility of the solute in the solvent, whereas it lowers the solvent viscosity, which causes an increase in the solvent diffusivity in the tissue matrix. Similar behavior was observed in the case of flavonoids extraction from the seeds of hawthorn [112], antioxidants from sijiaoling stem [103], pectin from the peels of grapefruit [102], and pomegranate [107].

2.3.2 IDENTIFICATION AND CHARACTERIZATION OF THE BIOACTIVE COMPOUNDS

While searching for the bioactive compounds is to find out a suitable method that can monitor the bioactivity of the source material such as antioxidant, antibacterial, or cytotoxicity [113]. The primary step involved in the isolation and characterization of the bioactive compounds is the choice and collection of the plant resources, followed by the next step involving the recovery of the ethnobotanical information to distinguish the possible available bioactive compounds [114]. Column chromatographic techniques can be employed for the isolation/extraction and purification of the bioactive compounds depending upon their properties. Instruments such as HPLC are widely recommended for the identification of the purification extent of the bioactive compounds. Other spectroscopic techniques such as UV-visible (UV-Vis), infrared (IR), mass spectroscopy (MS), and nuclear magnetic resonance (NMR) can also be used for the identification of the purified compound [115].

2.3.3 PURIFICATION OF THE BIOACTIVE COMPOUNDS

Column chromatography and thin-layer chromatography (TLC) are predominantly used chromatographic techniques due to their convenience, low cost, and most importantly their accessibility in a range of stationary phases. Plant materials contain a large number of complex phytochemicals that make their separation difficult. Thus, in these types of cases, increasing the polarity of the multiple mobile phases is considered useful for the highly valued separations. Structural explanation and examination of the desired bioactive compounds employ data attained from a wide variety of these spectroscopic techniques. The transmittance of the electromagnetic radiation from the molecules forms the basis of these techniques. A spectrum thus can be formed by calculating the amount of these absorbed radiations. The obtained spectra are specific to the specific bonds present in the molecule. Also, the structure of the organic compounds can be comparatively recognized as well as determined.

2.3.3.1 UV VISIBLE (UV-VIS) SPECTROSCOPY

It is used for the qualitative and quantitative identification, measurement, and characterization of the biomolecules in their both crude and pure form. It can be the most commonly used technique for the quantification of compounds because the aromatic ones are potent chromophores in the UV range. Phenolic compounds, including tannins, anthocyanidins, and phenols, can be easily identified by UV-Vis spectroscopy [116]. The specific UV-Vis spectra used for the quantification of anthocyanidins, phenolic acids, flavones, and phenolic extract are 520, 360, 320, and 280 nm, respectively. This technique was found less discriminating, timesaving, and provide complex information on the total polyphenol content than the other techniques.

2.3.3.2 FOURIER TRANSFORM INFRARED SPECTROSCOPY (FTIR)

FTIR is an important analysis tool for the identification and characterization of the bioactive compounds and their functional groups and/or chemical bonds present in the unidentified mixture of the plant extract. The spectra of the pure compounds are so unique that they are similar to the same kind of molecules like a fingerprint. In the case of the commonly found plant compounds, the spectrum of the unknown compound can be recognized by comparing them with the spectrum of the known compounds. Sample preparation varies among the solid and liquid state samples. In the case of the solid sample, the samples are crushed with the potassium bromide (KBr) and then pressed to form a thin tablet or pellet, which can be then analyzed. Another method employed for the solid sample is the solubilization of the sample into the solvent such as methylene chloride, and afterward, the solution prepared can be put upon the single salt plate. The solvent used is then evaporated to form a thin layer or film of the material on the plate. While, in the case of the liquid sample, the simple way is to put a drop of the sample sandwiched in between the two plates of sodium chloride. The drop then forms a thin layer, which is further analyzed. FTIR is a high-resolution analytical technique that is employed widely for the identification of the chemical components and to illuminate the structural compounds. It offers a quick and non-destructive analysis to fingerprint the herbal extracts or the powders [117].

2.3.3.3 INFRARED (IR) SPECTROSCOPY

When the IR light is transmitted from the sample, some of the light is absorbed rather than being absorbed. This technique deals with the vibrational changes that occur when the sample molecules are exposed to IR radiation. Different bonds have some definite vibrational frequencies. Thus, if any of this bond is present in the molecules, they can be thus detected by knowing the typical frequency absorption band in spectrum [118].

2.3.3.4 NUCLEAR MAGNETIC RESONANCE (NMR) SPECTROSCOPY

NMR technique is related to the magnetic properties of the atomic nuclei. The technique has enabled the researchers to study the molecules by collecting the differences among the variety of the magnetic nuclei and thereby provide an understandable image of where these nuclei are located in the specific molecule. Various attempts were made in the earlier time to separate the individual phenol by using TLC, LC, and CC, while their structures were resolved later by using NMR.

2.3.3.5 MASS SPECTROSCOPY

In MS, the molecules are bombarded with electrons or lasers, which are further converted to highly energetic charged ions. It generates a plot between the quantity of the fragmented ions and the mass: charge of ions. By employing this technique, relative molecular weight (MW) and the precise molecular formula of the compounds can be obtained using the information where the molecules have been fragmented [119]. Tandem MS provides a lot of useful information regarding the structural compounds. Therefore, both the HPLC and tandem MS give quick and precise identification of the chemical compounds in the plants. In addition, for the very first time, UPLC-PDA (ultra-performance liquid chromatography-photodiode array) and MALDI-TOF-MS (matrix-assisted laser desorption/ionization-time of flight-mass spectroscopy) techniques were used for the screening and identification of the phenolic compounds such as quinic acid, quercetin, crysin, chlorogenic acid, and kaempferol, in the hydromethanolic extract of *A. aspera* [120].

2.3.4 COMPARATIVE EVALUATION OF THE DIFFERENT EXTRACTION TECHNOLOGY

The bioactive compounds, such as anthocyanins, were extracted from the grape by-products. These compounds can be used as coloring and antioxidant agents. The combined effect of heat treatment (70°C), ultrasound (35 kHz), pulse electric field (3 kV cm^{-1}) and high hydrostatic pressure of 600 MPa was observed [121]. After 2 hr. of the extraction period, the total phenols of the samples subjected to the novel technologies were approximately 50% higher than the control samples. Using these technologies led to increased antioxidant activity such as with the ultrasonics two-fold, HHP threefold, and PEF fourfold increment was observed in contrast to the control extraction which is carried out in the water bath (70°C for 1 hr). The effect of various technologies on the extraction of the bioactive components such as zeaxanthin, α-, and β-carotene, lutein, naringenin, α-, and β-cryptoxanthin, and hesperetin from the orange juice was investigated [122]. For this, the juice was subjected to low pasteurization (LPT) at 70°C for 30 sec, PEF at 35 kV/cm for 750 ms, and high pressure (HPT) of 400 MPa at 40°C for 1 min. The study concluded that the HPT was the most effective method for the extraction of the bioactive compounds with their highest recovery from the orange juice, followed by PEF and LPT. The combined effect of the various techniques such as blanching, high-intensity PEF, and ultrasonic on the further enhancement of the lycopene bio-accessibility of the tomato fruits concluded that the blanching followed by high-intensity PEF had the significant release of trans and cis lycopene, respectively as 4.01±0.48 lg/g and 5.04±0.26 lg/g [123]. The extraction yield of the air-dried Agiorgitico red grape pomace by-products was compared for the three different extraction methods such as MAE, UAE, and Soxhlet extraction using water, ethanol, and ethanol: water (1:1) as respective solvents [124]. The study concluded that the UAE method with ethanol: water (1:1) showed higher contents of the phenolic contents, i.e., up to 4,38,984 ppm GAE in dry extract.

The effect of SFE on the total phenolic contents (TPC) extracted from the orange pomace at varying pressure levels (15–35 MPa) and temperatures (40–60°C), by employing the pure ethanol and ethanol: water (9:1) as solvents observed that the use of high pressure and 90% ethanol improved and enhanced the recovery of TPC [125]. In another study, lycopene and carotene were extracted from the Tunisian tomato peel using SFE with CO_2 and SE using ethanol, hexane, and ethyl acetate [126]. SE using overnight maceration with ethanol, hexane, and ethyl acetate resulted in 284.53 mg/kg, 608.94 mg/kg, and 320.35 mg/kg of dry tomato peels. On the other hand, the SFE with CO_2 resulted in 728.98 mg/kg dry tomato peels at 80°C, 400 bar pressure, 105 min at 4 g CO_2/min. Thus, the SFE with CO_2 resulted in higher lycopene extraction compared to the SE. Some of these methods, the yield of bioactive compounds, and the optimum condition required for the extraction from the various food waste are shown in Table 2.3.

TABLE 2.3 Comparison of Various Extraction Techniques Employed for Extracting the Bioactive Compounds

Sources	Bioactive Compounds	Extraction Solvent Used	Method	Yield	Optimum Conditions	References
Grape	Anthocyanin	Water	WE	7.93[a]	70°C	[121]
		Ethanol and water	PEF	14.05	3 kVcm^{-1}	
		Water and ethanol	Ultrasonics	7.76	600 MPa	
		Glycol and water	HHP	11.21	35 kHz	
Orange	Lutein	–	HPT	361.17[b]	400 mpA/40° C/1 min	[122]
			PEF	260.86	35 kV cm^{1}/750 ms	–
			LPT	226.42	70°C/30s	–
Orange pomace (dry)	Total phenolic content	Pure ethanol Ethanol:water – 9:1	SFE	21.2[c]	25 MPa and 60°C	[125]
Orange	α-Carotene	–	PEF	26.64	35 kVcm^{1}/750 ms	[122]
			HPT	38.06	400 MPa/40°C/1 min	
			LPT	25.04	70°C/30s	
Orange	β-carotene	–	PEF	33.74[c]	35 kVcm^{1}/750 ms	[122]
			HPT	53.78	400 MPa/40°C/1 min	
			LPT	32.72	70°C/30 s	
Orange	β-Cryptoxanthin	–	PEF	230.53[c]	35 kVcm^{1}/750 ms	[122]
			HPT	330.07	400 MPa/40°C/1 min	
			LPT	235.21	70°C/30s	
Orange	Zeaxanthin	–	PEF	278.70[c]	35 kVcm^{1}/750 ms	[122]
			HPT	408.56	400 MPa/40°C/1 min	
			LPT	259.95	70°C/30s	
Tomato waste	Lycopene	Ethanol	SE	284.53[c]	–	[126]
		Ethyl acetate	–	320.35	–	
		Hexane	–	608.94	–	
		Liquid CO_2	SFE	728.98	400 bar/80°C/4 g CO_2/min/105 min	
Orange	Hesperetin	–	PEF	11.09[f]	35 kVcm^{1}/750 ms	[122]
			HPT	13.34	400 MPa/40°C/1 min	
			LPT	11.56	70°C/30s	

TABLE 2.3 *(Continued)*

Sources	Bioactive Compounds	Extraction Solvent Used	Method	Yield	Optimum Conditions	References
Macleaya cordata	Alkaloid (sanguinarine)	NaOH and HCl	Percolation	6.14	–	[115]
		HCl	MAE	17.10	280 W/5 min	
		HCl	UAE	10.74	250 W/30 min	
		HCl	Maceration	16.87	100°C/30 min	

Abbreviations: WE: Water extraction; SE: Solvent extraction; HHP: High hydrostatic pressure; LPT: Low pasteurization treatment; PEF: Pulsed electric field; HPT: High pressure treatment; and MAE: Microwave assisted extraction.

[a]*mg Cy-3-glu eq.* g^{-1} *dry matter;*

[b]*µg/100 ml;*

[c]*GAE/g extract;*

[d]*ppm GAE in dry extract;*

[e]*mg/kg;*

[f]*mg/100 ml;*

[g]*mg/g sample.*

2.4 BIOACTIVITY OF BIOACTIVE COMPOUNDS AND NUTRACEUTICALS

Bioactivity or biological activity of the bioactive compound or a nutraceutical describes their adverse or beneficial effect on the living system [127]. When a bioactive compound is a complex mixture, then its activity is exerted by the compound's active ingredient but can be further modified with the other constituents. The activity of the compounds depends largely upon their dosage. Also, the effects, either beneficial or adverse, vary with the number of doses from low to high. Also, the biochemical system encountered by a nutraceutical or bioactive compound is extremely complex. Therefore, the factors affecting their bioactivity are manifold. These factors can be grouped into three categories:

2.4.1 SOLUBILITY

The solubility possesses great importance as they all relate to the transport of the compounds to the site of action. These compounds taken orally must travel through several membranes before reaching the required destination.

So, their efficiency depends upon their solubility behavior in the aqueous medium and towards the lipids. Compounds in their bound form tend to decrease their effectiveness, and while traveling through the various compartments, the molecule is subjected to various factors which tend to lower their concentration in the active form. Solubility is an important parameter to bioactivity, as solubility is directly related to its absorption by the body.

2.4.2 ELECTROLYTIC NATURE

All the chemical and bioactive compounds can be grouped into various categories depending upon their electrical conductivity in the aqueous medium. For instance, the inorganic salts, when dissolved, get dissociated into ions or charged particles. These compounds are referred to the as strong electrolyte, and the others which do not dissociate completely are nonelectrolytes. The un-ionized species are more lipid-soluble, and the ionized one has much lower lipid solubility. Thus, their passage through the membranes reaches insignificant levels and affects the capacity of the compounds to absorb.

2.4.3 STRUCTURE OF THE MOLECULES

The structure of the bioactive compounds strongly affects their absorption. For example, high MW compounds, such as complex lipids, are unable to pass through the intestinal cells until and unless they are hydrolyzed into smaller units [128]. The sugar moiety of the flavonoids is an important determinant in their absorption. Flavonoids attached to the β-glucosides are one of the most predominant forms in nature and can be absorbed to a smaller extent [129].

2.4.4 TRANSPORT MECHANISM

The transport mechanism in the intestinal lumen is an important factor that affects the bioavailability of the ingested food compounds. This includes passive diffusion, facilitated diffusion, and active transport. The first two mechanisms include the diffusion towards a concentration gradient through the intestinal cells to the blood circulation. The last mechanism work against the concentration gradient and can increase the compounds in the blood circulation [130]. Many bioactive compounds do not have the required physicochemical properties required for enhancing their permeability. Transport through the membrane is related to the two mechanisms of the compound permeability, uptake, and efflux. GLUT-family, organic anion transporters, vitamin transporters are involved in the uptake of these compounds and transport across the intestine. On the other hand, breast cancer resistance protein and ATP binding cassette family are examples of efflux mechanisms and can hinder the bioavailability of the bioactive compounds [131]. The activity of these mechanisms affects the disposition of the bioactive compounds, their absorption, secretion, and their distribution to the targeted tissues [129].

2.5 USE OF BIOACTIVE COMPOUNDS AS NUTRACEUTICAL AND BIOACTIVE COMPOUNDS

Natural bioactive compounds (NBCs) interact efficiently with the DNA, proteins, and with other biological molecules to obtain desired results. This then can be used for the designing of natural therapeutic agents. Also, there is now a growing interest of the consumers for the food bioactive which provides beneficial impacts on the human lives in terms of disease risk reduction and thereby promotion of good health. Bioactive or biologically active compounds are known for lowering the risk of various diseases such as cataracts, cancer, Alzheimer's, Parkinson's, and many others. These health-related benefits are attributed to the antioxidant and radical scavenging properties of these compounds that delay and inhibit the oxidation of lipids, proteins, and DNA. Moreover, these compounds are also known for their antimicrobial activity and preventing fruit against the pathogenic agents [3]. ROS can be characterized as free and non-radical. The molecules with one or more unpaired electrons are known as free radicals, while the non-radical forms are formed when two free radicals share their free unpaired electrons. Superoxide anion (O_2^-), hydrogen peroxide (H_2O_2), and hydroxyl radical (OH^-) [132]. Production of these free radicals in the excess of the defensive effects of the antioxidants and cofactors results in the oxidative damage that gets accumulated during the life cycle causing aging, and chronic diseases (neurodegenerative disorders, cardiovascular diseases (CVDs), and cancer) [133].

Normal metabolic functions of the body cause the generation of these free radicals, which affects the vital cellular functions and their structures, leading to various degenerative diseases. These free radicals formed are then deactivated by the antioxidants that catalyze the oxidative and reductive reactions and also control the redox state of the functional proteins [134]. Nutraceuticals are largely employed in pharmaceutical preparations in the form of capsules, pills, tablets, and powder. The use of mangiferin (bioactive xanthonoid) in single or in combination with the other known anticancer chemicals has shown significant benefits against lung, breast, cervix, brain, and prostate cancer, apart from their antioxidant and anti-inflammatory activities [135]. The addition of flavonoids and saponins from the black bean seed coats to the formulations of whole wheat bread resulted in the retention of 80% anthocyanins and 90% flavonoids after baking [136]. Application of these bread might have significant health-promoting effects. Similarly, the pate, an olive by-product, can be used as a bioactive compound [137]. This by-product was then characterized by the presence of luteolin, hydroxytyrosol, oleoside derivative, etc., as a potential ingredient in the preparation of nutraceuticals.

2.6 CONCLUSION

Bioactive compounds are a major source of therapeutic drugs in the health system worldwide. Food materials and their by-products could be used for the extraction of bioactive compounds using different approaches as discussed above. A suitable extraction method must be selected based on the outcome of the optimization process. Food waste forms a major source of bioactive compounds. Developing and using an extraction method with superior effectiveness will not only enhance the worth of the food waste but also help lower the cost of the formulated products. Also, this will help in reducing the use of synthetic chemicals in these products. The increase in the population and demands for healthy food products have led to the increase in the setup of newer food industries. However, the postharvest losses of the food materials such as fruits and vegetables have increased the amount of their waste. The utilization of this waste as a source of bioactive compounds has decreased the waste burden and increased the financial status of the farmers. Extraction technologies employing lesser or no solvent will possess a great significance towards sustainability. The present chapter ascertains how different technologies can be used for extracting bioactive compounds and can be used as nutraceuticals as well as in the form of dietary supplements.

KEYWORDS

- **anti-inflammatory activities**
- **bioactive compounds**
- **dietary supplements**
- **extraction technologies**
- **mangiferin**
- **nutraceuticals**
- **therapeutic drugs**

REFERENCES

1. Dhaval, A., Yadav, N., & Purwar, S., (2016). Potential applications of food-derived bioactive peptides in management of health. *International Journal of Peptide Research and Therapeutics, 22,* 377–398.
2. Correia, R. T., Borges, K. C., Medeiros, M. F., & Genovese, M. I., (2012). Bioactive compounds and phenoliclinked functionality of powdered tropical fruit residues. *Food Science and Technology International, 18,* 539–547.
3. Ayala-Zavala, J. F., & González-Aguilar, G. A., (2011). *Use of Additives to Preserve the Quality of Fresh-Cut Fruits and Vegetables* (pp. 231–254). CRC Press, Boca Raton.
4. Cragg, G. M., & Newman, D. J., (2007). Anticancer drug discovery and development from natural products. In: *Bioactive Natural Products* (pp. 337–384). CRC Press.
5. Donnez, D., Jeandet, P., Clément, C., & Courot, E., (2009). Bioproduction of resveratrol and stilbene derivatives by plant cells and microorganisms. *Trends in Biotechnology, 27,* 706–713.
6. Mapari, S. A., Nielsen, K. F., Larsen, T. O., Frisvad, J. C., Meyer, A. S., & Thrane, U., (2005). Exploring fungal biodiversity for the production of water-soluble pigments as potential natural food colorants. *Current Opinion in Biotechnology, 16,* 231–238.
7. Silveira, S. T., Daroit, D. J., & Brandelli, A., (2008). Pigment production by *Monascus purpureus* in grape waste using factorial design. *LWT-Food Science and Technology, 41,* 170–174.
8. Demain, A. L., (2000). Microbial biotechnology. *Trends in Biotechnology, 18,* 26–31.
9. Rodríguez-Meizoso, I., Jaime, L., Santoyo, S., Señoráns, F. J., Cifuentes, A., & Ibáñez, E., (2010). Subcritical water extraction and characterization of bioactive compounds from *Haematococcus pluvialis* microalga. *Journal of Pharmaceutical and Biomedical Analysis, 51,* 456–463.
10. Onofrejová, L., Vašíčková, J., Klejdus, B., Stratil, P., Mišurcová, L., Kráčmar, S., & Vacek, J., (2010). Bioactive phenols in algae: The application of pressurized-liquid and solid-phase extraction techniques. *Journal of Pharmaceutical and Biomedical Analysis, 51,* 464–470.

11. Deng, G. F., Shen, C., Xu, X. R., Kuang, R. D., Guo, Y. J., Zeng, L. S., Gao, L. L., et al., (2012). Potential of fruit wastes as natural resources of bioactive compounds. *International Journal of Molecular Science*, *13*, 8308–8323.
12. Wang, J., Sun, B., Cao, Y., Tian, Y., & Li, X., (2008). Optimization of ultrasound-assisted extraction of phenolic compounds from wheat bran. *Food Chemistry*, *106*, 804–810.
13. Arogba, S. S., (2000). Mango (*Mangifera indica*) kernel: Chromatographic analysis of the tannin, and stability study of the associated polyphenol oxidase activity. *Journal of Food Composition Analysis*, *13*, 149–156.
14. Puravankara, D., Boghra, V., & Sharma, R. S., (2000). Effect of antioxidant principles isolated from mango (*Mangifera indica* L) seed kernels on oxidative stability of buffalo ghee (butter-fat). *Journal of Science Food and Agriculture*, *80*, 522–526.
15. Noda, Y., Kaneyuki, T., Mori, A., & Packer, L., (2002). Antioxidant activities of pomegranate fruit extract and its anthocyanidins: Delphinidin, cyanidin, and pelargonidin. *Journal of Agricultural and Food Chemistry*, *50*, 166–171.
16. Zeyada, N. N., Zeitoum, M., & Barbary, O., (2008). Utilization of some vegetables and fruit waste as natural antioxidants. *Alexandria Journal of Food Science and Technology*, *5*, 1–11.
17. Oliveira, R., Oliveira, V., Aracava, K. K., & Da Costa, R. C. E., (2012). Effects of the extraction conditions on the yield and composition of rice bran oil extracted with ethanol—A response surface approach. *Food and Bioproducts Processing*, *90*, 22–31.
18. Wolfe, K. L., & Liu, R. H., (2003). Apple peels as a value-added food ingredient. *Journal of Agricultural and Food Chemistry*, *51*, 1676–1683.
19. Maier, T., Schieber, A., Kammerer, D. R., & Carle, R., (2009). Residues of grape (*Vitis vinifera* L.) seed oil production as a valuable source of phenolic antioxidants. *Food Chemistry*, *112*, 551–559.
20. Duan, X., Jiang, Y., Su, X., Zhang, Z., & Shi, J., (2007). Antioxidant properties of anthocyanins extracted from litchi (*Litchi chinenesis* Sonn.) fruit pericarp tissues in relation to their role in the pericarp browning. *Food Chemistry*, *101*, 1365–1371.
21. Chantaro, P., Devahastin, S., & Chiewchan, N., (2008). Production of antioxidant high dietary fiber powder from carrot peels. *LWT Food Science and Technology*, *41*, 1987–1994.
22. Sainvitu, P., Nott, K., Richard, G., Blecker, C., Jérôme, C., Wathelet, J. P., Paquot, M., & Deleu, M., (2012). Structure, properties and obtention routes of flaxseed lignin secoisolariciresinol: A review. *Biotechnology Agronomy and Society and Environment*, *16*, 115.
23. Carbonell-Capella, J. M., Buniowska, M., Barba, F. J., Esteve, M. J., & Frı´gola, A., (2014). Analytical methods for determining bioavailability and bioaccessibility of bioactive compounds from fruits and vegetables: A review. *Comprehensive Reviews in Food Science and Food Safety*, *13*, 155–171.
24. Duthie, G. G., Duthie, S. J., & Kyle, J. A., (2000). Plant polyphenols in cancer and heart disease: Implications as nutritional antioxidants. *Nutrition Research Reviews*, *13*, 79–106.
25. Marinova, D., Ribarova, F., & Atanassova, M., (2005). Total phenolics and total flavonoids in Bulgarian fruits and vegetables. *Journal of the University of Chemical Technology and Metallurgy*, *40*, 255–260.

26. Rice-Evans, C. A., Miller, N. J., & Paganga, G., (1996). Structure-antioxidant activity relationships of flavonoids and phenolic acids. *Free Radical Biology and Medicine, 20,* 933–956.
27. Yao, L. H., Jiang, Y., Shi, J., Tomas-Barberan, F., Datta, N., Singanusong, R., & Chen, S., (2004). Flavonoids in food and their health benefits. *Plant Foods for Human Nutrition, 59,* 113–122.
28. Merken, H. M., & Beecher, G. R., (2000). Measurement of food flavonoids by high-performance liquid chromatography: A review. *Journal of Agricultural and Food Chemistry, 48,* 577–599.
29. Ahuja, I., Kissen, R., & Bones, A. M., (2012). Phytoalexins in defense against pathogens. *Trends in Plant Science, 17,* 73–90.
30. Hooper, A. M., Hassanali, A., Chamberlain, K., Khan, Z., & Pickett, J. A., (2009). New genetic opportunities from legume intercrops for controlling *Striga* spp. parasitic weeds. *Pest Management Science: Formerly Pesticide Science, 65,* 546–552.
31. Corradini, E., Foglia, P., Giansanti, P., Gubbiotti, R., Samperi, R., & Lagana, A., (2011). Flavonoids: Chemical properties and analytical methodologies of identification and quantitation in foods and plants. *Natural Product Research, 25*(5), 469–495.
32. Aron, P. M., & Kennedy, J. A., (2008). Flavan-3-ols: Nature, occurrence and biological activity. *Molecular Nutrition & Food Research, 52,* 79–104.
33. Ricardo-da-Silva, J. M., Cheynier, V., Souquet, J. M., Moutounet, M., Cabanis, J. C., & Bourzeix, M., (1991). Interaction of grape seed procyanidins with various proteins in relation to wine fining. *Journal of the Science of Food and Agriculture, 57,* 111–125.
34. Cassidy, A., Hanley, B., & Lamuela-Raventos, R. M., (2000). Isoflavones, lignans and stilbeneseorigins, metabolism and potential importance to human health. *Journal of the Science of Food and Agriculture, 80,* 1044–1062.
35. Ralph, J., Lundquist, K., Brunow, G., Lu, F., Kim, H., Schatz, P. F., Marita, J. M., et al., (2004). Lignins: Natural polymers from oxidative coupling of 4-hydroxyphenylpropanoids. *Phytochemistry Reviews, 3,* 29–60.
36. Barapatre, A., Meena, A. S., Mekala, S., Das, A., & Jha, H., (2016). In vitro evaluation of antioxidant and cytotoxic activities of lignin fractions extracted from *Acacia nilotica. International Journal of Biological Macromolecules, 86,* 443–453.
37. Andrei, G., Lisco, A., Vanpouille, C., Introini, A., Balestra, E., Van, D. O. J., Cihlar, T., et al., (2011). Topical tenofovir, a microbicide effective against HIV, inhibits herpes simplex virus-2 replication. *Cell Host & Microbe, 10,* 379–389.
38. Norikura, T., Mukai, Y., Fujita, S., Mikame, K., Funaoka, M., & Sato, S., (2010). Lignophenols decrease oleate-induced apolipoprotein-B secretion in HepG2 cells. *Basic and Clinical Pharmacology and Toxicology, 107,* 813–817.
39. Fischer, B. M., Pavlisko, E., & Voynow, J. A., (2011). Pathogenic triad in COPD: Oxidative stress, proteaseeantiprotease imbalance, and inflammation. *International Journal of Chronic Obstructive Pulmonary Disease, 6,* 413.
40. Verma, S., Bahorun, T., Singh, R. K., Aruoma, O. I., & Kumar, A., (2013). Effect of *Aegle marmelos* leaf extract on N-methyl N-nitrosourea-induced hepatocarcinogensis in Balb/c mice. *Pharmaceutical Biology, 51,* 1272–1281.
41. Huie, C. W., (2002). A review of modern sample-preparation techniques for the extraction and analysis of medicinal plants. *Analytical and Bioanalytical Chemistry, 373,* 23–30.

42. Gan, C. Y., & Latiff, A. A., (2011). Optimization of the solvent extraction of bioactive compounds from *Parkia speciosa* pod using response surface methodology. *Food Chemistry, 124,* 1277–1283.
43. Zarnowski, R., & Suzuki, Y., (2004). Expedient Soxhlet extraction of resorcinolic lipids from wheat grains. *Journal of Food Composition and Analysis, 17,* 649–664.
44. Li, H., Pordesimo, L., & Weiss, J., (2004). High intensity ultrasound assisted extraction of oil from soybeans. *Food Research International, 37,* 731–738.
45. Luque-Garcia, J. L., & Luque De, C. M. D., (2004). Ultrasound-assisted Soxhlet extraction: An expeditive approach for solid sample treatment—Application to the extraction of total fat from oleaginous seeds. *Journal of Chromatography A, 1034,* 237–242.
46. Mamidipally, P. K., & Liu, S. X., (2004). First approach on rice bran oil extraction using limonene. *European Journal of Lipid Science and Technology, 106,* 122–125.
47. Wang, L., & Weller, C. L., (2006). Recent advances in extraction of nutraceuticals from plants. *Trends in Food Science and Technology, 17,* 300–312.
48. Giannuzzo, A. N., Boggetti, H. J., Nazareno, M. A., & Mishima, H. T., (2003). Supercritical fluid extraction of naringin from the peel of *Citrus paradisi. Phytochemical Analysis 14,* 221–223.
49. Ashraf-Khorassani, M., & Taylor, L. T., (2004). Sequential fractionation of grape seeds into oils, polyphenols, and procyanidins via a single system employing CO_2-based fluids. *Journal of Agricultural and Food Chemistry, 52,* 2440–2444.
50. Jung, G. W., Kang, H. M., & Chun, B. S., (2012). Characterization of wheat bran oil obtained by supercritical carbon dioxide and hexane extraction. *Journal of Industrial and Engineering Chemistry, 18,* 360–363.
51. Şanal, İ., Bayraktar, E., Mehmetoğlu, Ü., & Çalımlı, A., (2005). Determination of optimum conditions for SC-(CO_2 + ethanol) extraction of β-carotene from apricot pomace using response surface methodology. *Journal of Supercritical Fluids, 34,* 331–338.
52. Castro-Vargas, H. I., Rodríguez-Varela, L. I., Ferreira, S. R., & Parada-Alfonso, F., (2010). Extraction of phenolic fraction from guava seeds (*Psidium guajava* L.) using supercritical carbon dioxide and co-solvents. *Journal of Supercritical Fluids, 51,* 319–324.
53. Shi, J., Khatri, M., Xue, S. J., Mittal, G. S., Ma, Y., & Li, D., (2009). Solubility of lycopene in supercritical CO_2 fluid as affected by temperature and pressure. *Separation and Purification Technology, 66,* 322–328.
54. Chang, C. J., Chiu, K. L., Chen, Y. L., & Chang, C. Y., (2000). Separation of catechins from green tea using carbon dioxide extraction. *Food Chemistry, 68,* 109–113.
55. Bimakr, M., Rahman, R. A., Taip, F. S., Chuan, L., Ganjloo, A., Selamat, J., & Hamid, A., (2009). Supercritical carbon dioxide (SC-CO_2) extraction of bioactive flavonoid compounds from spearmint (*Mentha spicata* L.) leaves. *European Journal of Scientific Research, 33,* 679–690.
56. Ghafoor, K., Park, J., & Choi, Y. H., (2010). Optimization of supercritical fluid extraction of bioactive compounds from grape (*Vitis labrusca* B.) peel by using response surface methodology. *Innovative in Food Science and Emerging Technologies, 11,* 485–490.
57. Paes, J., Dotta, R., & Martínez, J., (2013). Extraction of phenolic compounds from blueberry (*Vaccinium myrtillus* L.) residues using supercritical CO_2 and pressurized water. *Journal of Supercritical Fluids.* doi: 10.1016/j.supflu.2014.07.025.

58. Nobre, B. P., Palavra, A. F., Pessoa, F. L., & Mendes, R. L., (2009). Supercritical CO_2 extraction of trans-lycopene from Portuguese tomato industrial waste. *Food Chemistry, 116,* 680–685.
59. Pereira, C. G., & Meireles, M. A. A., (2010). Supercritical fluid extraction of bioactive compounds: Fundamentals, applications, and economic perspectives. *Food and Bioprocess Technology, 3,* 340–372.
60. Papamichail, I., Louli, V., & Magoulas, K., (2000). Supercritical fluid extraction of celery seed oil. *Journal of Supercritical Fluids, 18,* 213–226.
61. Herrero, M., Cifuentes, A., & Ibañez, E., (2006). Sub-and supercritical fluid extraction of functional ingredients from different natural sources: Plants, food-by-products, algae, and microalgae: A review. *Food Chemistry, 98,* 136–148.
62. Zakaria, S. M., & Kamal, S. M. M., (2016). Subcritical water extraction of bioactive compounds from plants and algae: Applications in pharmaceutical and food ingredients. *Food Engineering Reviews, 1,* 23–34.
63. Tunchaiyaphum, S., Eshtiaghi, M., & Yoswathana, N., (2013). Extraction of bioactive compounds from mango peels using green technology. *International Journal of Chemical Engineering and Applications, 4,* 194.
64. Rangsriwong, P., Rangkadilok, N., Satayavivad, J., Goto, M., & Shotipruk, A., (2009). Subcritical water extraction of polyphenolic compounds from *Terminalia chebula* Retz. fruits. *Separation and Purification Technology, 66,* 51–56.
65. Singh, P. P., & Saldaña, M. D., (2011). Subcritical water extraction of phenolic compounds from potato peel. *Food Research International, 44,* 2452–2458.
66. Mayanga-Torres, P., Lachos-Perez, D., Rezende, C., Prado, J., Ma, Z., Tompsett, G., Timko, M., & Forster-Carneiro, T., (2017). Valorization of coffee industry residues by subcritical water hydrolysis: Recovery of sugars and phenolic compounds. *Journal of Supercritical Fluids, 120,* 75–85.
67. Ahmadian-Kouchaksaraie, Z., Niazmand, R., & Najafi, M. N., (2016). Optimization of the subcritical water extraction of phenolic antioxidants from *Crocus sativus* petals of saffron industry residues: Box-Behnken design and principal component analysis. *Innovative in Food Science and Emerging Technologies, 36,* 234–244.
68. Kronholm, J., Hartonen, K., & Riekkola, M. L., (2007). Analytical extractions with water at elevated temperatures and pressures. *TRAC Trends in Analytical Chemistry, 26,* 396–412.
69. Cheng, Y., & Li, S. M., (2004). Analytical method development of long-chain ketones in PM2. 5 aerosols using accelerated solvent extraction and Gc/FID/MSD. *International Journal of Environmental Analytical Chemistry, 84,* 367–378.
70. Yang, Y., Kayan, B., Bozer, N., Pate, B., Baker, C., & Gizir, A. M., (2007). Terpene degradation and extraction from basil and oregano leaves using subcritical water. *Journal of Chromatography A, 1152,* 262–267.
71. Kronholm, J., Revilla-Ruiz, P., Porras, S. P., Hartonen, K., Carabias-Martínez, R., & Riekkola, M. L., (2004). Comparison of gas chromatography-mass spectrometry and capillary electrophoresis in analysis of phenolic compounds extracted from solid matrices with pressurized hot water. *Journal of Chromatography A, 1022,* 9–16.
72. Kim, W., Kim, J., Veriansyah, B., Kim, J., Lee, Y., & Oh, S., (2009). Extraction of bioactive components from *Centella asiatica* using subcritical water. *The Journal of Supercritical Fluids, 48,* 211–216.

73. Khajenoori, M., Asl, A. H., Hormozi, F., Eikani, M., & Bidgoli, H. N., (2009). Subcritical water extraction of essential oils from *Zataria multiflora* Boiss. *Journal of Food Process Engineering, 32*, 804–816.
74. Wiboonsirikul, J., & Adachi, S., (2008). Extraction of functional substances from agricultural products or by-products by subcritical water treatment. *Food Science and Technology Research, 14,* 319–328.
75. Plaza, M., & Turner, C., (2015). Pressurized hot water extraction of bioactives. *TRAC Trends in Analytical Chemistry, 71*, 39–54.
76. Singh, G., Verma, A., & Kumar, V., (2016). Catalytic properties, functional attributes, and industrial applications of β-glucosidases. *3 Biotechology*, *6*, 3.
77. Puri, M., Sharma, D., & Barrow, C. J., (2012). Enzyme-assisted extraction of bioactives from plants. *Trends in Biotechnology, 30,* 37–44.
78. Zuorro, A., Fidaleo, M., & Lavecchia, R., (2011). Enzyme-assisted extraction of lycopene from tomato processing waste. *Enzyme and Microbial Technology*, *49*, 567–573.
79. Reshmitha, T., Thomas, S., Geethanjali, S., Arun, K., & Nisha, P., (2017). DNA and mitochondrial protective effect of lycopene rich tomato (*Solanum lycopersicum* L.) peel extract prepared by enzyme assisted extraction against H_2O_2 induced oxidative damage in L6 myoblasts. *Journal of Functional Foods, 28,* 147–156.
80. Gardossi, L., Poulsen, P. B., Ballesteros, A., Hult, K., Švedas, V. K., Vasić-Rački, Đ., Carrea, G., et al., (2010). Guidelines for reporting of biocatalytic reactions. *Trends in Biotechnology*, *28,* 171–180.
81. Afoakwah, A., Owusu, J., Adomako, C., & Teye, E., (2012). Microwave-assisted extraction (MAE) of antioxidant constituents in plant materials. *Global Journal of Biosciences and Biotechnology*, *1,* 132–140.
82. Kerem, Z., German-Shashoua, H., & Yarden, O., (2005). Microwave-assisted extraction of bioactive saponins from chickpea (*Cicer arietinum* L). *Journal of the Science of Food and Agriculture*, *85*, 406–412.
83. Padmapriya, K., Dutta, A., Chaudhuri, S., & Dutta, D., (2012). Microwave assisted extraction of mangiferin from *Curcuma amada. 3 Biotech, 2*, 27–30.
84. Filip, S., Pavlić, B., Vidović, S., Vladić, J., & Zeković, Z., (2017). Optimization of microwave-assisted extraction of polyphenolic compounds from *Ocimum basilicum* by response surface methodology. *Food Analytical Methods, 10,* 2270–2280.
85. Bandar, H., Hijazi, A., Rammal, H., Hachem, A., Saad, Z., & Badran, B., (2013). Techniques for the extraction of bioactive compounds from Lebanese *Urtica dioica. American Journal of Phytomedicine and Clinical Therapeutics*, *6*, 507–513.
86. Nour, A. H., Oluwaseun, A. R., Nour, A. H., Omer, M. S., & Ahmed, N., (2021). Microwave-assisted extraction of bioactive compounds. *Microwave heating. Electromagnetic Fields Causing Thermal and Non-thermal Effects*, 1–31.
87. Chen, L., Song, D., Tian, Y., Ding, L., Yu, A., & Zhang, H., (2008). Application of on-line microwave sample-preparation techniques. *TrAC Trends in Analytical Chemistry*, *27*, 151–159.
88. Chan, C. H., Yusoff, R., Ngoh, G. C., & Kung, F. W. L., (2011). Microwave-assisted extractions of active ingredients from plants. *Journal of Chromatography A*, *1218*(37), 6213–6225.
89. Tatke, P., & Jaiswal, Y., (2011). An overview of microwave-assisted extraction and its applications in herbal drug research. *Research Journal of Medicinal Plant*, *5*, 21–31.

90. Deo, S., Janghel, A., Raut, P., Bhosle, D., Verma, C., Kumar, S. S., Agrawal, M., et al., (2015). Emerging microwave assisted extraction (MAE) techniques as an innovative green technologies for the effective extraction of the active phytopharmaceuticals. *Research Journal of Pharmacy and Technology*, *8*, 655–666.
91. Alara, O. R., Abdurahman, N. H., Ukaegbu, C. I., & Azhari, N. H., (2018). *Vernonia cinerea* leaves as the source of phenolic compounds, antioxidants, and anti-diabetic activity using microwave-assisted extraction technique. *Industrial Crops and Products*, *122*, 533–544.
92. Terigar, B. G., Balasubramanian, S., Boldor, D., Xu, Z., Lima, M., & Sabliov, C. M., (2010). Continuous microwave-assisted isoflavone extraction system: Design and performance evaluation. *Bioresource Technology*, *101*, 2466–2471.
93. Chemat, F., & Cravotto, G., (2012). *Microwave-Assisted Extraction for Bioactive Compounds: Theory and Practice* (Vol. 4). Springer Science & Business Media.
94. Chan, C. H., Lim, J. J., Yusoff, R., & Ngoh, G. C., (2015). A generalized energy-based kinetic model for microwave-assisted extraction of bioactive compounds from plants. *Separation and Purification Technology*, *143*, 152–160.
95. Piñeiro, Z., Marrufo-Curtido, A., Serrano, M. J., & Palma, M., (2016). Ultrasound-assisted extraction of stilbenes from grape canes. *Molecules*, *21*, 784.
96. Aguiló-Aguayo, I., Walton, J., Viñas, I., & Tiwari, B. K., (2017). Ultrasound-assisted extraction of polysaccharides from mushroom by-products. *LWT Food Science and Technology*, *77*, 92–99.
97. Kumar, K., Srivastav, S., & Sharanagat, V. S., (2020). Ultrasound-assisted extraction (UAE) of bioactive compounds from fruit and vegetable processing by-products: A review. *Ultrasonics Sonochemistry*, 105325.
98. Palma, M., & Barroso, C. G., (2002). Ultrasound-assisted extraction and determination of tartaric and malic acids from grapes and winemaking by-products. *Analytica Chimica Acta*, *458*, 119–130.
99. Samaram, S., Mirhosseini, H., Tan, C. P., Ghazali, H. M., Bordbar, S., & Serjouie, A., (2015). Optimization of ultrasound-assisted extraction of oil from papaya seed by response surface methodology: Oil recovery, radical scavenging antioxidant activity, and oxidation stability. *Food Chemistry*, *172*, 7–17.
100. Zhang, Z. S., Wang, L. J., Li, D., Jiao, S. S., Chen, X. D., & Mao, Z. H., (2008). Ultrasound-assisted extraction of oil from flaxseed. *Separation and Purification Technology*, *62*(1), 192–198.
101. Wang, W., Ma, X., Xu, Y., Cao, Y., Jiang, Z., Ding, T., Ye, X., & Liu, D., (2015). Ultrasound-assisted heating extraction of pectin from grapefruit peel: Optimization and comparison with the conventional method. *Food Chemistry*, *178*, 106–114.
102. Xu, Y., Zhang, L., Bailina, Y., Ge, Z., Ding, T., Ye, X., & Liu, D., (2014). Effects of ultrasound and/or heating on the extraction of pectin from grapefruit peel. *Journal of Food Engineering*, *126*, 72–81.
103. Raza, A., Li, F., Xu, X., & Tang, J., (2017). Optimization of ultrasonic-assisted extraction of antioxidant polysaccharides from the stem of *Trapa quadrispinosa* using response surface methodology. *International Journal of Biological Macromolecules*, *94*, 335–344.
104. Maran, J. P., & Priya, B., (2014). Ultrasound-assisted extraction of polysaccharide from *Nephelium lappaceum* L. fruit peel. *International Journal of Biological Macromolecules*, *70*, 530–536.

105. Kazemi, M., Khodaiyan, F., & Hosseini, S. S., (2019). Eggplant peel as a high potential source of high methylated pectin: Ultrasonic extraction optimization and characterization. *LWT-Food Science and Technology, 105*, 182–189.
106. Hosseini, S. S., Khodaiyan, F., Kazemi, M., & Najari, Z., (2019). Optimization and characterization of pectin extracted from sour orange peel by ultrasound-assisted method. *International Journal of Biological Macromolecules, 125*, 621–629.
107. Moorthy, I. G., Maran, J. P., Muneeswari, S., Naganyashree, S., & Shivamathi, C. S., (2015). Response surface optimization of ultrasound-assisted extraction of pectin from pomegranate peel. *International Journal of Biological Macromolecules, 72*, 1323–1328.
108. Guandalini, B. B. V., Rodrigues, N. P., & Marczak, L. D. F., (2019). Sequential extraction of phenolics and pectin from mango peel assisted by ultrasound. *Food Research International, 119*, 455–461.
109. Leong, T., Ashokkumar, M., & Kentish, S. (2011). The Fundamentals of Power Ultrasound – A Review. *Acoustics Australia, 39*(2), 54–63.
110. Ramić, M., Vidović, S., Zeković, Z., Vladić, J., Cvejin, A., & Pavlić, B., (2015). Modeling and optimization of ultrasound-assisted extraction of polyphenolic compounds from *Aronia melanocarpa* by-products from filter-tea factory. *Ultrasonics Sonochemistry, 23*, 360–368.
111. Rodrigues, S., Fernandes, F. A., De Brito, E. S., Sousa, A. D., & Narain, N., (2015). Ultrasound extraction of phenolics and anthocyanins from *jabuticaba* peel. *Industrial Crops and Products, 69*, 400–407.
112. Pan, G., Yu, G., Zhu, C., & Qiao, J., (2012). Optimization of ultrasound-assisted extraction (UAE) of flavonoids compounds (FC) from hawthorn seed (HS). *Ultrasonics Sonochemistry, 19*, 486–490.
113. SaraJliJa, H., ČukelJ, N., Mršić, G. N. D., Brnčić, M., & Ćurić, D., (2012). Preparation of flaxseed for lignan determination by gas chromatography-mass spectrometry method. *Czech Journal of Food Sciences, 30*, 45–52.
114. Popova, I., Hall, C., & Kubatova,´A., (2009). Determination of lignans in flaxseed using liquid chromatography with time-of-flight mass spectrometry. *Journal of Chromatography A, 1216*, 217–229.
115. Zhang, F., Chen, B., Xiao, S., & Yao, S. Z., (2005). Optimization and comparison of different extraction techniques for sanguinarine and chelerythrine in fruits of *Macleaya cordata* (Willd) R. Br. *Separation and Purification Technology, 42*, 283–290.
116. Urbano, M., De Castro, M. D. L., Pe´rez, P. M., Garcı´a-Olmo, J., & Gomez-Nieto, M. A., (2006). Ultraviolet visible spectroscopy and pattern recognition methods for differentiation and classification of wines. *Food Chemistry, 97*, 166–175.
117. Hazra, K., Roy, R., Sen, S., & Laskar, S., (2007). Isolation of antibacterial pentahydroxy flavones from the seeds of *Mimusops elengi* Linn. *African Journal of Biotechnology, 6*(12).
118. Sasidharan, S., Chen, Y., Saravanan, D., Sundram, K., & Latha, L. Y., (2011). Extraction, isolation, and characterization of bioactive compounds from plants' extracts. *African Journal of Traditional, Complementary and Alternative Medicines, 8*(1).
119. Christophoridou, S., Dais, P., Tseng, L. H., & Spraul, M., (2005). Separation and identification of phenolic compounds in olive oil by coupling high-performance liquid chromatography with post-column solid-phase extraction to nuclear magnetic resonance spectroscopy (LC-SPE-NMR). *Journal of Agricultural and Food Chemistry, 53*, 4667–4679.

120. Narayan, C., & Kumar, A., (2013). Identification and characterization of phenolic compounds in hydro methanolic extract *Achyranthes aspera* (HMEA) by UPLC and MALDI-TOF-MS and in vivo antioxidant activity. *Oriental Pharmacy and Experimental Medicine, 13*, 51–59.
121. Corrales, M., Toepfl, S., Butz, P., Knorr, D., & Tauscher, B., (2008). Extraction of anthocyanins from grape by-products assisted by ultrasonics, high hydrostatic pressure, or pulsed electric fields: A comparison. *Innovative Food Science & Emerging Technologies, 9*, 85–91.
122. Plaza, L., Sánchez-Moreno, C., De Ancos, B., Elez-Martínez, P., Martín-Belloso, O., & Cano, M. P., (2011). Carotenoid and flavanone content during refrigerated storage of orange juice processed by high-pressure, pulsed electric fields and low pasteurization. *LWT – Food Science and Technology, 44,* 834–839.
123. Jayathunge, K., Stratakos, A. C., Cregenzán-Albertia, O., Grant, I. R., Lyng, J., & Koidis, A., (2017). Enhancing the lycopene in vitro bioaccessibility of tomato juice synergistically applying thermal and non-thermal processing technologies. *Food Chemistry, 221,* 698–705.
124. Drosou, C., Kyriakopoulou, K., Bimpilas, A., Tsimogiannis, D., & Krokida, M., (2015). A comparative study on different extraction techniques to recover red grape pomace polyphenols from vinification byproducts. *Industrial Crops and Products, 75*, 141–149.
125. Espinosa-Pardo, F. A., Nakajima, V. M., Macedo, G. A., Macedo, J. A., & Martínez, J., (2017). Extraction of phenolic compounds from dry and fermented orange pomace using supercritical CO_2 and cosolvents. *Food and Bioproducts Processing, 101*, 1–10.
126. Kehili, M., Kammlott, M., Choura, S., Zammel, A., Zetzl, C., Smirnova, I., Allouche, N., & Sayadi, S., (2017). Supercritical CO_2 extraction and antioxidant activity of lycopene and β-carotene-enriched oleoresin from tomato (*Lycopersicum esculentum* L.) peels by-product of a Tunisian industry. *Food and Bioproducts Processing, 102*, 340–349.
127. Manzoor, M., Singh, J., & Gani, A., (2022). Exploration of bioactive peptides from various origin as promising nutraceutical treasures: In vitro, in silico and in vivo studies. *Food Chemistry, 373*, 131395.
128. Appeldoorn, M. M., Vincken, J. P., Aura, A. M., Hollman, P. C., & Gruppen, H., (2009). Procyanidin dimers are metabolized by human microbiota with 2-(3, 4-dihydroxyphenyl) acetic acid and 5-(3, 4-dihydroxyphenyl)-γ-valerolactone as the major metabolites. *Journal of Agricultural and Food Chemistry, 57*, 1084–1092.
129. Rein, M. J., Renouf, M., Cruz-Hernandez, C., Actis-Goretta, L., Thakkar, S. K., & Da Silva, P. M., (2013). Bioavailability of bioactive food compounds: A challenging journey to bioefficacy. *British Journal of Clinical Pharmacology, 75*, 588–602.
130. Brand, W., Schutte, M. E., Williamson, G., Van, Z. J. J., Cnubben, N. H., Groten, J. P., Van, B. P. J., & Rietjens, I. M., (2006). Flavonoid-mediated inhibition of intestinal ABC transporters may affect the oral bioavailability of drugs, food-borne toxic compounds and bioactive ingredients. *Biomedicine & Pharmacotherapy, 60*, 508–519.
131. Di, L., & Kerns, E., (2015). *Drug-Like Properties: Concepts, Structure Design and Methods from ADME to Toxicity Optimization*. Academic press.
132. Birben, E., Sahiner, U. M., Sackesen, C., Erzurum, S., & Kalayci, O., (2012). Oxidative stress and antioxidant defense. *World Allergy Organization Journal, 5*, 9–19.
133. Rahman, K., (2007). Studies on free radicals, antioxidants, and co-factors. *Clinical Interventions in Aging, 2*, 219–236.

134. Yang, H. Y., & Lee, T. H., (2015). Antioxidant enzymes as redox-based biomarkers: A brief review. *BMB Reports, 48,* 200.
135. Núñez, S. A. J., Daglia, M., & Rastrelli, L., (2016). The potential role of Mangiferin in cancer treatment through its immunomodulatory, anti-angiogenic, apoptotic, and gene regulatory effects. *BioFactors, 42,* 475–491.
136. Chávez-Santoscoy, R. A., Lazo-Vélez, M. A., Serna-Sáldivar, S. O., & Gutiérrez-Uribe, J. A., (2016). Delivery of flavonoids and saponins from black bean (*Phaseolus vulgaris*) seed coats incorporated into whole wheat bread. *International Journal of Molecular Sciences, 17,* 222.
137. Lozano-Sánchez, J., Bendini, A., Di Lecce, G., Valli, E., Gallina, T. T., & Segura-Carretero, A., (2017). Macro and micro functional components of a spreadable olive by-product (pâté) generated by new concept of two-phase decanter. *European Journal of Lipid Science and Technology, 119,* 1600096.

CHAPTER 3

Chemical Characteristics, Profile, and Physicochemical Aspects of Bioactive Compounds and Nutraceuticals from Natural Sources

NARENDER K. CHANDLA,[1] SAKSHI SUKHIJA,[2] NITIN WAKCHAURE,[3] and SUNIL K. KHATKAR[1]

[1]*Department of Dairy Engineering, College of Dairy Science & Technology, Guru Angad Dev Veterinary & Animal Sciences University, Ludhiana, Punjab, India*

[2]*Department of Biotechnology Engineering and Food Technology, UIE, Chandigarh University, Mohali, Punjab, India*

[3]*Department of Dairy Economics and Business Management, College of Dairy Science & Technology, Guru Angad Dev Veterinary & Animal Sciences University, Ludhiana, Punjab, India*

ABSTRACT

Bioactive chemicals are different from essential nutrients, and they have an impact on promoting human health. These are additional nutritional components found in modest amounts in a variety of foods. All plants contain phenolic chemicals, particularly flavonoids, which have been extensively investigated in cereals, legumes, nuts, olive oil, vegetables, fruits, tea, and red wine. Plant-derived bioactive chemicals and milk peptides are considered promising sources of unique biochemical structures for the development of new natural products/drugs for human health promotion and their well-being. Many studies are interested in bioactive peptides as a health-promoting

Bioactive Compounds and Nutraceuticals from Dairy, Marine, and Nonconventional Sources: Extraction Technology, Analytical Techniques, and Potential Health Prospects.
Charanjit Singh Riar, Parmjit S. Panesar (Eds.)

functional diet. Preliminary positive effects of milk-derived bioactive peptides on target disorders are also being evaluated as chemotherapeutic drugs. The functional characteristics of bioactive components are strongly influenced by extraction, purification, and methods adopted. To convert these peptides' functional qualities into food applications, extraction, separation, and identification of these peptides, as well as their drug delivery mechanism, are required. Scientific research and industrial development aimed at discovering novel bioactive peptides which have the potential to generate a number of drugs and health-promoting functional foods for their use in food and non-food applications in the near future.

3.1 INTRODUCTION

Bioactive chemicals, which are found in minute amounts in many plants and foods, are something other than nutrition. Phytochemicals, which can be derived from plants, foods, and food byproducts, are the most common bioactive substances. Bioactive chemicals can regulate the metabolic operations of the human body, resulting in positive outcomes [1]. Bioactive substances strengthen the human immune system, aid in the treatment of chronic diseases, and lower the risk of life-threatening illnesses such as cancer, cardiovascular disease, and diabetes in diabetics [2]. A wide range of antioxidants, anti-proliferation, and anti-inflammatory bioactive substances are found in nutraceuticals. As a result of their health-promoting and disease-prevention properties, bioactive substances and nutraceuticals are given significant consideration. These bioactive components are produced either naturally (natural biosynthesis) or artificially (metabolic bioengineering). Many reputable universities have recently conducted research on bioactive compounds' impact on inflammation prevention, functional foods' role in chronic disease prevention, polyphenols' role in bone health, and the effect of bioactive active components on the neuro-protective system. NBCs have a wide range of bio-structures and functions, making them a great source of bioactive molecules for nutraceuticals, functional foods, and food additives in a wide range of food compositions. Nutraceuticals are concentrated versions of bioactive components found in herbs, plants, and food. Bioactive components and nutraceuticals elements are derived from a variety of traditional and non-conventional sources, including pseudocereals and by-products, and offer a great deal of promise in the treatment of a variety of chronic conditions. Functional foods, on the other hand, have a specialized nutraceutical ingredient as an active component for specific health advantages [3–5]. Nutraceuticals are phytochemicals found naturally in various plant parts such as roots, leaves, flowers, fruits, seeds, and stems, which protect the plant from a variety of diseases. Functional foods are rich in phytochemicals, which act as antioxidants, and dietary fiber, which provide health advantages when consumed in moderation. Herbal medicine has been utilized for millennia, particularly in traditional folk medicine. The use of therapeutic and aromatic herbs in human culture has a long history [4, 5].

It's worth noting that soluble dietary fiber can dissolve or absorb water and is excellent at binding toxins and cholesterol in the intestinal tract, so improving gut health and intestine absorption capability. Insoluble dietary fiber, on the other hand, does not dissolve in water and is useful for bulking up the intestine and speeding up the transit of food through it. Insoluble dietary fiber works by diluting possible carcinogens and reducing toxins and carcinogens' contact with the intestinal system by hastening their exit from the body. As a result, dietary fiber as a nutraceutical element aids in the proper functioning of the microbial gut flora and intestine. Apples, cranberries, mango, oranges, asparagus, broccoli, carrots, peanuts, walnuts, most legumes, oats, and psyllium are all high in soluble dietary fiber. Apples, bananas, berries, broccoli, green peppers, spinach, almonds, sesame seeds, most legumes, brown rice, whole wheat products, cereals, and pseudocereals are all high in insoluble dietary fiber. Nutraceuticals derived from natural sources are currently being utilized to prevent disease, including inflammation, hypochelostrolemia, anti-proliferation, antimicrobial, neurological disorders, and many other chronic diseases [4]. Figure 3.1 depicts the properties of bioactive functional foods. Isoflavones, or phytoestrogens, on the other hand, have estrogen antagonist activity, resulting in lower overall estrogen exposure in premenopausal women and a lower risk of breast cancer. Phytoestrogen-rich diets prevent hormone-sensitive increases in plasma cholesterol levels and bone loss in postmenopausal women [6–8].

There is a great demand for nutraceuticals from natural sources, and the drug bank database included 116 nutraceutical ingredients in June 2020, indicating that bioactive components have a wide range of applications around the world [9]. Furthermore, future research will focus on using nutraceuticals as pharmacological substances to treat a variety of chronic disorders. As a result, the future of drug development will be to assess the mechanism of action of certain nutraceutical components to perform medicinal acts in the treatment of diseases such as Alzheimer's, rheumatic, cancer, diabetes, and Parkinson's [10].

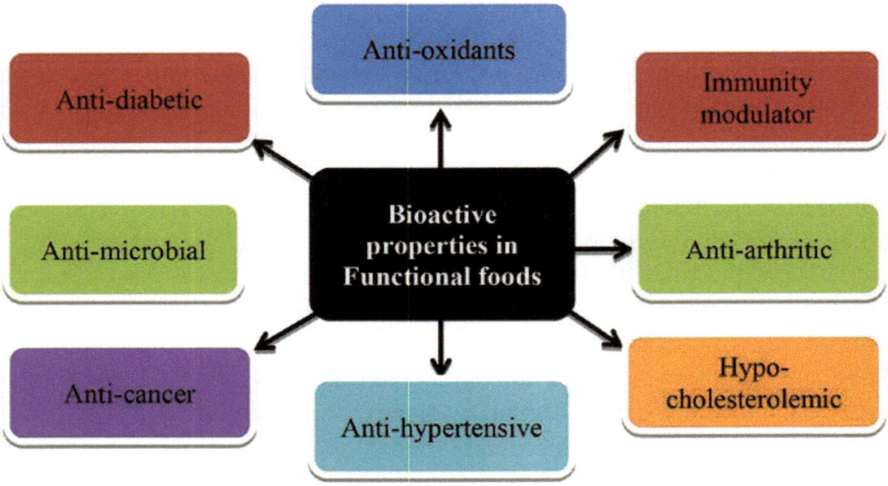

FIGURE 3.1 Bioactive properties of functional foods.

3.1.1 TYPES OF BIOACTIVE COMPOUNDS

Additional nutritional elements found in low concentrations in plant foods are known as food bioactive components. Figure 3.2 depicts the classification of functional meals. Polyphenols, phytosterols, tocopherols, carotenoids, organosulfur, tocotrienols, chemicals such as isothiocyanates and sulfide compounds, soluble and insoluble fiber, and fruto-ogliosaccharide are all bioactive dietary components found in plant-based foods and food products [4, 11]. Terpenes and terpenoids, phenolic compounds, and alkaloids are the most common bioactive chemicals. To extract these bioactive phytochemicals, most entire foods and plant sources are utilized. Polyphenols are an important class of bioactive chemicals found in plant sources. These phytochemicals are thought to be capable of influencing metabolic activity, resulting in positive benefits. Many experts throughout the world agree that carotenoids, tocopherols, phytosterols, phenolic acids, flavonoids, and saponins are among the bioactive substances that provide health benefits by reducing the risk of human diseases. The major bioactive elements taken from plant sources as nutraceutical supplements are polyphenols, also known as flavonoids, tannins, stillbenes, lignans, and phenolic acids, as shown in Figure 3.2.

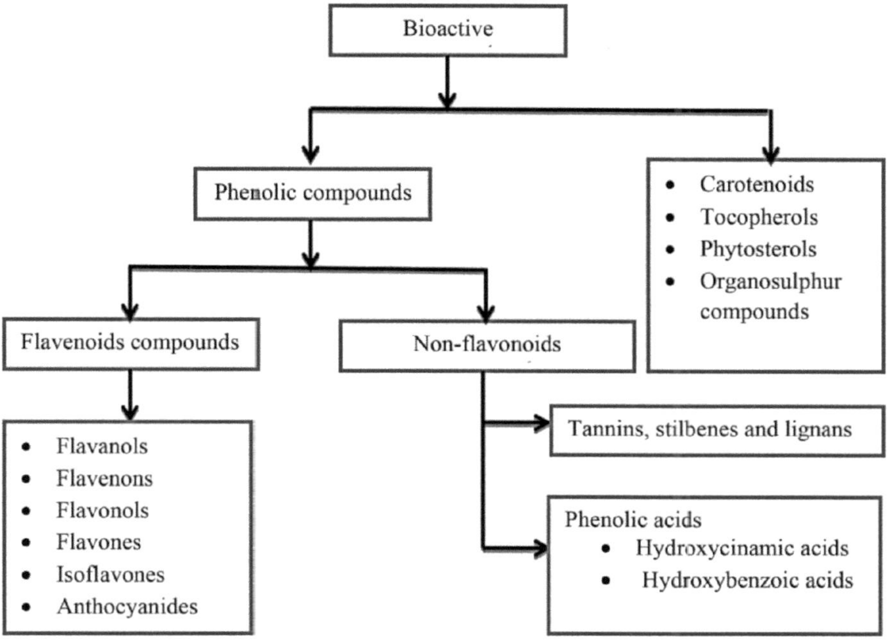

FIGURE 3.2 Classification of bioactive compounds.

Most of foods contain fats and oils; they are sensitive to oxidation reactions, resulting in oxidative damage to the overall food product during storage. Polyphenols can scavenge free radicals by removing reactive oxygen molecules from the environment. Free radicals are created during metabolism (aerobic cells) and are absorbed due to the bioactive substances' strong antioxidant qualities, which are mostly attributable to the hydroxyl group. This method has been shown to defend against cancer, degradation of the nervous system, cardiovascular disease, diabetes, and obesity. Coffee, tea, cocoa, fruits, vegetables, nuts, and most aromatic plants are some of the most common plant sources of polyphenols [11–17].

3.2 CHEMICAL CHARACTERISTICS AND PROFILING OF BIOACTIVE COMPONENTS

The chemical profile of a plant is determined by biosynthetic and physiological changes that occur during its growth. Although the chemical profile of bioactive components from various sources varies, the key bioactive components recovered showed comparable compound profiling. Terpenoids, carotenoids, phenylpropanoids, and alkaloids are bioactive components that are commercially extracted. The chemical profile of the extracted bioactive components is mostly determined by secondary metabolites (SMs), and extraction procedures have a substantial impact on the profile. SMs such as polyketides, isoprenoids, alkaloids, phenylpropanoids, and flavonoids are synthesized during biosynthesis in natural plant sources [18] and chemical compounds are synthesized during a specific growth period, at a specific stage, and most of them are seasonal [19]. Extracts of bioactive compounds such as ellagic acid, corilagin, vincristine, farnesene, and others are used to make nutraceutical ingredients for pharmaceutical applications, and they've also been used as food additives to improve food functionality [20, 21]. Chemical molecules are, however, derived from microbial/algal sources as well as metabolic engineering sources. They are considered as food extracts possessing various health benefits (Table 3.1). These extracted bioactive chemicals are often used as nutraceutical supplements, flavor enhancers, texturizers, bio-emulsifiers, acidulates, and thickeners in the food matrix.

3.2.1 CHEMICAL PROFILE OF BIOACTIVE COMPOUNDS IN FRUITS

The bioactive components are found in most fruits, but berries have a high concentration. Berries are considered great foods for a healthy diet since they contain many bioactive components. The main berries sources of bioactive compounds are blackberry, bilberry, blackcurrant, blueberry, chokeberry, cranberry, bayberry, raspberry, black raspberry, and strawberry, which are commonly taken either fresh or processed berries [22–24]. Phenolic acids, anthocyanins, tannins, stilbens, and carotenoids are all functional qualities of bioactive substances found in berries, and anthocyanins are an important group of polyphenols in berries. Berries such as bilberry, chokeberry, blackcurrant, honeyberry, grapes, blackberry, and berrycactus have high levels of anthocyanins [24, 25]. *S. cumini* is a traditional medicinal herb that contains a wide range of bioactive chemicals in all regions of the plant. Myricetin, oxalic acid, gallic acid, citronellol, cyanidin di-glucoside, hotrienol, phytosterols, flavonoids, carotenoids, and polyphenols, as well as minerals, are the main bioactive chemicals found in the edible section, which provide several health advantages. These bioactive chemicals have the potential to prevent/reduce a variety of ailments [26].

Citrus medica (Citron) is an unexplored fruit plant that has bioactive components in all its sections. Iso-limonene, citral, limonene, phenolics, flavonones, vitamin C, pectin, linalool, decanal, and nonanal are the main bioactive substances found in this fruit. These bioactive components account for a large portion of the health benefits. The anti-catarrhal, capillary protector, anti-hypertensive, diuretic, antibacterial, antifungal, anthelmintic, antimicrobial, analgesic, strong antioxidant, anti-cancerous, anti-diabetic, estrogenic, anti-ulcer, cardioprotective, and anti-hyperglycemic properties of this fruit have been proven in pharmacological studies [26].

Mango is a widely consumed fruit that contains fat, proteins, and beneficial substances. Mango kernels are not widely used, but studies show that they contain a wealth of nutrients and phytochemicals, including tocopherols, phytosterols, carotenoids, polyphenols (gallotannins, flavonols, benzophenone derivatives, mangiferin, homomangiferin, isomangiferin, anthocyanins, kaempferol, and quercetin), and phenolic acids (4-caffeoylquinic acids, caffeic, coumaric, ellagic, gallic, and ferulic acid (FA)). Antioxidant, antibacterial, anticancer, and antiproliferative properties are all present in phytochemicals. Total phenolic compounds in mango seed kernels range from 18.19 to 200.35 mg GAE/g db, depending on the cultivar and extraction method used. According to a study, there are 20.7 mg tannin, 20.2 mg vanillin, 12.6 mg coumarin, 11.2 mg cinnamic acid (CiA), 10.4 mg FA, 7.7 mg caffeic acid (CA), 6.0 mg gallic acid, and 4.2 mg mangiferin per 100 g of dry mango kernel [27]. Wastewater from the olive mill and pomace of wet olive is considered the materials of treatment, minimization, and prevention of many diseases as the prospect of being better utilized as an unconventional sources to recover bioactive compounds, likewise phenols and dietary fibers.

3.2.2 CHEMICAL PROFILE OF BIOACTIVE COMPOUNDS IN PSEUDO-FRUITS

The potential for emerging species with nutraceutical qualities of bioactive chemicals from Hovenia dulcis pseudo-fruits to be used in food applications is being assessed. Hovenia dulcis Thunberg (Rhamnaceae) is a hardy deciduous tree native to Japan, Korea, and China, although it has also been found to grow in northern Thailand and Vietnam's highland woods [28]. *H. dulcis*, sometimes known as the Japanese/Chinese raisin tree, is a pioneer tree. This pseudo-fruit contains substantial levels of (epi)-gallocatechin during maturation, which contributes to their antibacterial effects. During this stage of maturation, the tree's soluble sugar content increased, reducing the harshness of polyphenolic chemicals (e.g., tannins) and resulting in a pleasant-pear-scented fruit pulp. Many people around the world accept this fruit pulp for human consumption [29].

3.2.3 CHEMICAL PROFILE OF BIOACTIVE COMPOUNDS IN PLANTS

The doum palm is one of the most significant plants for weaving baskets since it produces fiber and leaflets. Antioxidants, tannins, phenols, sapocosides, flavonoids, tepenes, and terpenoids are all found in Doum nuts. The plant also has anti-inflammatory, anticancer, and antioxidant properties, as well as the ability to extract phenolics, flavonoids, and essential oils (EOs). When doum fruit was administered for 1 to 2 months, cholesterol, blood glucose, triglycerides (TAGs), and total lipids were considerably reduced [30].

Annona is utilized for medicinal and nutritional purposes in the form of stems, roots, seeds, pulp, and leaves, all of which have antibacterial characteristics. Anti-diabetic effects were found in A. Crassiflora. Anti-oxidant capabilities have been discovered in the Annona plant's fruit peel. Compounds isolated from Annona include phenols, alkanoids, acetogenins, and cyclopeptides [31]. Diabetes, hypercholesterolemia, hypertension, gastrointestinal problems, and cancer are all treated using different portions of this plant.

3.2.4 CHEMICAL PROFILE OF BIOACTIVE COMPOUNDS IN CEREALS

Cereals are high in protein, vitamins, minerals, fiber, and bioactive substances, among other things. These bioactive components are concentrated in the seed's outer layers, with phenolic compounds (phenolic acids and flavonoids) being one of the most important due to their health-promoting qualities. Although it is low in lysine and high in glutamic acid, the einkorn (ancient) wheat variety has been demonstrated to have a higher content of important amino acids (threonine, lysine, valine, methionine, leucine, isoleucine, and phenylalanine) than current species. Beta-glucans are polysaccharides made up of D-glucose molecules joined together by beta-glycosidic linkages. It is a component of dietary fibers, which are thought to provide a wide range of health advantages, including the treatment of gastrointestinal illnesses and immune system support. Many phytochemicals in colored wheat (*Triticum aestivum*) are responsible for a variety of health advantages. The outer aleurone layer of colored wheat (blue, black, purple, and red) includes a significant amount of anthocyanins and carotenoids [32]. Synthetic food colorants and dyes are being used less frequently by food regulatory and safety authorities and the food processing industry. High-value pigments are present in the bran layer (a milling industry co-product) may easily be removed and used as functional foods and natural colorants, making colored wheat essential for food processing companies. Extracted pigments, such as anthocyanin, can be used to replace synthetic colors in food, drugs, and cosmetics [33].

As a result, colored wheat has a high nutritious content as well as a high tensile strength, making it a popular bread ingredient. Carotenoids, total dietary fiber (TDF), total phenolic compounds (TPCs), and total anthocyanin compounds (TACs) are abundant in colored wheat products, which means they have strong antioxidant activity [33–36]. Black wheat has the most TPC, total flavonoid content (TFC), and antioxidant activity. According to Li et al. [94], while light purple wheat has the least. TPC and TFC concentration in black wheat were found to be 659.8 g gallic acid equivalents (GAEs) g1 and 319.3 g rutin equivalents g1, respectively, according to the study. Purple wheat has the highest TFC, with 21.59 to 102.95 mg of catechin equivalents per 100 g of wheat grain, for yellow, red, and common wheat was observed, for example, measured 13.44, 10.72, and 9.6 mg/100 g, respectively [37, 38].

Carotenoids, also known as tetraterpenoids, are a type of polyphenolic chemical that can be found in colored wheat as well as fruits and vegetables [39]. Carotene, cryptoxanthin, lutein, lutein dilinooleate, and zeaxanthin are the major carotenoids found in wheat. Antioxidant ability of colored wheat against free oxygen radicals and DPPH radicals is strong. Three assays are routinely used to determine antioxidant activity: DPPH, ABTS, and PCL (Photo-chemi-luminescence). Phenolic acids, flavones, flavonols, and anthocyanins are the key phenolic chemicals found in different colored wheat that have the most potent antioxidant properties. Anthocyanin content accounts for 69% of blue wheat's free radical scavenging activity, whereas phenolic acids account for 19%. The bran layer of cereals contains natural antioxidants and can be employed as a functional food ingredient or as a nutritional addition [40, 41].

3.2.5 CHEMICAL PROFILE OF BIOACTIVE COMPOUNDS IN MILLETS

Millets are coarse grains that are grown in many parts of the world to feed large populations in areas where wheat and rice cannot be grown owing to climate circumstances. Millets have been found to be high in polyphenols in several studies. Because millet is not a single plant species, the types and forms of bioactive chemicals isolated from diverse species, as well as their bioactivity and health-promoting characteristics, will vary significantly around the world [42, 43]. Millets offer a wide spectrum of bioactive phytochemicals that are helpful to human health, according to certain research. A research of finger millet found that it had hypoglycemic, hypocholesterolemic, and anti-ulcerative characteristics [44]. The majority of millet's bioactives are found in the seed coat, while the endosperm is mostly made up of carbohydrates and contains only a little amount of bioactives [42, 45].

3.2.6 CHEMICAL PROFILE OF BIOACTIVE COMPOUNDS IN PSEUDO-CEREALS

Psuedo-cereals (Buckwheat, Amaranth, and Quinoa) contain a variety of bioactive peptides from various families. Pseudo-cereals for nutraceutical extraction are being researched all over the world as an innovative and cost-effective non-conventional pseudo-cereal source. Pseudo-cereals are dicotyledonous plant species that are well-known for their nutraceutical properties. Pseudo-cereals are a good source of important nutrients and have the potential to provide health advantages due to the fiber and phytochemical content [5, 46]. Saponins, phenolic compounds, phytoecdysteroids, phytosterols, polysaccharides, betalains, and bioactive proteins and peptides are among the bioactive components found in pseudo-cereal grains [47]. The presence of phenolic and a wide variety of flavonoids (anthocyanins, isoflavonoids, flavonols, flavones, flavanones, and flavonoids) as well as derivatives of tyrosol was discovered in pseudocereals. Furthermore, squalene derived from amaranth grain flour oil inhibits the mRNA expression of 3-hydroxy-3-methylglutaryl coenzyme A, a major cholesterol biosynthesis reductase enzyme [48, 49]. Rutin is a glucoside flavonoid found in buckwheat, including its seed, and has a variety of pharmacological properties [47, 50]. Because of psuedocereals potential to scavenge free radicals, bind metal ions, and inhibit collagenase activity in the skin, pseudocereals have been postulated as promising bioactive compounds with anti-aging properties. In an insulin-resistant mouse model, buckwheat phagopyritols effectively suppressed increasing blood glucose, lowered fat levels, and improved insulin resistance *in vivo* [47]. Antioxidant, anti-inflammatory, and anti-diabetic properties are all found in phagopyritols. *Amaranth* spp. bioactive substances were found

TABLE 3.1 Bioactive Nutraceutical Extracted from Various Sources and Their Health Benefits

Sl. No.	Bioactive Ingredients	Source	Possible Bioactive
1.	Carotenoids: Alpha	Carrots, sweet potato, green vegetables, citrus fruits, tomatoes, and papaya, etc.	Radical, which accelerates the oxidation reaction and provides defense to cellular structure. Carotene, lutein, lycopene, curcumin shown anti-Alzheimer properties. However, lutein also shown anticancer activity in colon, zeaxanthin maintain the vision of eye, and lycopene is an antioxidant reduces the chance of prostate cancer.
2.	Polyphenolic compounds: quercetin, luteolin, apigenin, frulic acid, caffeic acids, flavanones, flavones, anthocyanins, resveratrol, curcumin, ellagic acid, tannins, and catechins.	Citrus fruits, fruit, vegetables, berries, turmeric roots (curcumin).	Curcumin: Antioxidants, reduce cardiovascular diseases, lower blood cholesterol and shown anti-Alzheimer properties antiproliferative (ellagic acid), protects skin from UV radiation (catechin).
3.	Dietary fiber: soluble fiber, insoluble fiber.	Beans, oats, barley, peas (soluble fiber sources) and wheat, corn bran (insoluble sources).	Reduce cardiovascular lower total cholesterol, and digestive stimulator (soluble fiber) however insoluble fiber benefits in reducing risk of breast and colon cancer and digestive stimulator.
4.	Fatty acids	Flaxseed, olive oils, canola oil milk, cheese, etc.	It possesses anti-inflammatory and cholesterol-lowering properties.
5.	Phytoestrogens	Soybean and legumes.	Antioxidants, cholesterol lowering and reduce menopause and improve the bone health.
6.	Terpenes: Tocopherols and tocotrienols	Cereals and palm oil.	Inhibit breast cancer in women and lower cholesterol.
7.	Phytosterols (sub-class of terpenes)	Green and yellow vegetables.	Anti-inflammatory, antineoplastic, antipyretic, possess immune modulating activities and facilitate the defecation of cholesterol in the body.
8.	Limonoids	Citrus fruits	Protect the lung tissue.
9.	Sulfides/thiols	Garlic and cruciferous vegetables.	Enhance the immunity

TABLE 3.1 *(Continued)*

Sl. No.	Bioactive Ingredients	Source	Possible Bioactive
10.	Fructo-oligosaccharides.	Whole grain, onions, artichokes.	Improve the microflora.
11.	Gallic, p-hydroxybenzoic, gentisic, vanillic, chlorogenic, sinapic, and ferulic acids, Ferulic, caffeic, p-coumaric, syringic acids, Quercetin, gallic, protocatechuic, Methyl vanillate, apigenin.	Millets (common millet, foxtail millet, finger millet, pearl millet, barnyard millet).	Anti-inflammatory, antiviral, and neuroprotective activities.
12.	Immunoglobulin, antibacterial peptides, antimicrobial proteins, oligosaccharides, and lipids. Conjugated linoleic acid, prebiotic, probiotic, bio-peptides and bioactive components along with amino acids and proteins molecules.	Milk and colostrum of bovine and other dairy species.	Antimicrobial, antihypertensive, anti-oxidative, anti-cytotoxic, immunomodulatory, opioid, digestive, endocrine, immune, and neurological activity and mineral carrying activities.
13.	Myricetin, oxalic acid, gallic acid, citronellol, cyanidin diglucoside, hotrienol, phytosterols, flavonoids, carotenoids, and polyphenols.	Herbal traditional medicinal plant plants (S. cumini).	Anticancer, antioxidant, antimicrobial, neuroprotective, antidiabetic, cardioprotective.

Source: Refs. [10, 26, 45, 49, 55, 57, 58, 60, 61].

to have antibacterial, gastroprotective, cardioprotective, hypolipidemic, anticancerous, antidiabetic, and antimicrobial effects [52]. Amaranth includes a considerable number of antihypertensive 11S and 7S globulin proteins. Bioactive peptides are produced when these proteins are broken down by digestive enzymes, and they have antihypertensive action up to 8 times higher than unmodified proteins, and they have a similar effect on decreasing blood pressure in rats as captopril. Quinoa produced from 11S globulin and other proteins contains 17 peptides with antioxidant potential [53]. Lunasin from amaranth can reach the cell nucleus and stop the transformation of fibroblast cancer cells, therefore avoiding cancer [18].

3.2.7 CHEMICAL PROFILE OF BIOACTIVE COMPOUNDS IN MEDICINAL PLANTS AND HERBS

Plants are a rich source of bioactive chemicals that have a variety of functions, including insect repellent action, antibacterial, and antifungal activity, and beneficial impacts on animals and humans. Polyphenols, natural colorants, essential oils (EOs), mineral compounds, vitamins, and many other phytochemicals with biological activity have been discovered in therapeutic plants. Medicinal herbs are typically described as any component of a plant that is consumed for its fragrant qualities. Herbs, on the other hand, have recently been found as sources of many phytochemicals, many of which have major biological activity and are of interest to various industries; today, medicinal herbs are of great importance and interest in food technology. Polyphenols are the most common bioactive chemicals found in medicinal and culinary herbs (especially phenolic acids and flavonoids). These chemicals are well-known for their biological activity, which includes anticancer, antioxidant, antibacterial, neuroprotective, antidiabetic, cardioprotective, and a variety of phenolic foam shown beneficial effects on the human body. Medicinal herbs, including oregano, sage, lavender, mint, lemon balm, and rosemary, have bioactive chemicals and sensory qualities (aroma, taste), making them useful in gastronomy and the food industry, particularly for developing high-value foods [62].

3.2.8 CHEMICAL PROFILE OF BIOACTIVE COMPOUNDS IN MILK

Milk has been more than a source of nutrients for any animal neonate, as well as for child growth and adult human nutrition. Milk is a complicated composition that is also a very nutritious food for humans. It's a fat globule emulsion in an aqueous environment. Lactose (milk sugar), protein (casein and whey), minerals, and certain vitamins can all be found in the aqueous portion of milk. It's also high in bio-peptides and other bioactive components, which are essential for body maintenance and the prevention of a variety of diseases. Milk is high in physiologically active compounds such as immunoglobulin, antibacterial peptides, antimicrobial proteins, oligosaccharides, and lipids, according to recent research. Milk contains conjugated linoleic acid (CLA), prebiotics, probiotics, biopeptides, and bioactive components, as well as amino acids and protein molecules. Cheese includes phosphor-peptides, which are proteolyzed during the ripening process [63]. Milk peptides are created by digestive enzymes breaking down milk proteins or by lactobacilli producing proteinase enzymes during milk fermentation [64]. Milk-derived bioactive peptides are normally 2–20 amino acids long and become active after being released from the precursor protein, where they are encrypted either *in vivo* or *in vitro* by digestion or proteolysis. Lipids, minerals, enzymes, vitamins, lactoferrin (LF), lactose, and oligosaccharides, immune-globulins, growth factors and cytokines, casein, and whey proteins, and derived peptides are all bioactive components of milk. Bioactive peptides can be found in a variety of foods, as well as dietary supplements, functional meals, and medical foods. Antimicrobial, antihypertensive, anti-oxidative, anti-cytotoxic, immune-modulatory, opioid, and mineral transporting capabilities are among the many biological activities and functions of these bioactive peptides [65, 66].

Natural bioactive components can be found in bovine and other dairy species' milk and colostrum. Biologically active peptides generated from milk are in an inactive form inside the sequence of the precursor molecules, and once liberated, they may impact a variety of physiological responses such as cardiovascular, digestive, endocrine, immunological, and neurological function. In the biological system, these bioactives and peptides have a variety of tertiary activities. The use of bio-peptides and the development of functional food products for human health is becoming increasingly important. Bioactive chemicals, such as alkaloids and flavonoids, are natural substances produced by plants and animals. Milk's diverse makeup and functionality can be found in just the protein components. Proteins are serum and glandular-derived substances with varying molecular sizes, concentrations, and functions [37, 67].

3.2.9 CHEMICAL PROFILE OF BIOACTIVE COMPOUNDS FROM MICROBIAL, ALGAL, AND METABOLIC SOURCES

Other than plant sources with significant extraction of bioactive components, microbial, algal, and metabolic sources are the principal natural sources for the extraction of valuable nutraceutical constituents, as illustrated in Figure 3.3.

3.2.9.1 MICROBIAL SOURCES

Microorganisms, particularly fungal microorganisms such as Penicillium, Aspergillus, and Streptomyces, are crucial in the synthesis of active molecules such as organic acids, enzymes, and antibodies [68]. Around 23,000 natural bioactive compounds (NBCs) have been found and isolated from microorganisms, which is rather remarkable [95]. Isoprenoids (carotene and lycopene), as well as phenylpropanoids such as stilbene derivatives (resveratrol), are separated from bacteria. However, only around 1.0% of microorganisms are cultured *in vitro*, leaving a vast number of microbes untapped for natural bioactive chemical extraction [69].

3.2.9.2 ALGAL SOURCES

Algae are an intriguing natural source for developing novel NBCs with antibacterial, antioxidant, and antiviral activities [70, 71]. Algae are an important part of the ecosystem in water, marine environments, and even barren sands, with more than 15,000 bioactive chemicals chemically synthesized from algae sources. Microalgae may produce a variety of useful components such as carotenoids, polyphenols, and flavonoids (quercetin, catechin, and others), and these chemicals are used in the food and pharmaceutical industries [72].

3.2.9.3 METABOLIC ENGINEERING

The creation of the target product can be hastened using biological reactions. Monitoring and controlling the biological process is critical, and metabolic/bio-engineering applications are an under-developed field that needs attention in order to discover novel paths for the production of NBSs.

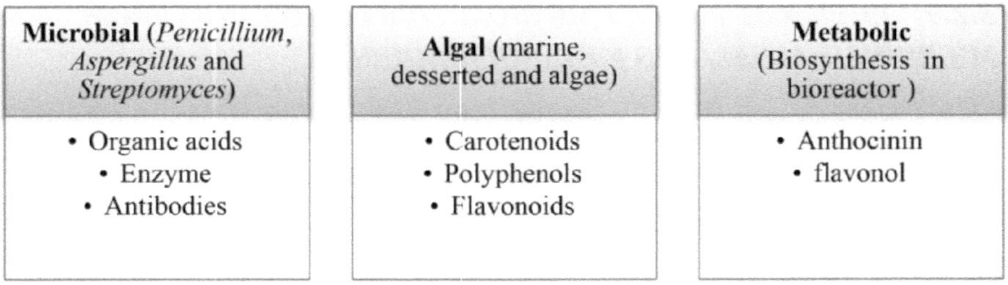

FIGURE 3.3 Microbial, algal, and metabolic sources of bioactive compounds production.

Many functional elements are altered for esthetic reasons, such as altering the anthocinin and flavonoid pathways to boost flavonol production in tomatoes [73]. SMs necessitate an understanding of the production route.

3.2.9.4 BIOACTIVE SUBSTANCES IN FUNGI

Bioactive compounds, also known as plant SMs, are low-molecular-weight compounds that are not required for life but are essential for the survival of the producing organism [71]. NMR, MS, and X-ray analysis have identified over 50,000 structures in plants. However, given that only around 20% of all plants have been investigated, the real number of SMs or bioactive substances in the plant kingdom is likely to exceed 1,00,000. [74, 75]. Fungi are affected by bioactive substances that interfere with molecular targets in their organs, tissues, and cells. Bio-membranes, proteins, and nucleic acids are the main targets. Bioactive chemicals and many volatile compounds presented anti-microbial properties and are still seen as a valuable resource for identifying new mechanisms of action. [76].

3.3 PHYSICOCHEMICAL ASPECTS OF BIOACTIVE COMPOUNDS AND NUTRACEUTICAL

3.3.1 EXTRACTION ASPECTS BIOACTIVE COMPONENTS

The global demand for nutraceuticals and functional meals with a wide range of chemical variety has prompted people to look for natural therapeutic solutions. Botanical-herbal formulations, which contain a variety of bioactive chemicals, are currently being studied extensively for medical applications. As a result, novel techniques, methodologies, and protocols were developed for the extraction, isolation, and characterization of bioactive components from natural resources. Some common difficulties and processing challenges must be re-examined with the advent of new isolation and extraction technologies due to their biochemical nature and aromatic volatility and bioreactivity. As extraction and the technique used to extract bioactive components are the most important unit operations that determine the yield, purity, and type of bioactive metabolites, as well as the magnitude or concentration of bioactive chemicals. Chemical analysis using phytochemical assays, chromatographic techniques such as thin layer chromatography (TLC) and high-performance chromatography (HPLC), immunoassay techniques, and Fourier transform infrared (IR) spectroscopy (FTIR) are all used to investigate bioactive mixtures found in plant-based extracts [80]. Several bioactive compounds are extracted from a variety of natural sources, including plants, animals, microorganisms, algae, and marine life, and these extracted nutraceuticals ingredients are used in a variety of functional foods, food additives, and preventive nutraceuticals-based new drug development. So, for the isolation/extraction of certain bioactive chemicals, solvent extraction (SE), ultrasonic extraction, liquid-liquid extraction, liquid-solid extraction, solid-liquid extraction, super critical extraction, and microwave extraction (MWE) techniques have been developed. Depending on the type of raw material used to extract the bioactive compound, the chemical sensitivity of bioactive ingredients, and the bio-economics of operational conditions, the required techniques, and mechanisms, as well as the bio-economics of operational conditions, are the parameters that are critical for deciding on isolation purposes [78, 79].

With the considerable use of byproducts from the agri-food-dairy business, sustainable food production can be achieved. Researchers are now focusing on valorizing waste (agro-industrial and dairy waste), which entails transforming byproducts and waste materials into value-added food and non-food items, as well as bioactive substances and biofuels, to improve the bio-economics of food production operations [80]. The type of chemical compounds and their reactivity toward suitable use for edible and non-edible is highly crucial to decide their employment, as are the concentrations of bioactive chemicals with the amount produced. These bioactive chemicals have demonstrated the capacity to affect a variety of metabolic processes as well as antioxidant capabilities. Certain bioactive components impacted gene expression by exhibiting inhibitory or induction capabilities toward enzymes. Baby foods with different kind of fruits used as ingredients showed higher antioxidant capacity. Among the commercial fruit-based baby foods after gentle steam cooking process presented high levels of bioactive compounds and antioxidant capacity [81, 82].

3.3.2 EFFECTS OF PROCESSING ON BIOACTIVE COMPONENTS

The processing method alters the content of bioactive food components, resulting in changes that may be advantageous or detrimental to health promotion. The extraction of bioactive components and, as a result, the features were shown to be strongly influenced by the processing method, processing time, and duration of post-harvest storage. It has been shown that procedures such as steam blanching and most of the thermal processing affect the bioactive retention in Cruciferae and allium vegetables [83, 84]. During processing, bioactive isothiocyanates have been lost in the range of 30 to 80% [83]. Key enzymes such as myrosinase in Cruciferae and allinase in allium plants, were shown to be damaged by temperatures of more than 100°C. The bioavailability of carotenoids and other lipid-soluble bioactives in food components, on the other hand, enhanced with processing. This could be owing to increases in surface area caused by size reduction (chopping and cutting) and heat treatments that break down the protein and carbohydrate food matrix that binds carotenoids [85, 86].

3.3.3 PHYSICOCHEMICAL ASPECTS OF BIOACTIVE COMPOUNDS FROM FRUITS

3.3.3.1 MANGO KERNELS

Mango kernels are high in calcium, potassium, magnesium, phosphorus, and vitamins A, E, K, and C, among other macronutrients and micronutrients. On a dry weight basis, mango kernels have a nutritional composition of 32.34% to 76.81% carbohydrate, 6% to 15.2% fat, 6.36% to 10.02% protein, 0.26% to 4.69% crude fiber, and 1.46% to 3.71% ash [87].

3.3.3.2 DRIED GOJI BERRIES

Dried goji berries 100-gram sample of contained 31.01.62 mg of ascorbic acid. The following are the mineral contents and percentage contribution to the RDA of 100 gram of dry goji berry fruits: Calcium is 49.0 mg (5.0%), phosphorus is 370.0 mg (67.2%), sodium is 1.32 mg (94.2%), potassium is 193.0 mg (>4,000%), magnesium is 120.0 mg (36.9%), iron is 0.04 mg (0.3%), copper is 0.01 mg (0.7%), and manganese is 0.008 mg (0.26%) [88–90].

3.3.4 PHYSICOCHEMICAL ASPECTS OF BIOACTIVE COMPOUNDS FROM CEREALS

3.3.4.1 ORYZANOL

Oryzanol in rice, avenanthramides, and saponins in oat, beta-glucans in oat and grain, alkyl resorcinols in rye, and anthocyanins and carotenoids in colored cereal grains were all found in cereals [91, 92]. Aleurone, germ, and bran layer make up a full grain of wheat. Phytochemicals such as phenolic acids, carotenoids, vitamin E, lignans, glucans, inulin, resistant starch, sterols, and phytates are abundant in these plants. Maximum bioactive mixtures are thus lacking in refined wheat products due to the evacuation of the grain and germ. Normal wheat includes 2.15% ash, 12.93% protein, 3.33% oil, 61.85% starch, and 2.29% cellulose. According to Zilic et al. [96], all of which are significantly lower than colored wheat values, both subjectively and numerically. Colored wheat had a protein content of 11.74%–18.17% greater, 7.31%–18.13% higher essential amino acids, and 8.88%–18.91% higher total amino acids than normal wheat [93].

3.3.4.2 AMARANTH

Amaranth seed, root, and leaves include a variety of functional chemicals such as fatty acids, steroids, lipids, amino acids, vitamins, minerals, and bioactives, as well as alkanoids, flavonoids, and glycoces, as well as phenolic acids, tannins, and saponins [82].

3.3.4.3 QUINOA

Quinoa grain is excellent nutritional source of well-balanced proteins, soluble fibers, and squalene. It contains exceptional sufficient quantity of nutritional constituents like high gluten-free proteins, several minerals, and fatty acids. This super grain pseudocereal also contains many health-promoting pharmacoactive constituents that could be utilized for the prevention and management of curing several diseases. Quinoa is a beneficial pseudo-cereal in particularly patients suffering with celiac disease. It is a gluten-free grain and hence can be easily digested by celiac patients. This grain presented very strong anti-hypercholesterolemic activity which could help in prevention of heart disease [59].

3.3.5 PHYSICOCHEMICAL ASPECTS OF BIOACTIVE COMPOUNDS FROM OIL SEEDS

Iodine value, 86.0–125.0 g I2/100 g oil; saponification value, 171.0–190.7 mg of KOH/g of oil; acid value, 1.1–12.9 mg of KOH/g of oil; free fatty acid, 0.6–6.5 g/100 g of oil were the physicochemical parameters of the oils extracted. Except for bittermelon, where eleostearic acid was the predominant fatty acid, palmitic, oleic, and linoleic acids were the major fatty acids in all the extracted seed oils. The extracted plant oils contained gallic, protocatechuic, p-hydroxybenzoic, vanillic, caffeic, syringic, p-coumaric, and ferulic acids (FAs). Vanillic acid was found in abundance in all the extracted oils [56].

3.3.6 PHYSICOCHEMICAL ASPECTS OF BIOACTIVE COMPOUNDS FROM MEDICINAL PLANTS

Dried plants and herbs are used to flavor and odor food, as well as infuse color into it. Primary metabolites (PMs) are carbohydrates, polysaccharides, amino acids, sugars, protein, and lipids, and SMs are carbohydrates, polysaccharides, amino acids, sugars, protein, and lipids. SMs such as phenolics, steroids, and lignins are found in a variety of species. A chemical in medicinal plants modulates benefits to the physiology of sick people and is utilized to achieve health benefits. Medicinal plants are valuable because of their distinct physico-chemical properties. Alkaloids, terpenoids, coumarins, flavonoids, nitrogen compounds, phenolics, and anti-inflammatory, immune-stimulatory, anticancer, antioxidant, and antibacterial characteristics were found in these plants [56].

3.3.7 PHYSICOCHEMICAL ASPECTS OF BIOACTIVE COMPOUNDS FROM MILK

Fat (lipod), proteins, sugars (lactose), minerals (ash), and minor additional elements such as phospolipds, cholesterol, pigments, enzymes (analase, lipase, phosphate, protease, peroxidase, and catalase) and vitamins A, D, E, and K, as well as water soluble vitamins of the B complex group Glycerides are fatty acid esters found in milk. Saturated fatty acids are more stable than unsaturated fatty acids, yet unsaturated fatty acids play a significant role in the physico-chemical properties of milk. Proteins are made up of a variety of amino acids, some of which are essential and others that aren't. The production of body proteins requires essential amino acids in the diet. Casein, beta-lactoglobuline, and alpha lacta-albumin are the proteins found in milk, and casein is only found in milk in the form of cesium caseinate phosphate complex. All of these nutrients are utilized to promote nutrition and diet, and their presence helps to prevent diseases caused by deficiency [11, 51].

3.4 CONCLUSION

Plants produce a wide range of bioactive metabolites that act as anti-pest defense mechanisms. Plants and many foods contain mostly antimicrobial bioactive chemicals likewise terpenoids and essential oils (EOs), phenolics, and polyphenols, alkaloids, polypeptides, and mixes (crude extract). Consumer desire for healthy products and this quench of functional foods will drive the through the development of bioactive based nutraceutical food products supplemented with medicinal herbs into the 21st century. However, due to a lack of advanced technology, enhanced goods, and molecular techniques, little work has been done in the field of bioactive based nutraceutical and functional foods. There is need to focus on developing novel facilities, such as advanced proteomics approaches, recombinant enzyme technologies, and microbial fermentation processes, to investigate the various effects of bioactive peptides on gene expression and to optimize the nutritional and health effects of these compounds. As a result, allergenicity, toxicity, and the stability of biological functions during gastrointestinal digestion should all be investigated in the formulation of bioactive peptide-containing products.

KEYWORDS

- **alkaloids**
- **beta-lactoglobulin**
- **bioactive peptide**
- **biological functions**
- **microbial fermentation**
- **polypeptides**
- **polyphenols**

REFERENCES

1. Galanakis, C. M., & Kotsiou, K., (2017). Recovery of bioactive compounds from olive mill waste. In: *Olive Mill Waste* (pp. 205–229). Academic Press.
2. Volp, A. C. P., Renhe, I. R. T., & Stringueta, P. C., (2009). Pigmentos naturais bioativos. *Alimentos e Nutrição Araraquara, 20*(1), 157–166.
3. Gulati, O. G., Ottaway, P. B., Jennings, S., Coppens, P., & Gulati, N., (2019). Chapter 20 – botanical nutraceuticals (food supplements and fortified and functional foods) and novel foods in the EU, with a main focus on legislative controls on safety aspects. *Nutraceutical and Functional Food Regulations in the United States and Around the World* (3rd edn., pp. 277–321.
4. Tanna, B., & Mishra, A., (2019). Nutraceutical potential of seaweed polysaccharides: Structure, bioactivity, safety, and toxicity. *Compr. Rev. Food Sci. Food Saf., 18*, 817–831.
5. Gutierrez-Del-Rıo, Fernandez, J., & Lombo, F., (2018). Plant nutraceuticals as antimicrobial agents in food preservation: Terpenoids, polyphenols and thiols. *Int. J. Antimicrob. Agents, 52*, 309–315.
6. Cassidy, A., Bingham, S., & Setchell, K. D., (1994). Biological effects of a diet of soy protein rich in isoflavones on the menstrual cycle of premenopausal women. *The American Journal of Clinical Nutrition, 60*(3), 333–340.
7. Potter, S. M., Baum, J. A., Teng, H., Stillman, R. J., Shay, N. F., & Erdman, Jr. J. W., (1998). Soy protein and isoflavones: Their effects on blood lipids and bone density in postmenopausal women. *The American Journal of Clinical Nutrition, 68*(6), 1375S–1379S.
8. Setchell, K. D., & Cassidy, A., (1999). Dietary isoflavones: Biological effects and relevance to human health. *The Journal of Nutrition, 129*(3), 758S–767S.
9. Drug Bank database. http://www.drugbank.ca/ (accessed on 13 September 2023).
10. Karthikeyan, K., Baskaran, A., & Rajasekaran, S., (2017). Antimicrobial and antifungal activity of *Barleria acuminata* Nees in ethanolic extract on different pathogens. *World J. Pharm. Res., 6*, 742–748.
11. https://www.encyclopedia.com/food/encyclopedias-almanacs-transcripts-and-maps/bioactive-food-components (accessed on 13 September 2023).
12. Mani, J. S., Johnson, J. B., Hosking, H., Ashwath, N., Walsh, K. B., Neilsen, P. M., Broszczak, D. A., & Naiker, M., (2021). Antioxidative and therapeutic potential of selected Australian plants: A review. *J. Ethnopharmacol., 268*, 113580.
13. Rodrıguez-Carrasco, Y., Gaspari, A., Graziani, G., Santini, A., & Ritieni, A., (2018). Fast analysis of polyphenols and alkaloids in cocoa-based products by ultra-high performance liquid chromatography and orbitrap high-resolution mass spectrometry (UHPLC-Q-Orbitrap-MS/MS). *Food Res. Int., 111*, 229–236.
14. Castaldo, L., Narvaez, A., Izzo, L., Graziani, G., & Ritieni, A., (2020). In vitro bioaccessibility and antioxidant activity of coffee silvers kin polyphenolic extract and characterization of bioactive compounds using UHPLC-Q-orbitrap HRMS. *Molecules, 25*, 2132.
15. Izzo, L., Castaldo, L., Narva´ ez, A., Graziani, G., Gaspari, A., Rodriguez- Carrasco, Y., & Ritieni, A., (2020). Analysis of phenolic compounds in commercial *Cannabis sativa* L. inflorescenses using UHPLC-Q-orbitrap HRMS. *Molecules, 25*, 631.
16. Mascoloti, S. R., Fernandes, A., Calhelha, R. C., Pereira, C., Pires, T. C. S. P., Alves, M. J., Canan, C., et al., (2020). Chemical and bioactive characterization of the aromatic

plant Levisticum officinale W.D.J. Koch: A comprehensive study. *Food Funct., 11*, 1292.

17. Maqsood, S., Adiamo, O., Ahmad, M., & Mudgil, P., (2020). Bioactive compounds from date fruit and seed as potential nutraceutical and functional food ingredients. *Food Chem., 308*, 125522.
18. Oksman-Caldentey, K. M., & Inzé, D., (2004). Plant cell factories in the post-genomic era: New ways to produce designer secondary metabolites. *Trends in Plant Science, 9*(9), 433–440.
19. Verpoorte, R., Contin, A., & Memelink, J., (2002). Biotechnology for the production of plant secondary metabolites. *Phytochemistry Reviews, 1*(1), 13–25.
20. Nobili, S., Lippi, D., Witort, E., Donnini, M., Bausi, L., Mini, E., & Capaccioli, S., (2009). Natural compounds for cancer treatment and prevention. *Pharmacological Research, 59*(6), 365–378.
21. Shahidi, F., (2009). Nutraceuticals and functional foods: Whole versus processed foods. *Trends in Food Science & Technology, 20*(9), 376–387.
22. Côté, J., Caillet, S., Doyon, G., Sylvain, J. F., & Lacroix, M., (2010). Bioactive compounds in cranberries and their biological properties. *Critical Reviews in Food Science and Nutrition, 50*(7), 666–679.
23. Garzón, G. A., Narváez-Cuenca, C. E., Vincken, J. P., & Gruppen, H., (2017). Polyphenolic composition and antioxidant activity of açai (*Euterpe oleracea* Mart.) from Colombia. *Food Chemistry, 217*, 364–372.
24. Jimenez-Garcia, S. N., Guevara-Gonzalez, R. G., Miranda-Lopez, R., Feregrino-Perez, A. A., Torres-Pacheco, I., & Vazquez-Cruz, M. A., (2013). Functional properties and quality characteristics of bioactive compounds in berries: Biochemistry, biotechnology, and genomics. *Food Research International, 54*(1), 1195–1207.
25. Paredes-López, O., Cervantes-Ceja, M. L., Vigna-Pérez, M., & Hernández-Pérez, T., (2010). Berries: Improving human health and healthy aging, and promoting quality life-a review. *Plant Foods for Human Nutrition, 65*(3), 299–308.
26. Chhikara, N., Kaur, R., Jaglan, S., Sharma, P., Gat, Y., & Panghal, A., (2018). Bioactive compounds and pharmacological and food applications of *Syzygium cumini*: A review. *Food & Function, 9*(12), 6096–6115.
27. Masibo, M., & He, Q., (2009). Mango bioactive compounds and related nutraceutical properties—A review. *Food Reviews International, 25*(4), 346–370.
28. Lim, J. C. W., Chan, T. K., Ng, D. S., Sagineedu, S. R., Stanslas, J., & Wong, W. F., (2012). Andrographolide and its analogues: Versatile bioactive molecules for combating inflammation and cancer. *Clinical and Experimental Pharmacology and Physiology, 39*(3), 300–310.
29. Maieves, H. A., López-Froilán, R., Morales, P., Pérez-Rodríguez, M. L., Ribani, R. H., Cámara, M., & Sánchez-Mata, M. C., (2015). Antioxidant phytochemicals of *Hovenia dulcis* Thunb. peduncles in different maturity stages. *Journal of Functional Foods, 18*, 1117–1124.
30. Hossam, S., Dhawi, F., & Amina, A., (2020). Chemical Compositions and Biological Activities of the Essential Oils from Gamma Irradiated Celery (*Apium graveolens* L.) Seed. Notulae Botanicae, Horti., *Agrobo. 48*(4); online first (DOI: 10.15835/nbha (12115).
31. Justino, A. B., Pereira, M. N., Peixoto, L. G., Vilela, D. D., Caixeta, D. C., De Souza, A. V., & Espindola, F. S., (2017). Hepatoprotective properties of a polyphenol-enriched

fraction from *Annona crassiflora* Mart. fruit peel against diabetes-induced oxidative and nitrosative stress. *Journal of Agricultural and Food Chemistry, 65*(22), 4428–4438.
32. Saini, P., Kumar, N., Kumar, S., Mwaurah, P. W., Panghal, A., Attkan, A. K., & Singh, V., (2021). Bioactive compounds, nutritional benefits and food applications of colored wheat: A comprehensive review. *Critical Reviews in Food Science and Nutrition, 61*(19), 3197–3210.
33. Gupta, R., Meghwal, M., & Prabhakar, P. K., (2021). Bioactive compounds of pigmented wheat (*Triticum aestivum*): Potential benefits in human health. *Trends in Food Science & Technology, 110*, 240–252.
34. Dhua, S., Kumar, K., Kumar, Y., Singh, L., & Sharanagat, V. S., (2021). Composition, characteristics and health promising prospects of black wheat: A review. *Trends in Food Science & Technology, 112*, 780–794.
35. Pasqualone, A., Bianco, A. M., Paradiso, V. M., Summo, C., Gambacorta, G., Caponio, F., & Blanco, A., (2015). Production and characterization of functional biscuits obtained from purple wheat. *Food Chemistry, 180*, 64–70.
36. Ficco, D. B. M., De Simone, V., De Leonardis, A. M., Giovanniello, V., Del Nobile, M. A., Padalino, L., & De Vita, P., (2016). Use of purple durum wheat to produce naturally functional fresh and dry pasta. *Food Chemistry, 205*, 187–195.
37. Li-Chan, E. C., (2015). Bioactive peptides and protein hydrolysates: Research trends and challenges for application as nutraceuticals and functional food ingredients. *Current Opinion in Food Science, 1*, 28–37.
38. Liu, F., Zhang, S., Li, J., McClements, D. J., & Liu, X., (2018). Recent development of lactoferrin-based vehicles for the delivery of bioactive compounds: Complexes, emulsions, and nanoparticles. *Trends in Food Science & Technology, 79*, 67–77.
39. Perez-Gregorio, R., & Simal-Gandara, J., (2017). A critical review of bioactive food components, and of their functional mechanisms, biological effects, and health outcomes. *Current Pharmaceutical Design, 23*(19), 2731–2741.
40. Zamaratskaia, G., Gerhardt, K., & Wendin, K., (2021). Biochemical characteristics and potential applications of ancient cereals – An underexploited opportunity for sustainable production and consumption. *Trends in Food Science & Technology, 107*, 114–123.
41. Van, H. P., (2016). Phenolic compounds of cereals and their antioxidant capacity. *Critical Reviews in Food Science and Nutrition, 56*(1), 25–35.
42. Akanbi, T. O., Timilsena, Y., & Dhital, S., (2019). Bioactives from millet: Properties and effects of processing on bioavailability. In: Wang, J., Sun, B., & Tsao, R., (eds.), *Bioactive Factors and Processing Technology for Cereal Foods* (pp. 171–183). Springer. https://doi.org/10.1007/978-981-13-6167-8_10.
43. Chethan, S., & Malleshi, N. G., (2007). Finger millet polyphenols: Characterization and their nutraceutical potential. *Am. J. Food Technol., 2*, 618–629.
44. Bhargava, A., & Srivastava, S., (2019). Cereals. In: *Participatory Plant Breeding: Concept and Applications* (pp. 129–173). Springer, Singapore.
45. Chandrasekara, A., & Shahidi, F., (2011). Determination of antioxidant activity in free and hydrolyzed fractions of millet grains and characterization of their phenolic profiles by HPLCDAD-ESI-MS. *J. Funct. Foods, 3*, 144–158.
46. Priego-Poyato, S., Rodrigo-Garcia, M., Escudero-Feliu, J., Garcia-Costela, M., Lima-Cabello, E., Carazo-Gallego, A., & Jimenez-Lopez, J. C., (2021). Current advances research in nutraceutical compounds of legumes, pseudocereals and cereals. In: *Grain and Seed Proteins Functionality*. IntechOpen.

47. Martínez-Villaluenga, C., Peñas, E., & Hernández-Ledesma, B., (2020). Pseudocereal grains: Nutritional value, health benefits and current applications for the development of gluten-free foods. *Food and Chemical Toxicology: An International Journal Published for the British Industrial Biological Research Association, 137*, 111178.
48. Ahumada, A., Ortega, A., Chito, D., & Benítez, R., (2016). Saponinas de quinua (*Chenopodium quinoa* Wild.): Un subproducto con alto potencial biológico. *Rev. Colomb. Cienc. Quím. Farm., 45*(3), 438–469.
49. Pradeep, P. M., & Sreerama, Y. N., (2015). Impact of processing on the phenolic profiles of small millets: Evaluation of their antioxidant and enzyme inhibitory properties associated with hyperglycemia. *Food Chem., 169*, 455–463.
50. Kumari, A., & Chaudhary, H. K., (2020). Nutraceutical crop buckwheat: A concealed wealth in the lap of Himalayas. *Critical Reviews in Biotechnology, 40*(4), 539–554.
51. Meng, F., Uniacke-Lowe, T., Ryan, A. C., & Kelly, A. L., (2021). The composition and physico-chemical properties of human milk: A review. *Trends in Food Science & Technology, 112*, 608–621.
52. Santiago-Silva, P., Soares, N. F., Nóbrega, J. E., Júnior, M. A., Barbosa, K. B., Volp, A. C. P., & Würlitzer, N. J., (2009). Antimicrobial efficiency of film incorporated with pediocin (ALTA® 2351) on preservation of sliced ham. *Food Control, 20*(1), 85–89.
53. Vilcacundo, R., Miralles, B., Carrillo, W., & Hernández-Ledesma, B., (2018). In vitro chemopreventive properties of peptides released from quinoa (*Chenopodium quinoa* Willd.) protein under simulated gastrointestinal digestion. *Food Research International (Ottawa, Ont.), 105*, 403–411.
54. Orona-Tamayo, D., Valverde, M. E., & Paredes-López, O., (2019). Bioactive peptides from selected Latin American food crops – A nutraceutical and molecular approach. *Critical Reviews in Food Science and Nutrition 59*(12), 1949–1975.
55. Devi, P. B., Vijayabharathi, R., Sathyabama, S., Malleshi, N. G., & Priyadarisini, V. B., (2014). Health benefits of finger millet (*Eleusine coracana* L.) polyphenols and dietary fiber: A review. *J. Food Sci. Technol., 51*, 1021–1040.
56. Nyam, K. L., Tan, C. P., Lai, O. M., Long, K., & Man, Y. C., (2009). Physicochemical properties and bioactive compounds of selected seed oils. *LWT-Food Science and Technology, 42*(8), 1396–1403.
57. Yazdani, D., Tan, Y. H., Zainal, A. M. A., & Jaganath, I. B., (2011). A review of bioactive compounds isolated from plants against plant pathogenic fungi. *Journal of Medicinal Plants Research, 5*(30), 6584–6589.
58. Park, Y. W., (2009). Bioactive components of goat milk. In: Park, Y. W., (ed.), *Bioactive Components in Milk and Dairy Products* (pp. 43–82). Wiley Blackwell Publishers, Ames, Iowa and Oxford, England.
59. Adetunji, C. O., Michael, O. S., Kadiri, O., Varma, A., Akram, M., Oloke, J. K., & Ubi, B. E., (2021). Quinoa: From farm to traditional healing, food application, and phytopharmacology. In: *Biology and Biotechnology of Quinoa* (pp. 439–466). Springer, Singapore.
60. Okwudili, U. H., Gyebi, D. K., & Obiefuna, J. A. I., (2017). Finger millet bioactive compounds, bioaccessibility, and potential health effects: A review. *Czech Journal of Food Sciences, 35*(1), 7–17.
61. Shahidi, F., & Zhong, Y., (2008). Bioactive peptides. *Journal of AOAC International, 91*(4), 914–931.

62. Hu, L., Zhou, Y., Zhang, M., & Liu, R., (2012). Characterization and properties of a lignosulfonate-based phenolic foam. *Bio Resources, 7*(1), 0554–0564.
63. Saito, T., Nakamura, T., Kitazawa, H., Kawai, Y., & Itoh, T., (2000). Isolation and structural analysis of antihypertensive peptides that exist naturally in Gouda cheese. *Journal of Dairy Science, 83*(7), 1434–1440.
64. Jauhiainen, T., & Korpela, R., (2007). Milk peptides and blood pressure. *The Journal of Nutrition, 137*(3), 825S–829S.
65. Kitts, D. D., & Weiler, K., (2003). Bioactive proteins and peptides from food sources. Applications of bioprocesses used in isolation and recovery. *Curr. Pharm. Des., 9*(16), 1309–1323.
66. Mohanty, D. P., Mohapatra, S., Misra, S., & Sahu, P. S., (2016). Milk derived bioactive peptides and their impact on human health: A review. *Saudi J. Biol. Sci., 23*(5), 577–583.
67. Regester, G. O., Smithers, G. W., Mitchell, I. R., McIntosh, G. H., & Dionysius, D. A., (1997). Bioactive factors in milk: Natural and induced. In: Welch, R., Burns, D., Davis, S., Popay, A., & Prosser, C., (eds.), *Milk Composition, Production and Biotechnology* (pp. 119–132). CAB International, UK.
68. Silveira, S. T., Daroit, D. J., & Brandelli, A., (2008). Pigment production by *Monascus purpureus* in grape waste using factorial design. *LWT-Food Science and Technology, 41*(1), 170–174.
69. Van, L. S. G., & Shen, B., (2006). Progress in combinatorial biosynthesis for drug discovery. *Drug Discovery Today: Technologies, 3*(3), 285–292.
70. Boziaris, I. S., (2014). Food ingredients from the marine environment. Marine biotechnology meets food science and technology. *Frontiers in Marine Science, 1*, 66.
71. Plaza, M., Santoyo, S., Jaime, L., Reina, G. G. B., Herrero, M., Señoráns, F. J., & Ibáñez, E., (2010). Screening for bioactive compounds from algae. *Journal of Pharmaceutical and Biomedical Analysis, 51*(2), 450–455.
72. El Gamal, A. A., (2010). Biological importance of marine algae. *Saudi Pharmaceutical Journal, 18*(1), 1–25.
73. Oksman-Caldentey, K. M., & Saito, K., (2005). Integrating genomics and metabolomics for engineering plant metabolic pathways. *Current Opinion in Biotechnology, 16*(2), 174–179.
74. Abidin, Z., (2011). A review of bioactive compounds isolated from plants against plant pathogenic fungi. *Journal of Medicinal Plants Research, 5*(30), 6584–6589.
75. Wink, M., (2006). Importance of plant secondary metabolites for protection against insects and microbial infections. *Advances in Phytomedicine, 3*, 251–268.
76. Hamamoto, E., Kimura, N., Nishino, S., Ishihara, A., Otani, H., & Osaki-Oka, K., (2021). Antimicrobial activity of the volatile compound 3, 5-dichloro-4-methoxybenzaldehyde, produced by the mushroom *Porostereum spadiceum*, against plant-pathogenic bacteria and fungi. *Journal of Applied Microbiology, 131*(3), 1431–1439.
77. Sasidharan, S., Chen, Y., Saravanan, D., Sundram, K. M., & Latha, L. Y., (2011). Extraction, isolation, and characterization of bioactive compounds from plants' extracts. *African Journal of Traditional, Complementary and Alternative Medicines, 8*(1).
78. Gill, B. S., & Qiu, F., (2020). Technologies for extraction and production of bioactive compounds. In: *Biotechnological Production of Bioactive Compounds* (pp. 1–36). Elsevier.
79. Mir, R. H., Shah, A. J., Mohi-Ud-Din, R., Pottoo, F. H., Dar, M., Jachak, S. M., & Masoodi, M. H., (2021). Natural anti-inflammatory compounds as drug candidates in Alzheimer's disease. *Current Medicinal Chemistry, 28*(23), 4799–4825.

80. Kumar, S., Kushwaha, R., & Verma, M. L., (2020). Recovery and utilization of bioactives from food processing waste. In: *Biotechnological Production of Bioactive Compounds* (pp. 37–68). Elsevier.
81. Carbonell-Capella, J. M., Buniowska, M., Esteve, M. J., & Frigola, A., (2015). Effect of *Stevia rebaudiana* addition on bioaccessibility of bioactive compounds and antioxidant activity of beverages based on exotic fruits mixed with oat following simulated human digestion. *Food Chemistry, 184*, 122–130.
82. Gandhi, P., Samarth, R. M., & Peter, K., (2020). Bioactive compounds of amaranth (genus Amaranthus). *Bioactive Compounds in Underutilized Vegetables and Legumes*, 1–37.
83. Wijngaard, H., Hossain, M. B., Rai, D. K., & Brunton, N., (2012). Techniques to extract bioactive compounds from food by-products of plant origin. *Food Research International, 46*(2), 505–513.
84. Song, K., & Milner, J. A., (2001). The influence of heating on the anticancer properties of garlic. *The Journal of Nutrition, 131*(3), 1054S–1057S.
85. Stahl, W., & Sies, H., (2005). Bioactivity and protective effects of natural carotenoids. *Biochimica et Biophysica Acta (BBA)-Molecular Basis of Disease, 1740*(2), 101–107.
86. Parker, S., May, B., Zhang, C., Zhang, A. L., Lu, C., & Xue, C. C., (2016). A pharmacological review of bioactive constituents of *Paeonia lactiflora* Pallas and *Paeonia veitchii* Lynch. *Phytotherapy Research, 30*(9), 1445–1473.
87. Mwaurah, P. W., Kumar, S., Kumar, N., Panghal, A., Attkan, A. K., Singh, V. K., & Garg, M. K., (2020). Physicochemical characteristics, bioactive compounds and industrial applications of mango kernel and its products: A review. *Comprehensive Reviews in Food Science and Food Safety, 19*(5), 2421–2446.
88. Agagunduz, D., Koseler-Beyaz, E., & Duman, S., (2021). Assessment of the physicochemical and antioxidant profile of dried goji berries. *Progress in Nutrition, 23*(4), e2021193.
89. Rodrigues, F. T., Koike, A. C. R., Da Silva, P. G., Negrão, B. G., De Alencar, S. M., Mancini, F. J., & Villavicencio, A. L. C., (2021). Effects of electron beam irradiation on the bioactive components of goji-berry. *Radiation Physics and Chemistry, 179*, 109144.
90. Pedro, A. C., Maurer, J. B. B., Zawadzki-Baggio, S. F., Ávila, S., Maciel, G. M., & Haminiuk, C. W. I., (2018). Bioactive compounds of organic goji berry (*Lycium barbarum* L.) prevent oxidative deterioration of soybean oil. *Industrial Crops and Products, 112*, 90–97.
91. Ciecierska, A., Drywien, M., Hamulka, J., & Sadkowski, T., (2019). Nutraceutical functions of beta-glucans in human nutrition. *Roczniki Pantwowego Zakladu Higieny, 70*(4), 315–324.
92. Ficco, D. B. M., De Simone, V., De Leonardis, A. M., Giovanniello, V., Del, N. M. A., Padalino, L., & De Vita, P., (2016). Use of purple durum wheat to produce naturally functional fresh and dry pasta. *Food Chemistry, 205*, 187–195.
93. Žilić, S., Basić, Z., Hadži-Tašković, Š. V., Maksimović, V., Janković, M., & Filipović, M., (2014). Can the sprouting process applied to wheat improve the contents of vitamins and phenolic compounds and antioxidant capacity of the flour? *International Journal of Food Science & Technology, 49*(4), 1040–1047.
94. Li, Y., Ma, D., Sun, D., Wang, C., Zhang, J., Xie, Y., & Guo, T. (2015). Total phenolic, flavonoid content, and antioxidant activity of flour, noodles, and steamed bread made from different colored wheat grains by three milling methods. *The Crop Journal, 3*(4), 328–334.

95. Joana Gil-Chavez, G., Villa, J. A., Fernando Ayala-Zavala, J., Basilio Heredia, J., Sepulveda, D., Yahia, E. M., & González-Aguilar, G. A. (2013). Technologies for extraction and production of bioactive compounds to be used as nutraceuticals and food ingredients: An overview. Comprehensive *Reviews in Food Science and Food Safety, 12*(1), 5–23.

PART II

Technology of Bioactive Compounds and Nutraceuticals From Dairy and Marine Sources

CHAPTER 4

Technological Aspects of Bioactive Compounds and Nutraceuticals from Milk/Dairy Products

LOULOUDA BOSNEA,[1] ANTONIA TERPOU,[2] VASILIKI KONTOGIANNI,[1] and MARIOS MATARAGAS[1]

[1]*ELGO-DIMITRA, Institute of Technology of Agricultural Products, Dairy Department, Ethnikis Antistaseos 3, Katsikas GR, Ioannina, Greece*

[2]*Department of Agricultural Development, Agri-Food and Natural Resources Management, School of Agri-Cultural Development, Nutrition and Sustainability, National and Kapodistrian University of Athens, Evripos Group GR, Greece*

ABSTRACT

Milk emerges as one of the most significant food components produced by the mammary gland of lactating females. The production of milk by mammalian females targets nourishment and immunological sealing of immature neonates. Beyond providing basic nutrition, several milks have been identified over the years of research to derive biologically active compounds. Likewise, intense scientific interest has focused on the identification of factors within milk as well as dairy products that may contribute to the improvement of human health. This chapter emphasizes on bioactive compounds and nutraceuticals provided *via* milk and dairy products and focuses on aspects that are linked to nutritional and technological matters.

Bioactive Compounds and Nutraceuticals from Dairy, Marine, and Nonconventional Sources: Extraction Technology, Analytical Techniques, and Potential Health Prospects.
Charanjit Singh Riar, Parmjit S. Panesar (Eds.)

4.1 INTRODUCTION

Milk is known as human's first source of nutrients and is one of the most nutritional foods. The traditional role of milk related to the growth of infants and newborn mammals; however, milk's role is more than the obvious. For centuries, milk has been used for the growth of children and as food for adults, either in raw form or in its modified versions [1]. Beyond providing basic nutrition, several milk-derived biologically active compounds such as milk proteins (caseins, lactoglobulins, lactoalbumins, immunoglobulins, peptides, etc.) seems to have significant effect on physiological and biochemical functions of human metabolism [2–4]. The last decades the term '*Bioactive Food Component*' was reported as a significant term of the scientific literature targeting to describe "*food components that can affect biological processes or substrates and hence have an impact on body function or condition and ultimately health*" [5, 6]. However, the term is a bit abstract, and there could be several interpretations. Lately, these components are extensively studied because it is believed to promote health and prevent diseases [5]. In international scientific databases many published studies focus on the role of bioactive components in milk, with milk proteins being the most well-studied and considered as the most important source of bioactive components [4, 7]. Specifically, protein-based bioactive compounds can contribute to the defense and healthy development of infants. However, the bioactivity of the milk protein is considered incomplete or absent in native proteins and the bioactivity is achieved *via* proteolytic digestion to release the bioactive peptides from the native protein [4, 8]. Through milk fermentation, cheese maturation, or enzymatic hydrolysis through digestive enzymes in the gastrointestinal transportation, several bioactive components are produced [6, 9]. Although bioactive peptides can be found in foods it is believed that their majority are encoded within the parental proteins [4, 6, 10, 11]. Also, the high levels of calcium in mammalian milk play an important role in the development of children and in the prevention of osteoporosis-related diseases in elderly people, reduce cholesterol absorption, and control body weight and blood pressure [12]. Figure 4.1 shows the potential bioactive compounds in milk and dairy products.

Recent studies, research reports, and advanced compositional identification of many dairy bioactive compounds have led to the discovery of specific physiological, biochemical, and nutritional functionalities and characteristics that may positively influence human metabolism and health [4, 6, 12, 13]. A representative category of milk bioactive compounds is microbial metabolites such as short-chain fatty acids (SCFAs). These metabolites are mainly derived from bacteria (LAB) and can be produced in the gastrointestinal tract (GIT) upon fermentation of dietary fiber [6, 12]. Moreover, milk is known to contain prebiotic bioactives, with lactose being the predominant disaccharide, while human milk contains approx. 12 g/L of total oligosaccharides that could stimulate the growth of beneficial bacteria in the GIT [14, 15]. Subsequently, the of biological compounds in milk contribute *via* a wide spectrum while biochemical mechanisms of action that optimize infant nutrition substitutes still remain unknown with new data constantly being discovered.

4.2 BIOACTIVE COMPONENTS IN MILK

4.2.1 CALCIUM

Calcium is an essential mineral and one the major minerals in the human body mainly responsible for bone growth and health, while also participates in other significant human biological functions [16]. Calcium exists in milk in a dynamic equilibrium with two forms, micellar calcium, and soluble calcium [17]. In micellar calcium state, calcium is usually found as colloidal calcium phosphate (CCP) in the casein micelles, and as calcium ions bonded to phosphoserine residues. The soluble calcium in the whey phase is mostly in the form of calcium citrate, calcium phosphate or calcium ions [18]. The loss of calcium is required to be supplemented from the daily diet; therefore, dairy products remain the main source of calcium in the human diet. During the past 20 years, research has proven that calcium in dairy products has a beneficial role (Table 4.1) and the level of consumption has been associated to 21st century diseases such as obesity and cardiovascular diseases (CVDs) [19]. Calcium plays an important role in the control of blood glucose and energy metabolism through insulin-dependent and independent routes and regulates the blood lipid profile by enabling fecal fat excretion [19–21].

Recently calcium protective effect against obesity-related metabolic complications is gaining attention mainly because of the minimum side effects and cost [21, 23, 27]. A study on adolescent girls revealed that an increment of milk consumption improves the bone mineral acquisition, influence bone mass and reported an association between high intake of calcium with reduced fat mass in young women [22]. Other studies also reported that dietary calcium supplementation reduces weight gain and metabolic complications associated with obesity [28]. Umesawa et al. [25]

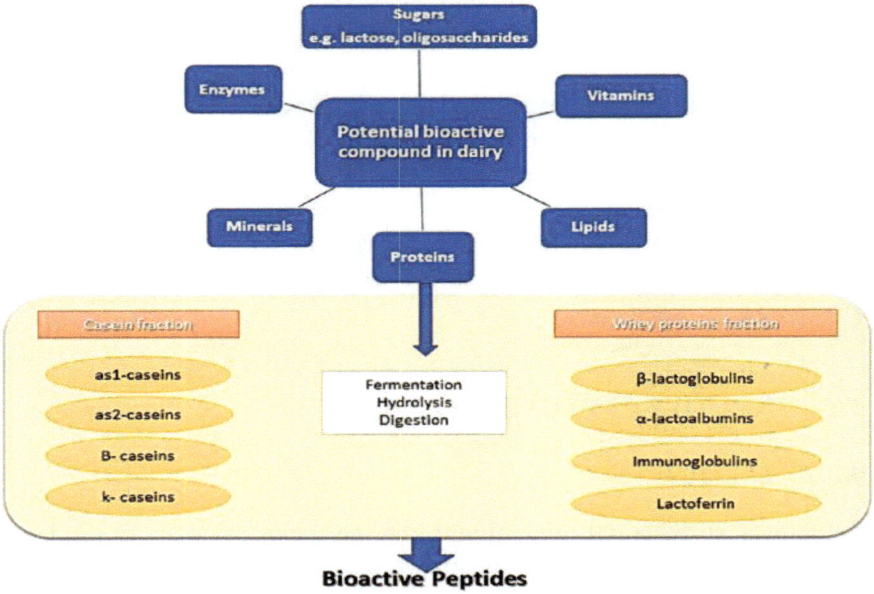

FIGURE 4.1 Milk and dairy bioactive compounds.

TABLE 4.1 Bioactivity and Suggested Health Effects of Calcium from Milk and Dairy Products

Bioactivity and Health Effects of Calcium Intake	References
Reducing fat mass/anti-obesity effects/weight regulation/obesity complications reduction.	[22–24]
Hepatoprotective effect against high-fat diet.	[24]
Increased bone mass.	[16]
Reduced mortality from cardiovascular disease.	[25]

managed to correlate the dietary calcium intake from dairy products with mortality reduction from stroke for Japanese men and women [25]. Calcium supplementation in rats has showed a hepatoprotective effect against high-fat diet by regulating oxidative-induced inflammatory response and lipogenesis activity [24]. Also, in a recent systematic review, the effects of dietary calcium supplement on circulating lipoprotein concentrations and atherogenic indices in overweight and obese individuals are summarized. The authors concluded that most studies evaluated co-supplementation of calcium and vitamin D and reported the positive effects of calcium supplementation with or without vitamin D in serum levels of total cholesterol, low-density lipoprotein (LDL-C), high-density lipoprotein (HDL-C) cholesterol, and triglyceride (TG) concentrations. Furthermore, they reported that several studies demonstrated the beneficial effects of calcium supplementation on at least one atherogenic index. Also, it is reported that calcium supplementation may not be helpful to reduce serum levels of TC and TG in overweight and obese individuals, however, it may modulate LDL-C and HDL-C concentration [29]. The mechanisms associated with the beneficial effects of calcium can be attributed to various reasons. In some cases, it has been suggested that calcium can bind fatty acids to calcium fatty acid soaps and increase fecal fat excretion by interacting with fatty acids; thus, plasma triglyceride concentration decreases as fatty acid absorption is inhibited [20]. Heshmati et al. [29] reported that the ratio of HDL-C to LDL-C is thought to improve with calcium intake [29].

4.2.2 MILK SUGARS

Milk is an important source of energy, high-quality protein, vitamins, and minerals. Lactose, the major carbohydrate in milk, is a disaccharide consisting of a molecule of galactose and molecule of glucose with numerous applications in foods and pharmaceutical products. Lactose concentration may vary among different milk origins from 4% up to 7% w/w [30–32]. Lactose's nutritional value besides being an energy source, resides in favoring calcium, magnesium, and phosphorous intestinal absorption along with the utilization of vitamin C [33]. Recently, lactose content in cow's milk has been claimed as a biomarker of health status and reproductive success [31]. Hydrolyzed lactose can enhance transportation and absorption of calcium ions in the form of calcium gluconate in the intestine. On the other hand, lactose malabsorption and/or intolerance has been associated with bone mass reduction and fractures [34]. Unhydrolyzed lactose can act as a prebiotic by stimulating the growth of lactic acid bacteria and more importantly probiotic strains [15, 35]. Lactose has also been reported to enhance the formation of non-digestible oligosaccharides (NDOs) such as trans-galacto-oligosaccarides through growth of lactic acid bacteria [15, 35].

Oligosaccharides and especially human milk oligosaccharides (HMOs) pose as significant bioactive compounds while most recently studies are applied for the delivery of these bioactives from biowaste material [15, 36–39]. HMOs consist of 3 to 14 different monosaccharides while the structure of these oligosaccharides is based on five structural sugars including glucose, galactose, N-acetylglucosamine, fucose, or sialic acid residues (N-acetyl-d-neuraminic acid, N-acetyl-d-glucosamine) [33, 40]. Even though d-galactose has been associated with inducing aging, oxidative stress, and inflammation, it is known that it can form oligosaccharides (galactooligosaccharides, transgalactooligosaccharides) that stimulate calcium absorption in humans and improve bone health [36]. Human milk oligosaccharides can be found in human milk either as free molecules or conjugated to lipids and proteins [39, 41].

Oligosaccharide concentrations in farm animals (OSs) are significantly lower in concentrations than HMOs, and differ in composition where there is lower number of different OSs, a higher abundance of sialylated and a lower abundance of fucosylated OSs but have β-glycosidic linkage of Gal and *N* acetylhexosamine to lactose as HMOs [33]. Bioactivity of OSs from bovine milk is similar with HMOs and can be applied as substitutes in milk products. Caprine milk, on the other hand, is rich in complex lactose-derived OSs with prebiotic and pathogen-binding abilities. Moreover, caprine milk contains large amounts of different sialylated and neutral lactose-derived OSs similar to human milk therefore it could provide very promising functional ingredients for human nutrition applications such as infant formulae supplementation [33].

4.2.3 MILK FATS

The nutritional quality of dairy products is highly associated with milk fat composition because of the high concentration of fat-soluble vitamins and n–3 fatty acids, as well as high content of CLA. The concentration of fat in milk may vary because it depends on several factors such as the animal species, breed, nutrition, individual traits of the animal, and period of lactation [42]. Over the years, milk and dairy products have been blamed for CVDs, mainly because of their cholesterol and saturated fatty acids content even though there is not a clear correlation between dietary cholesterol and LDL cholesterol [43].

Milk and other dairy products constitute a significant source of fat and liposoluble compounds, vitamins such as A and E, and beta-carotene [44]. Vitamin E, especially tocopherols, is important for the human diet and health, with numerous applications in food production. Tocopherols have also, an important task in maintaining the oxidative stability of dairy products [45].

Minor sterols that could be included in bioactive compounds can be found in dairy products and are mainly intermediate compounds of cholesterol synthesis. Phytosterols that derive from the diet of animals and lanosterol, which is an animal-origin sterol that can be found in sheep milk, have bioactive effects on human health [46]. The chemical structure of phytosterols is known to be very similar to cholesterol. Phytosterol intake is reported to reduce the levels of low-density lipoprotein-cholesterol, thus minimizing the risks associated with CVDs [46].

Fatty acids have high nutritional value and are considered as important bioactive dairy components as they can influence the growth of mammals and the development of the nervous system [48]. Milk fat also contains monounsaturated (MUFAs) and polyunsaturated fatty acids (PUFAs) that are generally regarded as good fat as they have a beneficial impact on cholesterol levels. Several scientific reports on some PUFAs have demonstrated that they can prevent cardiovascular disease, improve the immune response, and have anticancer and anti-atherogenic properties, while others have shown the ability to reduce the incidence of type 2 diabetes [49]. However, milk and dairy products have significant amounts of fatty acids (Table 4.2) such as lauric (C12:0), myristic (C14:0) and palmitic (C16:0) acids that have been associated with increased risk of atherosclerosis, obesity, and heart diseases [50]. Milk and dairy products have also some amounts of bioactive CLA such as cis-9, trans-11 isomers also commonly known as rumenic acid. *In vitro* and *in vivo* studies have demonstrated the various health properties of CLA in milk such as anticarcinonegic effect, regeneration of the nervous system, reduction of heart disease, inhibition of both benign and malignant tumors while also posing anti-oxidative, anti-atherosclerotic, and anti-inflammatory effects [51–54]. Bioactive MUFAs, PUFAs, and CLA are generally higher in sheep milk and cheese products than bovine and their health effects have been reported by researchers [55]. Also, Megalemou et al. [56]; and Poutzalis et al. [57] have studied the anti-throbotic and anti-atherogenic activities of traditional Greek yogurts and shown that the most active inhibitors of platelet-activating factor (PAF) were found in yogurt made from sheep's milk and ewe's milk and are related with the polar lipids of milk [56, 57].

4.2.4 MILK PROTEINS AND BIOACTIVE PEPTIDES

Proteins in milk are complexes of caseins and whey protein. The majority of the proteins are caseins (as1, as2-, β-, γ-, and k-casein) and depending

TABLE 4.2 Fatty Acid Composition Expressed as Percent by Weight of Total Fatty Acids in Bovine Milk

Fatty Acid	Common Name	Systematic Name	Percentage (% w/v)
		Saturated Fatty Acids Total	67.1–74.4
4:0	Butyric	Butanoic acid	4.0–5.1
6:0	Caproic	Caproic acid	2.1–2.9
8:0	Caprylic	Octanoic acid	1.2–1.9
10:0	Capric	Decanoic acid	2.4–3.5
12:0	Lauric	Dodecanoic acid	3.0–4.1
14:0	Myristic	Tetradecanoic acid	10.0–12.1
15:0	Pentadecylic	Pentadecanoic acid	0.8–1.1
16:0	Palmitic	Hexadecanoic acid	28.7–34.1
17:0	Margaric	Heptadecanoic acid	0.4–0.5
18:0	Stearic	Octadecanoic acid	10.3–13.3
20:0	Arachidic acid	Eicosanoic acid	0.2
10:1	Caproleic acid	Dec-9-enoic acid	0.2–0.4
		Mono-Unsaturated Fatty Acids, Cis, Total	22.2–26.7
14:1	Myristoleic acid	(Z)-tetradec-9-enoic acid	0.4–1.3
16:1	Palmitoleic	(Z)-hexadec-9-enoic acid	0.9–1.8
18:1	Oleic	(Z)-octadec-9-enoic acid	19.7–24.7
		Poly-Unsaturated Fatty Acids, Cis, Total	2.0–2.5
18:2	linoleic	(9Z,12Z)-octadeca-9,12-dienoic acid	1.4–1.8
18:3	a-Linolenic	(9Z,12Z,15Z)-octadeca-9,12,15-trienoic acid	0.6–0.9

Source: Adapted from: Lindmark [47].

on the milk origin can reach up to 80% of total proteinic content. The whey protein fraction includes α-lactalbumins, β-lactoglobulins, immunoglobulins, lactoperoxidase (LP), lysozyme, and lactoferrin (LF) posing various biofunctional properties [58]. Significantly, there has been observed by many different studies that milk proteins are composed of bioactive peptides (Table 4.3) that are inactive or latent in the raw protein form but can be activated by enzymatic proteolysis lead by proteolytic enzymes during cheese production, during milk fermentation from starter cultures, or during passage through the GIT from proteolytic enzymes [59, 60]. Caseins can be the source of bioactive milk peptides that are known for their functional properties. Such peptides can be produced by the proteolytic enzymes in digestive system that can break casein in oligopeptides that may release bioactive peptides. Casein phosphopeptides (CPP) derived from beta caseins of bovine milk are known to enhance the absorption of calcium by increasing the soluble calcium concentration in the small intestinal lumen [61]. Glycomacropeptide (GMP) is a fragment of the C-terminal from κ-casein during coagulation in the making of cheese by the action of the chymosin proteases that remain in the whey fraction after the curd formation. GMP is an acidic, hydrophilic peptide with a molecular weight (MW) of approximately 8,000 Daltons. Several research groups have demonstrated the bioactivity of GMP such as reduction of bone loss in ovariectomized rats and antimicrobial activity *in vitro* by inhibiting production of several toxins produced by bacteria [62, 63].

Alpha-lactalbumin (α-La) is the main whey protein in bovine and human milk that is rich with sulfur-containing amino acids such as cysteine, methionine, and tryptophan in its molecule [64]. α-La participates in lactose biosynthesis and can be bound with calcium. Somu et al. [65] has demonstrated that α-lactalbumin have reactive oxygen species (ROS)-dependent cytotoxicity in many tumor cells such as MCF-7, MDA-MB231, HeLa, and MG 63 [65]. Beta-Lactoglobulin (β-Lg) is the second whey protein that can be found in four genetic variants (A, B, C, and D) [64, 66]. Researchers have proved through numerous studies that β-Lg has a wide range of bioactivities, such as an effect on blood cholesterol since it can influence serum cholesterol levels and have hypocholesterolemic activity while it is reported to potentially have activity as an inhibitor of Angiotensin-converting enzyme (ACE) and therefore help in relaxation of the veins and arteries to lower blood pressure and prevent the narrowing of blood vessels [66, 67].

LF is a milk Fe-binding glycoprotein with several physiological factions and bioactive functions, that consists of a single chain polypeptide with two globular lobes [68]. LF has been reported to have antimicrobial, antioxidant, anti-inflammatory, and antibacterial properties, iron binding properties, potential anticancer properties, bone formation stimulating, osteoblasts stimulation, inhibition of osteoclasts and bone resorption cells [69, 70]. A recent study demonstrated that LF could mitigate hippocampus' oxidative stress [71]. LF also has potential antiviral activity against rotavirus, HIV, hepatitis C virus and most recently inhibition of SARS-CoV-2 variants has been detected [72–74]. Lysozyme as well as α-lactalbumin has shown ROS-dependent cytotoxicity in many tumor cells such as MCF-7, MDA-MB231, HeLa, and MG 63 [65]. Milk also contains proline, an amino acid with bioactive properties, which plays an important role in collagen synthesis. It has been reported that high levels of proline and hydroxyproline have been found in ewes milk proteins. A study showed an indirect effect of proline on the proliferation of MCF-7 and MDA-MB-231 breast cancer cells. On the contrary, proline-rich polypeptide (PRP) participates in the humoral immune response and is responsible for the maturation of regulatory T lymphocytes (Tregs) [75]. PRP has shown protection against Alzheimer's disease and other similar amyloidogenic diseases [76]. Finally, immunoglobulins (Ig), such as IgA, IgD, IgE, IgG, and IgM are antibodies present in the milk and colostrums that protect offsprings from pathogens and infection. Over the years, several researchers have demonstrated the antimicrobial and anti-cholestorelemic activity of Ig [67, 77].

TABLE 4.3 Dairy Bioactive Compounds and the Reported Bioactivity

Bioactive Peptides	Bioactivity and Suggested Health Effect	References
Glycomacropeptide (GMP)	Promoting growth of bifidobacterial	[78–80]
	Immunomodulatory actions	
Immunoglobulin	Lowering cholesterol levels.	[77, 81, 82]
	Protection against Shigella, Norovirus, and various other intestinal pathogens.	
α-Lactalbumin	Protection against gastric mucosal injury.	[65, 83–87]
	Binds calcium molecules.	
	Increases tryptophan in the plasma.	
Lactoferrin	Inhibition of SARS-CoV-2 variants.	[72, 74, 88, 89]
	Antimicrobial, anticancer.	
β-Lactoglobulin	hypocholesterolemic activity.	[87, 90–92]
	Antitumor effects.	
	Inhibition of angiotensin-I converting enzyme (ACE).	

4.3 DAIRY BIOACTIVE COMPOUNDS PHYSIOLOGICAL FUNCTIONALITIES

The main areas of bioactivity of milk components include probiotic activity and antioxidant activity, antimicrobial, anticancer, and anti-inflammatory effects. Figure 4.2 demonstrates the health benefits of bioactive dairy compounds

4.3.1 ANTIOXIDANT ACTIVITY OF BIOACTIVE MILK COMPOUNDS

Antioxidants are chemical compounds that can neutralize and scavenge the free radicals naturally produced in the body since oxidation is an indispensable biological process directly connected with energy production. It is known that oxidative stress can cause serious damage to biological systems by inducing atherosclerosis, diabetes, accelerated aging, CVDs and break down of vital biochemical compounds [93]. The consumption of compounds with antioxidative activity as supplements or foods rich in antioxidants may alleviate the effects of oxidative stress and damage in the human body. Dietarians have associated several metabolic diseases with the modern Western lifestyle and dietary habits. Milk and dairy products provides some antioxidant activity that is associated with sulfur amino acids such as cysteine, vitamins A, E, carotenoids, enzyme systems, superoxide dismutase (SOD), catalase, and glutathione peroxidase [94].

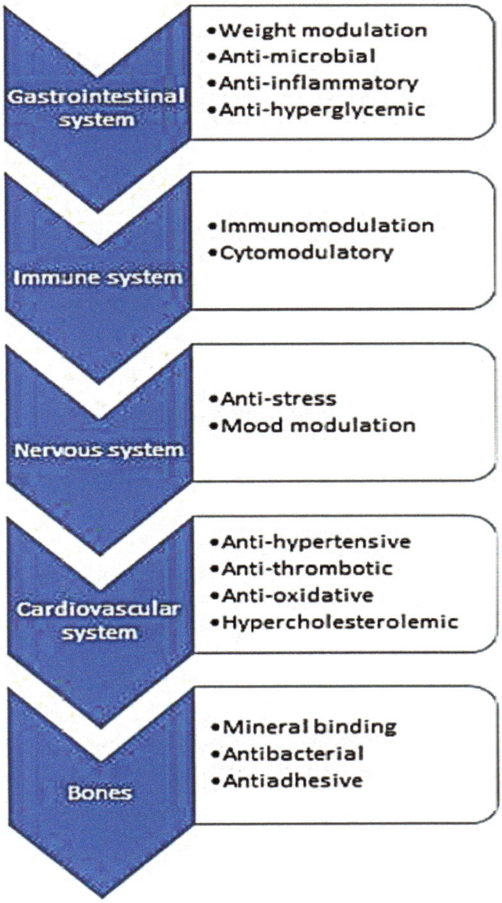

FIGURE 4.2 Health benefits of bioactive compounds in milk and dairy products.

Milk proteins have shown antioxidant activity for the scavenging of ROS. Suetsuna et al. have demonstrated that casein molecules may act as scavenger to hider the lipoxygenase-catalyzed lipid autoxidation [95]. Other studies showed that casein molecules can bind the non-heme iron and inhibit lipoxygenase activity [96]. Several casein hydrolysate isolates demonstrated antioxidant activities in human T cells that is independent of degree of hydrolysis by increasing cellular catalase activity and amount of reduced glutathione but without any effect on SOD activity. Also, whey proteins antioxidant activity is reported and has been associated with several biochemical mechanisms as the chelation of transition metals by LF and scavenging of free radicals by sulfur containing amino acids, inhibition of lipid peroxidation, generation of peroxide radicals, thiobarbituric acid reacting substances, uptake of oxygen and iron oxide free radicals [97]. Carotenoids in milk and dairy products are also important antioxidants that act as scavengers of singlet oxygen and peroxyl radicals [98, 99]. Milk also contains small quantities of ascorbic acid, which is known as one of the strongest and least toxic natural antioxidants and is the major water-soluble antioxidant in milk that can act as a strong free radical scavenger [100]. Other significant antioxidant compounds in milk are vitamins A and E that protect the PUFAs and are associated bio-chemical compounds from peroxidation. α-tocopherol presences in milk fat globule membrane act as a preventative, chain breaking antioxidant and quencher of singlet oxygen in milk [101]. γ-tocopherol is of high functional value as it can trap the nitrogen oxide species and shows protective effects in patients with kidney diseases and asthma [102]. Finally, other antioxidants in milk include zinc, selenium, and glutathione that improve the functionality and antioxidant capacity of milk [103].

4.3.2 ANTIHYPERTENSIVE EFFECT OF BIOACTIVE MILK COMPOUNDS

Hypertension prevailed over modern diseases in western world therefore, researchers have focused on the evaluation of antihypertensive peptides and their intervention in CVDs [6, 104]. Also, consumer's interest for more healthy foods leads to food choice of natural occurring peptides rather than synthetic [6, 105, 106]. An increasing worldwide interest towards the use of food proteins/peptides as novel alternatives is adapting for current therapeutic protocols in the prevention and treatment of high blood pressure [106, 107]. The potential effect of these anti-hypertensive peptides in therapeutic protocols has been widely researched [107–109]. It has been reported that several peptides, such as ACE inhibitory peptides, when released from the protein, have the ability to modulate renin-angiotensin systems and regulate blood pressure [109]. These peptides have been used to produce drugs and are usually isolated from several sources including milk. Two main approaches have been used to release of antihypertensive peptides from the parent milk proteins; hydrolyzation of milk proteins by proteolytic enzymes and milk fermentation. Angiotensin-converting enzyme (ACE) is a peptidyl dipeptidase enzyme able to cleave the carboxyl-end of substrate which upon conversion from angiotensin I into angiotensin II that control blood pressure. The role of ACE inhibitory peptides is to inhibit the conversion to Angiotensin II, hence ensuing the dilation of blood vessels and controlling the hypersensitivity. In addition, ACE degrades bradykinin that has vasco dilatory properties.

Milk proteins have been characterized as a range of inhibitory ACE peptides and effective ACE inhibitors of milk casein (casokinins) and whey proteins (lactokinins) [110]. A study demonstrated the bioavailability of the β-casein inhibitory peptide (LHLPLP) which was noted to be hydrolyzed to HLPLP and was further stable to different proteases resulting in maintaining the blood pressure. VPP and IPP are well-known dairy-originated antihypertensive tripeptides derived from milk fermentation by *Lactobacillus helveticus* and *Saccharomyces cerevisiae* [111]. Milk LF derived peptides RRWQWR, LIWKL, and RPYL significantly reduced blood pressure in SHRs and were also found to reduce Ang II induced vasoconstriction in isolated rabbit carotid arterial segments [112]. In addition, RYLGY and AYFYPEL peptides derived from bovine casein hydrolysation have demostrated antioxidant activity and *in vivo* ACE inhibitory effects [112]. Chen et al. [113] studied the fermentation activity of 59 strains of *Lb. helveticus* in cow milk and selected among them three strains with the highest ACE-I activity, namely, IMAU80872, IMAU80852, and IMAU80851 (76.64%, 81.68%, and 81.71, respectively). It was also reported the increasing ACE-I activity after *in vitro* digestion of the fermented milks, indicating the thermostability and resistance of the ACE-I constituents to gastrointestinal proteases and that the peptides are prodrugs [113]. Fermentation of goat milk by Mahdi et al. [114] revealed three potential antihypercholesterolemia peptides composed of 16 amino acid residues (LYQEPVLGPVRGPFPI, YQEPVLGPVRGFPIL, and VQSWMHQPPQPLSPT) able to reduce the production of malondialdehyde and cholesterol level in hypercholesterolemia rats [114]. Finally, α-lactorphine and β-lactorphine are known ACE inhibitory peptides that are released from whey proteins [110]. Garcia-Tejedor et al. [115] have examined dairy yeast strains and showed the feasibility for the production of milk protein-derived antihypertensive hydrolysates [115].

4.3.3 ANTIMICROBIAL PEPTIDES

Antimicrobial peptides are the peptides which provide protection against several species such as bacteria, fungi, viruses, and protozoans as they have toxicity towards bacterial pathogens without impairing the host tissue. The overall antibacterial activity in milk is higher than would be expected from the amount of the various immunoglobulin and non-immunoglobulin protein contributions. Early studies have demonstrated the antimicrobial activity of milk as lactoferricin that has potential antimicrobial activity while is the main precursor for the formation of bioactive peptides. Several antimicrobial peptides have been isolated from caseins, lactoalbumins, and lactoglobulins. Recently, there is a renewed interest in exploiting bioactive peptides from the pharmaceutical industry as some of them have provided an inhibitory action against several pathogenic Gram-positive and Gram-negative bacterial strains [110]. Alpha lactoferrampin fragment released from the digestion of milk LF has bactericidal action against *E. coli* and *Pseudomonas aeruginosa* [116]. Likewise, LF and its derivatives display antimicrobial activity (Table 4.3) against several pathogens, such as *Salmonella enteritidis*, *Haemophilus influenzae*, *Clostridium perfringens*, *Pseudomonas aeruginosa*, *Listeria monocytogenes*, *Staphylococcus aureus*, *C. albicans*, *Vibrio cholerae*, *Helicobacter pylori*, and also inhibition of viruses such as poliovirus, hepatitis C, B, and G, herpes simplex virus, HIV-1, rotavirus, and SARS-CoV-2 [72, 74, 88, 89, 117].

4.3.4 MINERAL BINDING

Several peptides with great anionic activity have the ability to form soluble complexes with minerals such as calcium, phosphorus, etc., in a solution at intestinal pH and subsequently increasing the bioavailability of the minerals and their resistance to the proteolytic activity of intestines [118]. Milk caseins are well known for stabilizing calcium and phosphate ions in their molecule; thus several phosphopeptides have been identified from the enzymatic digestion of milk proteins. During the formation of casein micelle, casein proteins are subjected to tryptic digestion releasing caseinophosphopeptides (CPPs) from the N-terminus polar domain containing a cluster of phosphorylated seryl residues that are responsible for interaction between calcium phosphate and casein. Also, these phosphorylated peptides bind to calcium and phosphate ions, improving their bioavailability [118].

4.3.5 ANTITHROMBOTIC

Antithrombotic means the prevention of development or augmentation of blood clots in the body. The major role of an antithrombotic drug is to increase fibrinolysis and reduce platelet aggregation. Reports revealed that usually antithrombotic peptides contain 3 to 20 amino acids residues that possess a certain homology of the γ-chain of fibrinogen as they can compete for the receptors of the thrombocytes and subsequently inhibiting the development or reduce platelet aggregation. For the moment, the majority of food-derived antithrombotic peptides are obtained by gastrointestinal digestion of milk casein or enzymatic hydrolysis or digestion of blue mussel, legume, amaranth grain, and brewer's spent grain proteins [118]. Casoplatelins which are k-casein fragments isolated from tryptic hydrolysates may inhibit the activity of fibrinogen binding platelet further showing antithrombotic action during gastrointestinal digestion and are easily assimilated into the blood. Erdmann et al. showed that bovine k-casein f106–f116, with amino acid sequence MAIPPKKNQDK, hindered platelet aggregation and binding with the receptor site thus preventing fibrinogen binding with blood platelets. Yogurt prepared from skim milk fortified with WPC using a yogurt starter culture and a combination with highly proteolytic co culture showed a high antithrombotic activity ranging from 42% to 74% [119]. The high antithrombotic activity was attributed to the structural similarity of milk LF to the fibrinogen α-chain that has been known to constrain platelet aggregation. Even though, the probable physiological effects of these antithrombotic peptides have not been well documented and integrated in therapheutic protocols, it is reported that such peptides have been identified in the plasma of infants after breastfeeding or on ingestion of cow milk-based infant formula.

4.4 ANTICANCER/ANTIPROLIFERATIVE ACTIVITY

Cancer is a disease caused through growth of abnormal cells. Despite the many years of investigation and scientist efforts, no clear results have been jet reached regarding the pathway affecting tumor progression. However, general mechanisms of cancer inhibition have been well-identified. Having in mind that "*prevention is better than cure*," food-derived antiproliferative peptides show potential on supporting chemoprevention compounds and drug adjuncts by reducing the chances of cancer occurrence, slowing cancer progression, and sensitizing tumors to drug treatment. Antiproliferative peptides, usually received through our diet, have provided several advantages such as elevated specificity to tumor cells, high affinity and permeability to tissues, and most importantly, low toxicity. Research investigating milk-derived proteins has brought to light the potential for their use as novel anticancer agents. Evaluation of evaluated LF, α-lactalbumin, and its complexes, such as HAMLET, BAMLET, and lactalbumin-oleic acid complexes, as well as LP, whey, and casein provided evidence on their effectiveness in the treatment of head and neck cancer cells lines as it was found that these compounds can inhibit tumor growth, modulate cancer gene expression, and have cytotoxic effects on cancer cells [120]. Also, in preliminary studies, HMP-S7, which is a human milk-derived peptide, showed cytotoxic activity against four distinct leukemic cell lines in a dose-dependent manner. HMP-S7 induced leukemic cell death by penetrating the plasma membrane to enter the cytoplasm and cause the leakage of lactate dehydrogenase, thus acting in a membranolytic manner [121]. Breast cancer has a diverse etiology characterized by the heterogeneous expression of hormone receptors and signaling molecules, resulting in varied sensitivity to chemotherapy. The adverse side effects of chemotherapy, coupled with the development of drug resistance, have prompted the exploration of natural products to combat cancer. Lactoferricin B (LfcinB) is a bovine milk-derived peptide that exhibits anticancer properties. LfcinB was considered as a promising and safe candidate that can be considered for the treatment of breast cancer [122].

4.5 BIOAVAILABILITY OF BIOACTIVE PEPTIDES

Bioactivity of different peptides is different while in observation *in vitro* than the real-life applications as there are concerns for its bioavailability, absorption, and susceptibility to breakdown into inactive fragments [6, 15, 123–125]. For effective health promoting activity from bioactive peptides, resistance against gastrointestinal proteases and easy absorption into the blood serum without any degradation is necessary. This stability makes certain that the ingested peptide sequences which showed *in vitro* bioactivity have safely reached to the desired cellular sites of action [124, 125]. It is reported that MAP1 and MAP2, milk proteins derived peptides, showed *in vitro* ACE inhibitory activity for hypertensivity but only MAP1 could show potent inhibitory effect and reduction in blood pressure (Table 4.4). Intestinal-expressed peptide transporters help in carrying the smaller peptides across the interests, whereas oligopeptides are transported passively through hydrophobic regions of epithelial membrane. Also, small-sized peptides have demonstrated more *in vivo* bioactivity and can be easily absorbed into blood circulation.

TABLE 4.4 Dairy Products Bioactive Components and Their Potential Health Effects

Dairy Product	Bioactivity and Suggested Health-Effect	Study Design	References
Gouda cheese	Beneficial effect on abdominal adipose.	Animal study (rats).	[126]
Cheese	ACE-inhibitory	–	[127]
"Festivo" cheese	ACE-inhibitory	*In vitro*	[128]
Italian cheese (Mozzarella, Italico, Crescenza, and Gorgonzola)	ACE-inhibitory	*In vitro*	[129]
Yoghurt	ACE-inhibitory	*In vitro*	[130]
Sour milk	Reduce hypertension	Placebo-controlled study	[131]
High-fat dairy products	Lower risk of colorectal cancer	Prospective cohort study	[26]

4.6 FUTURE PROSPECTS

Bioactive compounds provide numerous health benefits to consumers, as has been well documented in recent literature. The notion that components within milk and milk products can modulate physiological functions and provide various beneficial effects on human health is more and more highlighted in late research. With today's accurate analytical and biochemical research tools, the presence of various less studied compounds posing biological activity can be demonstrated in both milk and dairy products. Recent scientific achievements have overcome prior obstacles in separation, isolation as well as concentration of bioactive compounds, so that their application in functional foods and nutraceuticals can be more easily achieved. On the other hand, bioactivity mainly depends on the biochemical state of bioactive compounds reaching the bloodstream. As a result, research is now to target the protection and enhancement of these bioactive. Thus, research efforts have subsequently focused not only on the identification of factors within milk and dairy products that may be relevant to improving human health but as well to the sustainability of these compounds. Moreover, in future applications, milk and dairy bioactive compounds can be applied as food enrichment to produce novel functional foods, food preservatives, food colorants, anti-cancer agents, or anti-inflammatory agents.

KEYWORDS

- **bioactive compounds**
- **dairy**
- **food preservatives**
- **hypertensivity**
- **milk**
- **novel functional foods**
- **nutraceutical**
- **peptides**

REFERENCES

1. Lawrence, R. A., (2022). 4 – biochemistry of human milk. In: Lawrence, R. A., & Lawrence, R. M., (eds.), *Breastfeeding* (9th edn., pp. 93–144), Elsevier: Philadelphia.
2. Kilic-Akyilmaz, M., Ozer, B., Bulat, T., & Topcu, A., (2022). Effect of heat treatment on micronutrients, fatty acids, and some bioactive components of milk. *International Dairy Journal, 126*, 105231. doi: https://doi.org/10.1016/j.idairyj.2021.105231.
3. Akkurt, S., & Tomasula, P. M., (2022). Bioactivity of milk components. In: McSweeney, P. L. H., & McNamara, J. P., (eds.), *Encyclopedia of Dairy Sciences* (3rd edn., pp. 314–326). Academic Press: Oxford.
4. Hendricks, G. M., & Guo, M. R., (2021). 3 – Bioactive components in human milk. In: Guo, M., (ed.), *Human Milk Biochemistry and Infant Formula Manufacturing Technology* (2nd edn., pp. 61–80). Woodhead Publishing.
5. Sharma, R., (2021). Bioactive food components for managing cellular senescence in aging and disease: A critical appraisal and perspectives. *Pharma Nutrition, 18*, 100281. doi: https://doi.org/10.1016/j.phanu.2021.100281.
6. Terpou, A., & Rai, A. K., (2022). 2 – microbial transformation for improving food functionality. In: Rai, A. K., Singh, S. P., Pandey, A., Larroche, C., & Soccol, C. R.,

(eds.), *Current Developments in Biotechnology and Bioengineering* (pp. 31–45). Elsevier.

7. Qu, J., Zhang, L., Yin, L., Liu, J., Sun, Z., & Zhou, P., (2021). Changes in bioactive proteins and serum proteome of human milk under different frozen storage. *Food Chemistry, 352*, 129436. doi: https://doi.org/10.1016/j.foodchem.2021.129436.
8. Lönnerdal, B., (2017). Bioactive proteins in human milk—potential benefits for preterm infants. *Clinics in Perinatology, 44*, 179–191. doi: https://doi.org/10.1016/j.clp.2016.11.013.
9. Terpou, A., (2020). Selected ethnic fermented foods of Greece. In: Sankaranarayanan, A., Amaresan, N., & Dhanasekaran, D. (Eds.). *Fermented Food Products* (1st ed.). CRC Press. https://doi.org/10.1201/9780429274787.
10. Puchalska, P., Marina, A. M. L., & García, L. M. C., (2015). Isolation and characterization of peptides with antihypertensive activity in foodstuffs. *Critical Reviews in Food Science and Nutrition, 55*, 521–551. doi: 10.1080/10408398.2012.664829.
11. Li, S., Ye, A., & Singh, H., (2021). Impacts of heat-induced changes on milk protein digestibility: A review. *International Dairy Journal, 123*, 105160. doi: https://doi.org/10.1016/j.idairyj.2021.105160.
12. Perrella, S., Gridneva, Z., Lai, C. T., Stinson, L., George, A., Bilston-John, S., & Geddes, D., (2021). Human milk composition promotes optimal infant growth, development and health. *Seminars in Perinatology, 45*, 151380. doi: https://doi.org/10.1016/j.semperi.2020.151380.
13. Cakir, B., & Tunali-Akbay, T., (2021). Potential anticarcinogenic effect of goat milk-derived bioactive peptides on HCT-116 human colorectal carcinoma cell line. *Analytical Biochemistry, 622*, 114166. doi: https://doi.org/10.1016/j.ab.2021.114166.
14. Urashima, T., Katayama, T., Sakanaka, M., Fukuda, K., & Messer, M., (2022). Evolution of milk oligosaccharides: Origin and selectivity of the ratio of milk oligosaccharides to lactose among mammals. *Biochimica et Biophysica Acta (BBA) – General Subjects, 1866*, 130012. doi: https://doi.org/10.1016/j.bbagen.2021.130012.
15. Terpou, A., Papadaki, A., Lappa, I. K., Kachrimanidou, V., Bosnea, L. A., & Kopsahelis, N., (2019). Probiotics in food systems: Significance and emerging strategies towards improved viability and delivery of enhanced beneficial value. *Nutrients, 11*, 1591.
16. Ilesanmi-Oyelere, B. L., & Kruger, M. C., (2020). The role of milk components, pro-, pre-, and synbiotic foods in calcium absorption and bone health maintenance. *Frontiers in Nutrition, 7*. doi: 10.3389/fnut.2020.578702.
17. Huang, M., Wang, C., Cheng, M., Zhang, X., Jiang, H., & Wang, J., (2022). Effects of quantity and source of calcium on the behavior of goat milk after heating and acidification. *LWT, 153*, 112535. doi: https://doi.org/10.1016/j.lwt.2021.112535.
18. Deeth, H. C., & Lewis, M. J., (2015). Practical consequences of calcium addition to and removal from milk and milk products. *International Journal of Dairy Technology, 68*, 1–10. doi: https://doi.org/10.1111/1471-0307.12188.
19. Daley, D. K., & Myrie, S. B., (2021). Chapter 1 – extra-skeletal effects of dietary calcium: Impact on the cardiovascular system, obesity, and cancer. In: Eskin, N. A. M., (ed.), *Advances in Food and Nutrition Research* (Vol. 96, pp. 1–25). Academic Press.
20. Baspinar, B., & Güldaş, M., (2021). Traditional plain yogurt: A therapeutic food for metabolic syndrome? *Crit. Rev. Food Sci. Nutr., 61*, 3129–3143. doi: 10.1080/10408398.2020.1799931.

21. Kjølbæk, L., Lorenzen, J. K., Larsen, L. H., & Astrup, A., (2017). Calcium intake and the associations with fecal fat and energy excretion, and lipid profile in a free-living population. *J. Nutr. Sci.*, *6*, e50–e50. doi: 10.1017/jns.2017.55.
22. Lin, Y. C., Lyle, R. M., McCabe, L. D., McCabe, G. P., Weaver, C. M., & Teegarden, D., (2000). Dairy calcium is related to changes in body composition during a two-year exercise intervention in young women. *J. Am. Coll. Nutr.*, *19*, 754–760. doi: 10.1080/07315724.2000.10718075.
23. Zhang, F., Ye, J., Zhu, X., Wang, L., Gao, P., Shu, G., Jiang, Q., & Wang, S., (2019). Anti-Obesity effects of dietary calcium: The evidence and possible mechanisms. *Int. J. Mol. Sci.*, *20*, 3072. doi: 10.3390/ijms20123072.
24. Das, S., & Choudhuri, D., (2020). Calcium supplementation shows a hepatoprotective effect against high-fat diet by regulating oxidative-induced inflammatory response and lipogenesis activity in male rats. *J. Tradit. Complement Med.*, *10*, 511–519. doi: 10.1016/j.jtcme.2019.06.002.
25. Umesawa, M., Iso, H., Date, C., Yamamoto, A., Toyoshima, H., Watanabe, Y., Kikuchi, S., Koizumi, A., Kondo, T., Inaba, Y., et al., (2006). Dietary intake of calcium in relation to mortality from cardiovascular disease. *Stroke*, *37*, 20–26. doi: doi: 10.1161/01.STR.0000195155.21143.38.
26. Larsson, S. C., Bergkvist, L., & Wolk, A., (2005). High-fat dairy food and conjugated linoleic acid intakes in relation to colorectal cancer incidence in the Swedish mammography cohort. *The American Journal of Clinical Nutrition*, *82*, 894–900. doi: 10.1093/ajcn/82.4.894.
27. Das, S., & Choudhuri, D., (2020). Dietary calcium regulates the insulin sensitivity by altering the adipokine secretion in high fat diet induced obese rats. *Life Sciences*, *250*, 117560. doi: https://doi.org/10.1016/j.lfs.2020.117560.
28. Shidfar, F., Moghayedi, M., Jafarian, K. S. R., Hosseini, S., & Shidfar, S., (2010). Effects of a calcium supplement on serum lipoproteins, apolipoprotein B, and blood pressure in overweight men. *International Journal of Endocrinology and Metabolism (IJEM)*, *8*, 194–200.
29. Heshmati, J., Sepidarkish, M., Namazi, N., Shokri, F., Yavari, M., Fazelian, S., Khorshidi, M., & Shidfar, F., (2019). Impact of dietary calcium supplement on circulating lipoprotein concentrations and atherogenic indices in overweight and obese individuals: A systematic review. *Journal of Dietary Supplements*, *16*, 357–367. doi: 10.1080/19390211.2018.1440685.
30. Terpou, A., Bosnea, L., Kanellaki, M., Plessas, S., Bekatorou, A., Bezirtzoglou, E., & Koutinas, A. A., (2018). Growth capacity of a novel potential probiotic *Lactobacillus paracasei* K5 strain incorporated in industrial white brined cheese as an adjunct culture. *J. Food Sci.*, *83*, 723–731. doi: 10.1111/1750-3841.14079.
31. Televičius, M., Juozaitiene, V., Malašauskienė, D., Antanaitis, R., Rutkauskas, A., Urbutis, M., & Baumgartner, W., (2021). Inline milk lactose concentration as a biomarker of the health status and reproductive success in dairy cows. *Agriculture*, *11*, 38.
32. Coppa, G., Pierani, P., Zampini, L., Carloni, I., Carlucci, A., & Gabrielli, O., (1999). Oligosaccharides in human milk during different phases of lactation. *Acta Paediatrica*, *88*, 89–94. doi: https://doi.org/10.1111/j.1651-2227.1999.tb01307.x.
33. Oliveira, D. L., Wilbey, R. A., Grandison, A. S., & Roseiro, L. B., (2015). Milk oligosaccharides: A review. *International Journal of Dairy Technology*, *68*, 305–321. doi: https://doi.org/10.1111/1471-0307.12209.

34. Hodges, J. K., Cao, S., Cladis, D. P., & Weaver, C. M., (2019). Lactose intolerance and bone health: The challenge of ensuring adequate calcium intake. *Nutrients, 11*, 718. doi: 10.3390/nu11040718.
35. Zhang, Q., Ni, Y., & Kokot, S., (2016). Competitive interactions between glucose and lactose with BSA: Which sugar is better for children? *Analyst, 141*, 2218–2227. doi: 10.1039/C5AN02420J.
36. Whisner, C. M., Martin, B. R., Schoterman, M. H., Nakatsu, C. H., McCabe, L. D., McCabe, G. P., Wastney, M. E., et al., (2013). Galacto-oligosaccharides increase calcium absorption and gut bifidobacteria in young girls: A double-blind cross-over trial. *Br. J. Nutr., 110*, 1292–1303. doi: 10.1017/s000711451300055x.
37. Awasthi, M. K., Tarafdar, A., Gaur, V. K., Amulya, K., Narisetty, V., Yadav, D. K., Sindhu, R., Binod, P., Negi, T., Pandey, A., et al., (2022). Emerging trends of microbial technology for the production of oligosaccharides from biowaste and their potential application as prebiotic. *International Journal of Food Microbiology, 368*, 109610. doi: https://doi.org/10.1016/j.ijfoodmicro.2022.109610.
38. Coppa, G. V., & Gabrielli, O., (2008). Human milk oligosaccharides as prebiotics. In: *Therapeutic Microbiology* (pp. 131–146).
39. Bode, L., (2009). Human milk oligosaccharides: Prebiotics and beyond. *Nutrition Reviews, 67*, S183–S191. doi: https://doi.org/10.1111/j.1753-4887.2009.00239.x.
40. Wang, Y., Zhou, X., Gong, P., Chen, Y., Feng, Z., Liu, P., Zhang, P., et al., (2020). Comparative major oligosaccharides and lactose between Chinese human and animal milk. *International Dairy Journal, 108*, 104727. doi: https://doi.org/10.1016/j.idairyj.2020.104727.
41. Li, W., Wang, J., Lin, Y., Li, Y., Ren, F., & Guo, H., (2021). How far is it from infant formula to human milk? A look at the human milk oligosaccharides. *Trends in Food Science & Technology, 118*, 374–387. doi: https://doi.org/10.1016/j.tifs.2021.09.021.
42. Markiewicz-Kęszycka, M., Czyżak-Runowska, G., Lipińska, P., & Wójtowski, J., (2013). Fatty acid profile of milk – a review. *Journal of Veterinary Research, 57*, 135–139. doi: doi: 10.2478/bvip-2013–0026.
43. Zhu, Z., Wu, F., Lu, Y., Wang, Z., Zang, J., Yu, H., Guo, C., et al., (2018). The association of dietary cholesterol and fatty acids with dyslipidemia in Chinese metropolitan men and women. *Nutrients, 10*, 961.
44. Panfili, G., Manzi, P., & Pizzoferrato, L., (1994). High-performance liquid chromatographic method for the simultaneous determination of tocopherols, carotenes, and retinol and its geometric isomers in Italian cheeses. *Analyst, 119*, 1161–1165. doi: 10.1039/AN9941901161.
45. Cervinkova, B., Krcmova, L. K., Solichova, D., Melichar, B., & Solich, P., (2016). Recent advances in the determination of tocopherols in biological fluids: From sample pretreatment and liquid chromatography to clinical studies. *Analytical and Bioanalytical Chemistry, 408*, 2407–2424. doi: 10.1007/s00216-015-9214-0.
46. Martini, M., Salari, F., Buttau, L., & Altomonte, I., (2021). Natural content of animal and plant sterols, alpha-tocopherol and fatty acid profile in sheep milk and cheese from mountain farming. *Small Ruminant Research, 201*, 106419. doi: https://doi.org/10.1016/j.smallrumres.2021.106419.
47. Lindmark, M. H., (2008). Fatty acids in bovine milk fat. *Food & Nutrition Research, 52*(1), doi: 10.3402/fnr.v52i0.1821.

48. Martysiak-Żurowska, D., Kiełbratowska, B., & Szlagatys-Sidorkiewicz, A., (2018). The content of conjugated linoleic acid and vaccenic acid in the breast milk of women from Gdansk and the surrounding district, as well as in infant formulas and follow-up formulas. Nutritional recommendation for nursing women. *Dev. Period. Med.*, *22*, 128–134. doi: 10.34763/devperiodmed.20182202.128134.
49. Hanuš, O., Samková, E., Křížová, L., Hasoňová, L., & Kala, R., (2018). Role of fatty acids in milk fat and the influence of selected factors on their variability: A review. *Molecules*, *23*, 1636. doi: 10.3390/molecules23071636.
50. Haug, A., Høstmark, A. T., & Harstad, O. M., (2007). Bovine milk in human nutrition – a review. *Lipids in Health and Disease*, *6*, 25. doi: 10.1186/1476-511X-6-25.
51. Savoini, G., Farina, G., Dell'Orto, V., & Cattaneo, D., (2016). Through ruminant nutrition to human health: Role of fatty acids. *Advances in Animal Biosciences*, *7*, 200–207. doi: 10.1017/S2040470016000133.
52. Benjamin, S., & Spener, F., (2009). Conjugated linoleic acids as functional food: An insight into their health benefits. *Nutr. Metab. (Lond)*, *6*, 36. doi: 10.1186/1743-7075-6-36.
53. Saba, F., Sirigu, A., Pillai, R., Caria, P., Cordeddu, L., Carta, G., Murru, E., et al., (2019). Downregulation of inflammatory markers by conjugated linoleic acid isomers in human cultured astrocytes. *Nutr. Neurosci.*, *22*, 207–214. doi: 10.1080/1028415x.2017.1367130.
54. Moutsioulis, A. A., Rule, D. C., Murrieta, C. M., Bauman, D. E., Lock, A. L., Barbano, D. M., & Carey, G. B., (2008). Human breast milk enrichment in conjugated linoleic acid after consumption of a conjugated linoleic acid–rich food product: A pilot study. *Nutrition Research*, *28*, 437–442. doi: https://doi.org/10.1016/j.nutres.2008.04.001.
55. Tsorotioti, S., Nasopoulou, C., Detopoulou, M., Sioriki, E., Demopoulos, C., & Zabetakis, I., (2014). In vitro anti-atherogenic properties of traditional Greek cheese lipid fractions. *Dairy Science & Technology*, *94*, 269–281. doi: 10.1007/s13594-014-0161-x.
56. Megalemou, K., Sioriki, E., Lordan, R., Dermiki, M., Nasopoulou, C., & Zabetakis, I., (2017). Evaluation of sensory and in vitro anti-thrombotic properties of traditional Greek yogurts derived from different types of milk. *Heliyon*, *3*, e00227. doi: https://doi.org/10.1016/j.heliyon.2016.e00227.
57. Poutzalis, S., Anastasiadou, A., Nasopoulou, C., Megalemou, K., Sioriki, E., & Zabetakis, I., (2016). Evaluation of the in vitro anti-atherogenic activities of goat milk and goat dairy products. *Dairy Science & Technology*, *96*, 317–327. doi: 10.1007/s13594-015-0266-x.
58. Caboni, P., Murgia, A., Porcu, A., Manis, C., Ibba, I., Contu, M., & Scano, P., (2019). A metabolomics comparison between sheep's and goat's milk. *Food Research International*, *119*, 869–875. doi: https://doi.org/10.1016/j.foodres.2018.10.071.
59. Meisel, H., & Bockelmann, W., (1999). Bioactive peptides encrypted in milk proteins: Proteolytic activation and thropho-functional properties. *Antonie Van Leeuwenhoek*, *76*, 207–215.
60. Korhonen, H., (2009). Milk-derived bioactive peptides: From science to applications. *Journal of Functional Foods*, *1*, 177–187. doi: https://doi.org/10.1016/j.jff.2009.01.007.
61. Iwatani, S., & Yamamoto, N., (2019). Functional food products in Japan: A review. *Food Science and Human Wellness*, *8*, 96–101. doi: https://doi.org/10.1016/j.fshw.2019.03.011.
62. Kruger, M., Poulsen, R., Schollum, L., Haggarty, N., Ram, S., & Palmano, K., (2006). A comparison between acidic and basic protein fractions from whey or milk for reduction of bone loss in the ovariectomised rat. *International Dairy Journal*, *16*, 1149–1156.

63. Rong, Y., Lu, Z., Zhang, H., Zhang, L., Song, D., & Wang, Y., (2015). Effects of casein glycomacropeptide supplementation on growth performance, intestinal morphology, intestinal barrier permeability and inflammatory responses in *Escherichia coli* K88 challenged piglets. *Animal Nutrition, 1*, 54–59. doi: https://doi.org/10.1016/j.aninu.2015.05.006.
64. Wang, Q., Pan, M. H., Chiou, Y. S., Li, Z., Wei, S., Yin, X., & Ding, B., (2022). Insights from alpha-lactoalbumin and beta-Lactoglobulin into mechanisms of nanoliposome-whey protein interactions. *Food Hydrocolloids, 125*, 107436. doi: https://doi.org/10.1016/j.foodhyd.2021.107436.
65. Somu, P., & Paul, S., (2019). Supramolecular nanoassembly of lysozyme and α-lactalbumin (apo α-LA) exhibits selective cytotoxicity and enhanced bioavailability of curcumin to cancer cells. *Colloids and Surfaces B: Biointerfaces, 178*, 297–306. doi: https://doi.org/10.1016/j.colsurfb.2019.03.016.
66. Sunde, H., Ryder, K., Bekhit, A. E. D. A., & Carne, A., (2021). Analysis of peptides in a sheep beta-lactoglobulin hydrolysate as a model to evaluate the effect of peptide amino acid sequence on bioactivity. *Food Chemistry, 365*, 130346. doi: https://doi.org/10.1016/j.foodchem.2021.130346.
67. Mehra, R., Kumar, H., Kumar, N., Ranvir, S., Jana, A., Buttar, H. S., Telessy, I. G., Awuchi, C. G., Okpala, C. O. R., Korzeniowska, M., et al., (2021). Whey proteins processing and emergent derivatives: An insight perspective from constituents, bioactivities, functionalities to therapeutic applications. *Journal of Functional Foods, 87*, 104760. doi: https://doi.org/10.1016/j.jff.2021.104760.
68. Plaut, A. G., & Geme, J. S., (2013). Chapter 805 – lactoferrin. In: Rawlings, N. D., & Salvesen, G., (eds.), *Handbook of Proteolytic Enzymes* (3rd edn., pp. 3635–3640). Academic Press.
69. Chaneton, L., Bontà, M., Pol, M., Tirante, L., & Bussmann, L. E., (2013). Milk lactoferrin in heifers: Influence of health status and stage of lactation. *Journal of Dairy Science, 96*, 4977–4982. doi: https://doi.org/10.3168/jds.2012-6028.
70. Guo, H. Y., Jiang, L , Ibrahim, S. A., Zhang, L., Zhang, H., Zhang, M., & Ren, F. Z., (2009). Orally administered lactoferrin preserves bone mass and microarchitecture in ovariectomized rats. *J. Nutr., 139*, 958–964. doi: 10.3945/jn.108.100586.
71. Zheng, J., Xie, Y., Li, F., Zhou, Y., Qi, L., Liu, L., & Chen, Z., (2020). Lactoferrin improves cognitive function and attenuates brain senescence in aged mice. *Journal of Functional Foods, 65*, 103736. doi: https://doi.org/10.1016/j.jff.2019.103736.
72. Drago-Serrano, M. E., Campos-Rodríguez, R., Carrero, J. C., & De La Garza, M., (2017). Lactoferrin: Balancing ups and downs of inflammation due to microbial infections. *Int. J. Mol. Sci., 18*, 501.
73. Niaz, B., Saeed, F., Ahmed, A., Imran, M., Maan, A. A., Khan, M. K. I., Tufail, T., et al., (2019). Lactoferrin (LF): A natural antimicrobial protein. *International Journal of Food Properties, 22*, 1626–1641. doi: 10.1080/10942912.2019.1666137.
74. Wotring, J. W., Fursmidt, R., Ward, L., & Sexton, J. Z., (2022). Evaluating the in vitro efficacy of bovine lactoferrin products against SARS-CoV-2 variants of concern. *Journal of Dairy Science, 105*, 2791–2802. doi: https://doi.org/10.3168/jds.2021-21247.
75. Singh, J., Singh, R., Gupta, P., Rai, S., Ganesher, A., Badrinarayan, P., Sastry, G. N., et al., (2017). Targeting progesterone metabolism in breast cancer with l-proline derived new 14-azasteroids. *Bioorganic & Medicinal Chemistry, 25*, 4452–4463. doi: https://doi.org/10.1016/j.bmc.2017.06.031.

76. Bharadwaj, P., Head, R., Martins, R., Raussens, V., Sarroukh, R., Jegasothy, H., Waddington, L., & Bennett, L., (2013). Modulation of amyloid-β 1-42 structure and toxicity by proline-rich whey peptides. *Food & Function, 4*, 92–103. doi: 10.1039/C2FO30111C.
77. Mehra, R., Marnila, P., & Korhonen, H., (2006). Milk immunoglobulins for health promotion. *International Dairy Journal, 16*, 1262–1271. doi: https://doi.org/10.1016/j.idairyj.2006.06.003.
78. Wu, Y., Zhang, X., Tao, S., Pi, Y., Han, D., Ye, H., Feng, C., et al., (2020). Maternal supplementation with combined galactooligosaccharides and casein glycomacropeptides modulated microbial colonization and intestinal development of neonatal piglets. *Journal of Functional Foods, 74*, 104170. doi: https://doi.org/10.1016/j.jff.2020.104170.
79. Requena, P., González, R., López-Posadas, R., Abadía-Molina, A., Suárez, M. D., Zarzuelo, A., De Medina, F. S., & Martínez-Augustin, O., (2010). The intestinal anti-inflammatory agent glycomacropeptide has immunomodulatory actions on rat splenocytes. *Biochemical Pharmacology, 79*, 1797–1804. doi: https://doi.org/10.1016/j.bcp.2010.02.008.
80. Mikkelsen, T. L., Rasmussen, E., Olsen, A., Barkholt, V., & Frøkiær, H., (2006). Immunogenicity of κ-casein and glycomacropeptide. *Journal of Dairy Science, 89*, 824–830. doi: https://doi.org/10.3168/jds.S0022-0302(06)72145-2.
81. Labayo, H. K. M., Pajuelo, M. J., Tohma, K., Ford-Siltz, L. A., Gilman, R. H., Cabrera, L., Mayta, H., Sanchez, G. J., Cornejo, A. T., Bern, C., et al., (2020). Norovirus-specific immunoglobulin A in breast milk for protection against norovirus-associated diarrhea among infants. *EClinicalMedicine, 27*, 100561. doi: https://doi.org/10.1016/j.eclinm.2020.100561.
82. Bíró, A., Cervenak, L., Balogh, A., Lőrincz, A., Uray, K., Horváth, A., Romics, L., et al., (2007). Novel anti-cholesterol monoclonal immunoglobulin G antibodies as probes and potential modulators of membrane raft-dependent immune functions. *Journal of Lipid Research, 48*, 19–29. doi: https://doi.org/10.1194/jlr.M600158-JLR200.
83. Dowlati, Y., Ravindran, A. V., Maheux, M., Steiner, M., Stewart, D. E., & Meyer, J. H., (2015). No effect of oral l-tryptophan or alpha-lactalbumin on total tryptophan levels in breast milk. *European Neuropsychopharmacology, 25*, 779–787. doi: https://doi.org/10.1016/j.euroneuro.2015.03.005.
84. Beulens, J. W. J., Bindels, J. G., De Graaf, C., Alles, M. S., & Wouters-Wesseling, W., (2004). Alpha-lactalbumin combined with a regular diet increases plasma Trp–LNAA ratio. *Physiology & Behavior, 81*, 585–593. doi: https://doi.org/10.1016/j.physbeh.2004.02.027.
85. Chenal, A., Vernier, G., Savarin, P., Bushmarina, N. A., Gèze, A., Guillain, F., Gillet, D., & Forge, V., (2005). Conformational states and thermodynamics of α-lactalbumin bound to membranes: A case study of the effects of pH, calcium, lipid membrane curvature and charge. *Journal of Molecular Biology, 349*, 890–905. doi: https://doi.org/10.1016/j.jmb.2005.04.036.
86. Jiang, Y., Liu, X. C., Ahrné, L. M., & Skibsted, L. H., (2021). Enthalpy-entropy compensation in calcium binding to acid-base forms of glycine tyrosine dipeptides from hydrolysis of α-lactalbumin. *Food Research International, 149*, 110714. doi: https://doi.org/10.1016/j.foodres.2021.110714.
87. Abedin, M. M., Chourasia, R., Chiring, P. L., Singh, S. P., & Kumar, R. A., (2022). Characterization of ACE inhibitory and antioxidant peptides in yak and cow milk hard

chhurpi cheese of the Sikkim Himalayan region. *Food Chemistry: X*, *13*, 100231. doi: https://doi.org/10.1016/j.fochx.2022.100231.
88. Bertuccini, L., Russo, R., Iosi, F., & Superti, F., (2018). Lactobacilli and lactoferrin: Biotherapeutic effects for vaginal health. *Journal of Functional Foods*, *45*, 86–94. doi: https://doi.org/10.1016/j.jff.2018.03.033.
89. Santos-Pereira, C., Andrés, M. T., Fierro, J. F., Rodrigues, L. R., & Côrte-Real, M., (2022). A review on lactoferrin as a proton pump inhibitor. *International Journal of Biological Macromolecules*, *202*, 309–317. doi: https://doi.org/10.1016/j.ijbiomac.2022.01.075.
90. Xu, X., Feng, Y., Chen, X., Wang, Q., Meng, T., & Liu, A., (2019). Antitumor effects of seleno-β-lactoglobulin on human breast cancer MCF-7 and MDA-MB-231 cells in vitro. *Toxicology in Vitro*, *61*, 104607. doi: https://doi.org/10.1016/j.tiv.2019.104607.
91. Nagaoka, S., Futamura, Y., Miwa, K., Awano, T., Yamauchi, K., Kanamaru, Y., Tadashi, K., & Kuwata, T., (2001). Identification of novel hypocholesterolemic peptides derived from bovine milk β-lactoglobulin. *Biochemical and Biophysical Research Communications*, *281*, 11–17. doi: https://doi.org/10.1006/bbrc.2001.4298.
92. Hernández-Ledesma, B., Recio, I., Ramos, M., & Amigo, L., (2002). Preparation of ovine and caprine β-lactoglobulin hydrolysates with ACE-inhibitory activity. Identification of active peptides from caprine β-lactoglobulin hydrolyzed with thermolysin. *International Dairy Journal*, *12*, 805–812. doi: https://doi.org/10.1016/S0958-6946(02)00080-8.
93. Kris-Etherton, P. M., Hecker, K. D., Bonanome, A., Coval, S. M., Binkoski, A. E., Hilpert, K. F., Griel, A. E., & Etherton, T. D., (2002). Bioactive compounds in foods: Their role in the prevention of cardiovascular disease and cancer. *The American Journal of Medicine*, *113*, 71–88. doi: https://doi.org/10.1016/S0002-9343(01)00995-0.
94. Usta, B., & Yilmaz-Ersan, L., (2013). Antioxidant enzymes of milk and their biological effects. *Ziraat Fakültesi Dergisi*, Uludağ Üniversitesi, *27*(2), 123–130.
95. Suetsuna, K., Ukeda, H., & Ochi, H., (2000). Isolation and characterization of free radical scavenging activities peptides derived from casein. *The Journal of Nutritional Biochemistry*, *11*, 128–131. doi: https://doi.org/10.1016/S0955-2863(99)00083-2.
96. Rival, S. G., Fornaroli, S., Boeriu, C. G., & Wichers, H. J., (2001). Caseins and casein hydrolysates. 1. lipoxygenase inhibitory properties. *Journal of Agricultural and Food Chemistry*, *49*, 287–294. doi: 10.1021/jf000392t.
97. Chen, J., Lindmark-Månsson, H., & Åkesson, B., (2000). Optimization of a coupled enzymatic assay of glutathione peroxidase activity in bovine milk and whey. *International Dairy Journal*, *10*, 347–351. doi: https://doi.org/10.1016/S0958-6946(00)00057-1.
98. Young, A. J., & Lowe, G. M., (2001). Antioxidant and prooxidant properties of carotenoids. *Archives of Biochemistry and Biophysics*, *385*, 20–27. doi: https://doi.org/10.1006/abbi.2000.2149.
99. Nozière, P., Graulet, B., Lucas, A., Martin, B., Grolier, P., & Doreau, M., (2006). Carotenoids for ruminants: From forages to dairy products. *Animal Feed Science and Technology*, *131*, 418–450. doi: https://doi.org/10.1016/j.anifeedsci.2006.06.018.
100. Choe, E., & Min, D. B., (2009). Mechanisms of antioxidants in the oxidation of foods. *Comprehensive Reviews in Food Science and Food Safety*, *8*, 345–358. doi: https://doi.org/10.1111/j.1541-4337.2009.00085.x.
101. Jensen, S. K., & Nielsen, K. N., (1996). Tocopherols, retinol, beta-carotene and fatty acids in fat globule membrane and fat globule core in cows' milk. *J. Dairy Res*, *63*, 565–574. doi: 10.1017/s0022029900032106.

102. Jiang, Q., Im, S., Wagner, J. G., Hernandez, M. L., & Peden, D. B., (2022). Gamma-tocopherol, a major form of vitamin E in diets: Insights into antioxidant and anti-inflammatory effects, mechanisms, and roles in disease management. *Free Radical Biology and Medicine, 178*, 347–359. doi: https://doi.org/10.1016/j.freeradbiomed.2021.12.012.
103. Zhang, S., Wu, Z., Heng, J., Song, H., Tian, M., Chen, F., & Guan, W., (2020). Combined yeast culture and organic selenium supplementation during late gestation and lactation improve preweaning piglet performance by enhancing the antioxidant capacity and milk content in nutrient-restricted sows. *Animal Nutrition, 6*, 160–167. doi: https://doi.org/10.1016/j.aninu.2020.01.004.
104. Mohanty, D. P., Mohapatra, S., Misra, S., & Sahu, P. S., (2016). Milk derived bioactive peptides and their impact on human health: A review. *Saudi Journal of Biological Sciences, 23*, 577–583. doi: https://doi.org/10.1016/j.sjbs.2015.06.005.
105. Ganatsios, V., Nigam, P., Plessas, S., & Terpou, A., (2021). Kefir is a functional beverage gaining momentum towards its health-promoting attributes. *Beverages, 7*, 48.
106. Depta, J., Małkowska, P., Wysokińska, M., Todorska, K., Sierawska, O., Hrynkiewicz, R., Bębnowska, D., & Niedźwiedzka-Rystwej, P., (2022). Therapeutic role of antimicrobial peptides in diabetes mellitus. *Biologics, 2*, 92–106.
107. León-López, A., Pérez-Marroquín, X. A., Estrada-Fernández, A. G., Campos-Lozada, G., Morales-Peñaloza, A., Campos-Montiel, R. G., & Aguirre-Álvarez, G., (2022). Milk whey hydrolysates as high value-added natural polymers: Functional properties and applications. *Polymers, 14*, 1258.
108. Coscueta, E. R., Batista, P., Gomes, J. E. G., Da Silva, R., & Pintado, M. M., (2022). Screening of novel bioactive peptides from goat casein: In silico to in vitro validation. *Int. J. Mol. Sci., 23*, 2439.
109. Ahtesh, F. B., Stojanovska, L., Mishra, V., Donkor, O., Feehan, J., Bosevski, M., Mathai, M., & Apostolopoulos, V., (2022). Identification and effects of skim milk-derived bioactive antihypertensive peptides. *Biologics, 2*, 1–14.
110. Punia, H., Tokas, J., Malik, A., Sangwan, S., Baloda, S., Singh, N., Singh, S., Bhuker, A., Singh, P., Yashveer, S., et al., (2020). Identification and detection of bioactive peptides in milk and dairy products: Remarks about agro-foods. *Molecules, 25*, 3328.
111. Majumder, K., & Wu, J., (2015). Molecular targets of antihypertensive peptides: Understanding the mechanisms of action based on the pathophysiology of hypertension. *Int. J. Mol. Sci., 16*, 256–283.
112. Fernández-Musoles, R., Castelló-Ruiz, M., Arce, C., Manzanares, P., Ivorra, M. D., & Salom, J. B., (2014). Antihypertensive mechanism of lactoferrin-derived peptides: Angiotensin receptor blocking effect. *Journal of Agricultural and Food Chemistry, 62*, 173–181. doi: 10.1021/jf404616f.
113. Chen, Y., Li, C., Xue, J., Kwok, L. Y., Yang, J., Zhang, H., & Menghe, B., (2015). Characterization of angiotensin-converting enzyme inhibitory activity of fermented milk produced by *Lactobacillus helveticus*. *Journal of Dairy Science, 98*, 5113–5124. doi: https://doi.org/10.3168/jds.2015–9382.
114. Mahdi, C., Untari, H., & Padaga, M. C., (2018). Identification and characterization of bioactive peptides of fermented goat milk as a source of antioxidants as a therapeutic natural product. *IOP Conference Series: Materials Science and Engineering, 299*, 012014. doi: 10.1088/1757-899x/299/1/012014.

115. García-Tejedor, A., Padilla, B., Salom, J. B., Belloch, C., & Manzanares, P., (2013). Dairy yeasts produce milk protein-derived antihypertensive hydrolysates. *Food Research International, 53*, 203–208. doi: https://doi.org/10.1016/j.foodres.2013.05.005.
116. Kraan, M. I. A. V. D., Nazmi, K., Teeken, A., Groenink, J., Hof, W. V. T., Veerman, E. C. I., Bolscher, J. G. M., & Amerongen, A. V. N., (2005). *Lactoferrampin, an Antimicrobial Peptide of Bovine Lactoferrin, Exerts its Candidacidal Activity by a Cluster of Positively Charged Residues at the C-Terminus in Combination with a Helix-Facilitating N-Terminal Part, 386*, 137–142. doi: 10.1515/BC.2005.017.
117. Farnaud, S., & Evans, R. W., (2003). Lactoferrin—A multifunctional protein with antimicrobial properties. *Molecular Immunology, 40*, 395–405. doi: https://doi.org/10.1016/S0161-5890(03)00152-4.
118. Chai, K. F., Voo, A. Y. H., & Chen, W. N., (2020). Bioactive peptides from food fermentation: A comprehensive review of their sources, bioactivities, applications, and future development. *Comprehensive Reviews in Food Science and Food Safety, 19*, 3825–3885. doi: https://doi.org/10.1111/1541-4337.12651.
119. Abd El-Fattah, A., Sakr, S., El-Dieb, S., & Elkashef, H., (2018). Developing functional yogurt rich in bioactive peptides and gamma-aminobutyric acid related to cardiovascular health. *LWT, 98*, 390–397. doi: https://doi.org/10.1016/j.lwt.2018.09.022.
120. Wang, T., Liu, X., Ng, Y. Y., Tarleton, K., Tran, A., Tran, T., Xue, W. Y., Youssef, P., Yuan, P., Zhang, D., et al., (2022). Milk-derived proteins and peptides in head and neck carcinoma treatment. *Biomolecules, 12*, 290.
121. Chiangjong, W., Panachan, J., Vanichapol, T., Pongsakul, N., Pongphitcha, P., Siriboonpiputtana, T., Lerksuthirat, T., Nuntnarumit, P., Supapannachart, S., Srisomsap, C., et al., (2021). HMP-S7 is a novel anti-leukemic peptide discovered from human milk. *Biomedicines, 9*, 981. doi: 10.3390/biomedicines9080981.
122. Rahman, R., Fonseka, A. D., Sua, S. C., Ahmad, M., Rajendran, R., Ambu, S., Davamani, F., et al., (2021). Inhibition of breast cancer xenografts in a mouse model and the induction of apoptosis in multiple breast cancer cell lines by lactoferricin B peptide. *J. Cell Mol. Med., 25*, 7181–7189. doi: 10.1111/jcmm.16748.
123. Bosnea, L. A., Moschakis, T., Nigam, P. S., & Biliaderis, C. G., (2017). Growth adaptation of probiotics in biopolymer-based coacervate structures to enhance cell viability. *LWT, 77*, 282–289. doi: https://doi.org/10.1016/j.lwt.2016.11.056.
124. Shivanna, S. K., & Nataraj, B. H., (2020). Revisiting therapeutic and toxicological fingerprints of milk-derived bioactive peptides: An overview. *Food Bioscience, 38*, 100771. doi: https://doi.org/10.1016/j.fbio.2020.100771.
125. Manzoor, M., Singh, J., & Gani, A., (2022). Exploration of bioactive peptides from various origin as promising nutraceutical treasures: In vitro, in silico and in vivo studies. *Food Chemistry, 373*, 131395. doi: https://doi.org/10.1016/j.foodchem.2021.131395.
126. Higurashi, S., Kunieda, Y., Matsuyama, H., & Kawakami, H., (2007). Effect of cheese consumption on the accumulation of abdominal adipose and decrease in serum adiponectin levels in rats fed a calorie-dense diet. *International Dairy Journal, 17*, 1224–1231. doi: https://doi.org/10.1016/j.idairyj.2007.03.012.
127. Meisel, H., (1998). Overview of milk protein-derived peptides. *International Dairy Journal, 8*, 363–373. doi: https://doi.org/10.1016/S0958-6946(98)00059-4.
128. Ryhänen, E. L., Pihlanto-Leppälä, A., & Pahkala, E., (2001). A new type of ripened, low-fat cheese with bioactive properties. *International Dairy Journal, 11*, 441–447. doi: https://doi.org/10.1016/S0958-6946(01)00079-6.

129. Smacchi, E., & Gobbetti, M., (1998). Peptides from several Italian cheeses inhibitory to proteolytic enzymes of lactic acid bacteria, pseudomonas fluorescens ATCC 948 and to the angiotensin I-converting enzyme. *Enzyme and Microbial Technology*, *22*, 687–694.
130. Shakerian, M., Razavi, S. H., Ziai, S. A., Khodaiyan, F., Yarmand, M. S., & Moayedi, A., (2015). Proteolytic and ACE-inhibitory activities of probiotic yogurt containing non-viable bacteria as affected by different levels of fat, inulin, and starter culture. *J. Food Sci. Technol.*, *52*, 2428–2433. doi: 10.1007/s13197-013-1202-9.
131. Hata, Y., Yamamoto, M., Ohni, M., Nakajima, K., Nakamura, Y., & Takano, T., (1996). A placebo-controlled study of the effect of sour milk on blood pressure in hypertensive subjects. *Am. J. Clin. Nutr.*, *64*, 767–771. doi: 10.1093/ajcn/64.5.767.

CHAPTER 5

Technological and Analytical Aspects of Bioactive Compounds and Nutraceuticals from Marine Algae

VIDA ŠIMAT,[1] ROBERTA FRLETA,[2] MARTINA ČAGALJ,[1] and DANIJELA SKROZA[3]

[1]*University Department of Marine Studies, University of Split, Split, Croatia*

[2]*Center of Excellence for Science and Technology-Integration of Mediterranean Region (STIM), Faculty of Science, University of Split, Split, Croatia*

[3]*Department of Food Technology and Biotechnology, Faculty of Chemistry and Technology, University of Split, Split, Croatia*

ABSTRACT

The high-value compounds from marine sources show promising potential for biotechnological applications, so the interest in their exploitation is growing. Among marine organisms, marine microalgae and seaweeds are highly diverse sources of a wide range of valuable molecules with bioactive potential, such as pigments, lipids, and bioactive compounds. The compounds isolated from algae are used as nutraceuticals, superfood, and functional food ingredients but also pharma and cosmetic industry. Along with the rise of the demand for such molecules/products on the market, the use of algal biomass is sustainable and renewable, reducing the negative impact on the environment and meeting the market demand. This chapter summarizes the recent trend and challenges in the technical procedures

Bioactive Compounds and Nutraceuticals from Dairy, Marine, and Nonconventional Sources: Extraction Technology, Analytical Techniques, and Potential Health Prospects.
Charanjit Singh Riar, Parmjit S. Panesar (Eds.)

for the cultivation of microalgae, various pre-treatments and extraction methods for micro- and macroalgae, and analytical methods currently used for the identification and quantification of their compounds.

5.1 INTRODUCTION

Nutraceutical is a term coined from two words, "nutrient" and "pharmaceutical," referred to bioactive molecules that are found in raw foods, functional foods or dietary supplements. These molecules have physiological benefit and the ability to protect human health either by preventing chronic disease or treating a certain disease/condition [1, 2]. The human health benefits from marine-originated bioactives in several ways; through direct consumption of organisms such as fish, invertebrates, shellfish or algae; through uptake of supplements or natural drugs derived from marine organisms, or indirectly through their use in feed and agriculture [3, 4]. The vast diversity of marine ecosystems and the amount of unexplored sources that might carry bioactives has drawn the attention of researchers to organisms that might contribute to the growth of the blue bioeconomy. Some of the bioactives from marine organisms, such as omega-3 long-chain PUFAs, amino acids and peptides from fish and fish by-products, collagen, and chitin from seafood processing by-products have been studied for years and their technological and analytical aspects, as well as their health benefits and possible application in cosmetic, food and pharma industry are well known and documented [some recent work: 2, 4–9]. However, as the fishing effort grows, so is the pressure on natural marine populations. Overfishing is depleting wildlife populations, and with the growth of the world population, the demand for nutritive-rich foods and protective supplements is rising. The only way to reply to this challenge is to find the renewable and sustainable sources of bioactives and nutraceutical. These molecules should have strong bioactive properties, and their cultivation should benefit the human race without further depletion of the seas and oceans. In regard to this, in the last decades there is a great interest in the research of micro- and macroalgae (also known as seaweeds) as a source of bioactive compounds. According to Science Direct, the number of publications with key words microalgae/macroalgae and bioactive has grown from 10–20 in 2000 to >1,000 in 2021 (Figure 5.1).

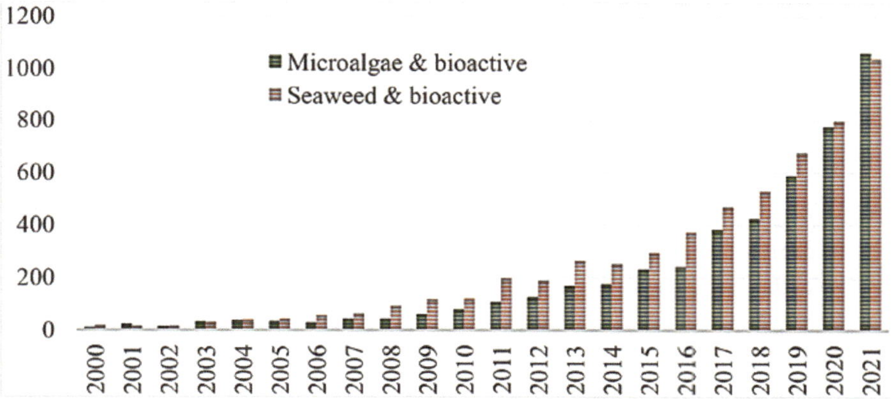

FIGURE 5.1 An overview of research publication with key words "microalga and bioactive" and "seaweed and bioactive."

Source: From Science Direct (data accessed in September 2021).

This chapter describes the two renewable and sustainable marine sources of bioactives and nutraceutical, micro-, and macroalgae, and recent technological and analytical challenges of their cultivation and extraction of valuable compounds.

5.2 TECHNOLOGICAL ASPECTS

5.2.1 MICROALGAE

The marine ecosystem contains a vast biodiversity of microalgae, with over 7,000 species of these free living microorganisms that present a bases of food chain in marine ecosystem [10, 11]. The most important classes of microalgae in terms of abundance are diatoms (Bacillariophyceae), green algae (Chlorophyceae) and golden algae (Chrysophyceae). In addition, cyanobacteria or blue-green algae (Cyanophyceae) are also referred to as microalgae [12]. They have the ability to grow rapidly in very harsh environment and tolerate exposure to heat, cold, wide range of salinity and ultra-violet light radiation. They can grow in various closed (controlled) or open systems such as, photobioreactors or artificial ponds and lakes. In these conditions, microalgae produce molecules that might be used for different applications in white (industry and energy), green (agriculture and environment), and red (medical and health care) biotechnology [13]. As mentioned, the field of red biotechnology is growing rapidly, as is the market demand for microalgae-derived products. Consumers are growing awareness of the importance of a balanced, high-nutritive, and healthy diet. With that, the consumption of foods that carry value beyond nutritive and food supplements containing bioactive substances that fall between pharmaceuticals and nutrition is increasing from year to year [14].

Microalgae are unicellular organisms that are capable of generating chemical energy through the process of photosynthesis. The chemical energy in microalgae is found in the form of various compounds. The uniqueness of microalgae is their ability to synthesize high-value compounds, such as carotenoids (especially astaxanthin and lutein), PUFA, sterols, phenols, and others, with low energy investment. For this reason they are becoming an interesting field of biotechnology, in pharmaceutical, nutraceutical, cosmetic, and food industries as natural therapeutics, food additives and UV-light protective agents [11]. In general, the consumption of various nutraceuticals of microalgal origin has many health benefits since they show antioxidant, antiviral, and anticancer properties, play a role in protection of the cardiovascular system, weight loss, improved immune system and brain development, healthier gut and digestive system [15]. The main challenge in the biotechnological process is to ensure sustainable, renewable, eco-friendly, and profitable production of high-value compounds. The compounds mainly accumulate inside the cell, which requires the constant development of efficient technologies for their extraction, recovery, and purification [16].

Depending on the growing conditions, microalgae change the composition of the metabolites. Besides those locked in the cell, some metabolites are excreted in the culture medium. Under stressful conditions, they synthesize organic components from products of primary metabolism, so-called secondary metabolites (SMs). Although they are not involved in the pathways necessary for growth, development, and reproduction, they are important for performing other physiological functions. They are classified on the basis of synthesis, function, and structure. The five main classes of SMs are terpenoids and steroids, substances derived from fatty acids and polyketides, alkaloids, non-ribosomal polypeptides and enzyme cofactors [17, 18]. The primary purpose of these compounds is to ensure the organism's survival in adverse environmental conditions. SMs have a significant potential for the use in various industries such as pharmaceuticals, food, and agriculture. Today, between 3,000 and 5,000 compounds are known in microalgae, of which 1,500 have some kind of biological activity and almost 1,000 compounds have antibiotic activity [19]. The structural specificity of these compounds is often species dependent [20].

So far, *Haematococcus pluvialis*, *Chlorella vulgaris*, *Spirulina*, *Nannochloropsis oculata*, *Arthrospira platensis*, are the most investigated microalgae [21]. Microalgae *Chlorella* sp. proved to be an excellent source of steroid components. These compounds showed pronounced pharmaceutical potential as anticancer and antitumor agents [22]. On the other hand, in three genera of diatoms, *Chaetoceros*, *Skeletonema*, and *Thalassiosira*, a high content of phenolic compounds was recorded and, consequently good antioxidant and antibacterial activity [23]. Generally, the main features of SMs are cheap and easy scale-up for higher yields, it can be the starting material for obtaining the product, and the main advantage is that they can serve as a template in the pharmaceutical industry [18]. Microalgae can be isolated from a fresh sample of seawater, but there is also a culture collection from which microalgae stains can be purchased. Isolation methods include micromanipulation, serial dilution, streak plating, and density centrifugation. Sterile conditions are required for cultivation to avoid contamination issues. The preparation of the culture medium includes filtration and sterilization of seawater, and enrichment with various nutrients such as phosphate, nitrate, trace metals, vitamins, and others. The choice of medium for cultivation determines the growth performance of the microalgae. The media formulation such as "f/2 medium" is most often used for marine species cultivation in the laboratory. Cultivation of microalgae starts with inoculation of the selected stain in the culture medium (Figure 5.2).

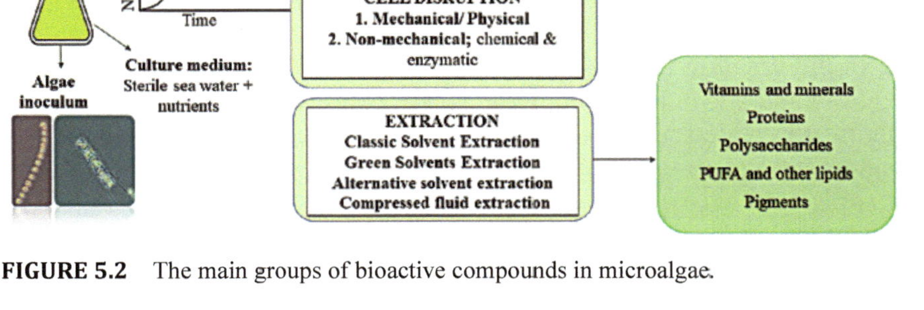

FIGURE 5.2 The main groups of bioactive compounds in microalgae.

Parameters, such as temperature and photoperiod, directly affect the growth of the culture. Different strains have different nutritional needs and growing conditions (temperature, photoperiod, salinity, etc.). The growth of microalgae takes place in several stages. The first phase is the lag phase, when the culture is at the lowest concentration, and it lasts the shortest. The second phase is the logarithmic or exponential growth phase, in which the culture has the fastest growth. Growth then slows down and reaches the stationary phase when no growth is recorded, followed by the death phase. Harvesting of microalgae is done when the growth reaches the late logarithmic phase. It has been shown that in this phase, the cells possess the largest number of bioactive compounds. Selection of harvesting methods are based on the six important criteria, i.e., biomass quantity, biomass quality, cost, processing time, species related criteria, and toxicity. The collection methods include centrifugation, flocculation, flotation, and filtration [16].

The process of collecting microalgae by centrifugation involves the separation of cells from the culture medium using centrifugal force. At the industrial level different types of centrifugal systems such as decanters, perforated basket centrifuges, disk stack centrifuges, imperforated basket centrifuges, and hydro-cyclones are used [24]. Although this procedure is quite effective, it requires high consumption of energy and time. The high cost of maintenance and cell damage makes this method less suitable for large-scale application. However, for small-scale lab production, this method is suitable and ensures high yields. The flocculation process encompasses the addition of flocculation agents in the medium, resulting in aggregation of microalgae cells in larger particles. This process is efficient and cheap, thus suitable for large-scale production. The use of chemical flocculants, such as toxic iron or aluminum, requires additional treatments of collected biomass for their removal. To resolve this issue, alternative in form of bio-flocculants are available. The most common bio-flocculants in these processes are acrylic acid and chitosan. In addition to being safer and more eco-friendly, they are also effective at lower concentrations, and their use does not require additional treatments, avoiding the possible contamination of algal biomass [25].

In the flotation process, the microalgae cells in the culture medium get exposed and attach to gas bubbles. The bubbles, carrying the cells, eventually float to the surface, resulting in easy biomass collection. The three most common flotation systems are dissolved air flotation system, dispersed air flotation, and electro-flotation [26–28]. This procedure is simple and very effective and requires low time and space consumption. However, high energy consumption makes it unprofitable for application at the industrial level. Filtration is the process of microalgae harvesting using a semipermeable membrane. Different forms of filtering, such as microfiltration (MF) (0.1–10 μm), macrofiltration (10 μm), dead end filtration, ultrafiltration (UF) (0.02–0.2 μm), tangential flow filtration (TFF), vacuum filtration, and pressure filtration, are applied depending on the cell size. Filters are made of different materials like polyvinylidene fluoride (PVDF), polyacrylonitrile (PAN), polyether sulfone (PES) and polytetrafluoroethylene (PTFE), polyether sulfone polyvinyl-pyrrolidone (PES-PVP), polyvinyl chloride (PVC) [24]. The filtration method is effective and recovers a high yield of biomass without chemical treatment and expensive equipment. The cost of the filtration increases because of the high price of synthetic filters and time need for filtration of large amounts of medium. Production of cheaper filters from organic material is a good alternative, increasing the profitability of the process; for example, the harvesting efficiency of stretch cotton filters was estimated at 66%–93% [29]. The maintenance costs of the larger filtration systems make this method unprofitable for large-scale production. The harvesting techniques are commonly combined to enhance the process and modified to specific characteristics of the selected species, desired concentration and quality of the final product [24]. In this process, flocculation and flotation generally precede centrifugation or filtration to maximize harvesting yield and reduce the cost and time [30].

5.2.2 MACROALGAE

Macroalgae (seaweeds) are divided in three classes: green (Chlorophyta), red (Rhodophyta), and brown (Phaeophyta) [31]. The collection of wild seaweeds did not increase in the last 50 years and accounts only for 3% of the world seaweed production in 2019 [32]. Cultivated seaweeds contributed to almost 30% of 120 million tons produced by world aquaculture in 2019. Red and brown seaweeds were second (15.20%) and third (13.65%) species groups by quantity in world aquaculture production [33]. From more than 11,000 macroalgal species, only 27 were cultivated in 2019, mostly in Asia for human consumption or hydrocolloids (carrageenan, alginate, and agar) production [33]. Only 10 species are cultivated intensively, *Saccharina japonica*, *Undaria pinnatifida* and *Sargassum fusiforme* from brown seaweeds, *Enteromorpha clathrata*, *Monostroma nitidum*, and *Caulerpa* spp. from green seaweeds and *Phorphyra* spp., *Euchema* spp., *Kappaphycus alvarezii* and *Gracilaria* spp. from red seaweeds [34]. The major advantages of seaweeds' cultivation are that seaweeds don't need feed or elaborate setups, plus their cultivation does very little harm to environment, fishery resources and seabed. Sometimes, it could be integrated within a farming system where the seaweeds absorb dissolved nutrients and remove waste from feed. For example, *S. japonica* is cultivated with abalone and sea cucumber in China, seaweeds are cultivated within shrimp cages in Brazil, and *Gracilaria* are cultivated on salmon farms in Chile. Seaweed farming areas are in intertidal zones which makes them easily accessible, cultivation is relatively simple, initial capital is low and the crop can be harvested very fast, in about six weeks [34]. Seaweeds are used as food for the last 2000 years, but they are only at the beginning of exploitation for their SMs in the form of bioactive compounds and nutraceuticals [35]. The molecules from seaweeds demonstrated various biological activities *in vitro* and *in vivo*, including antioxidant, antibacterial, antiviral, antitumor, anticancer, anti-inflammatory, anticoagulant, cardioprotective, and antidiabetic activities with beneficial effects on health [31, 36, 37]. Seaweeds are recently brought to focus of research precisely because of their divergent bioactivity.

Bioactive compounds and nutraceuticals from seaweeds include polysaccharides, proteins, and amino acids, lipids, and fatty acids, vitamins, and phytochemicals. Seaweeds' polysaccharides (fucoidan, laminarin, alginate, carrageenan, agar, ulvan) are biological macromolecules with different degrees of sulfation [38], bound in algal cell wall or acting as storage polysaccharides. They have shown great antioxidant potential through their scavenging of reactive oxygen substances (ROS) [39]. There is a lack of research about seaweed proteins' functional properties that is impeding their nutraceutical use. The amount of proteins in seaweeds vary from low amounts in brown seaweeds, 3±15% of dry algal weight, and moderate amounts in green seaweeds, 9±26% of dry algal weight, to high amounts in red seaweeds, max. 47% of dry algal weight [40] and the seaweeds' amino acid fraction is mostly comprised of glutamic and aspartic acids [40]. Lipid content of the commonly used seaweeds doesn't account for more than 5% of dry algal weight. However, omega-3 and omega-6 PUFA are the significant part of seaweed lipids [41]. It is well known that PUFA have important nutritional properties and act beneficial on human health. Many factors influence lipids composition in seaweeds (species, season, location, and other environmental factors), however, glycoglycerolipid class is the most commonly found. Other lipid classes, phospholipids, triacylglycerols, sterols, and pigments are also found in seaweeds [42]. Seaweeds consumption can improve overall health as they are an important source of both hydro- and liposoluble vitamins [43].

Among phytochemicals present in seaweeds, gallic acid, quercetin, phloroglucinol, phlorotannin, and carotenoids are the most studied compounds [36]. Seaweeds' phenolics range from complex compounds (phlorotannin) to simple molecules (phenolic acids) [2] and are found in algal cell wall or dissolved in cytoplasm. Phlorotannins are phloroglucinol oligomers or polymers, found only in brown seaweeds [44]. They are reported to have high antioxidant potential, scavenging free radicals (superoxide, peroxyl, and nitric radicals) and chelating ferrous ions [31]. As their structure is very complex, their characterization and identification is rather difficult [45]. Carotenoids are, along with chlorophylls and phycobiliproteins, seaweeds natural pigments. They provide seaweeds with the protection against photo-oxidative processes as they efficiently scavenge singlet oxygen and peroxyl radicals [45]. The most abundant carotenoids in seaweeds are fucoxanthin and astaxanthin [2]. Seaweeds can also be used for other non-food applications such as nutraceutical, pharmaceutical, and cosmetic applications, production of animal feed and biofuel or bio-packaging. However, these applications encounter many challenges and constraints (technical and economic) [32]. Composition of nutraceuticals and bioactive compounds in seaweeds vary due to many factors, for instance different species, age, location of growth, harvesting season and other environmental conditions [31]. Besides, there is a lack of standard procedures for drying and extraction of algal material [46]. The availability of seaweeds used for extraction is often unknown [47] as most scientists harvest seaweeds for research purposes and only few procure cultivated seaweeds from seaweed farms. Numerous studies evaluated non-food applications of seaweed, but there is a need for research regarding seaweed cultivation, processing, and extraction of bioactive compounds and nutraceuticals along with their toxicity and safety for commercial scale-up [36].

5.3 PRETREATMENT METHODS

The recovery of targeted bioactive compounds can be enhanced by pretreatment or cell disruption techniques, depending on the type of targeted compounds. Cell wall represents a physical barrier to solvent entry, especially in microalgae that possess a multilayered membrane structure. Cell disruption is a process releasing of intracellular content rich in bioactive compounds caused by rupture of the cell wall. The choice of the adequate method provides effective releasing of cellular compounds. For this purpose, various mechanical and non-mechanical methods are used [16]. For microalgae, pretreatment methods (physical and chemical) are quite important and necessary to ensure the effective extraction of the components from the cells. On the other hand, pretreatment of macroalgae usually include drying and grounding/pulverization prior to extraction, and use of a specific solvent.

5.3.1 MICROALGAE

Non-mechanical cell disruption methods include various chemical methods, osmotic shock and enzymatic methods. During the interaction of cellular components of microalgae with various chemicals such as acids, alkalis, hypochlorite, and others, deformations occur and consequently the cell wall breaks [16]. The choice of non-toxic solvent helps removing the barriers of further application of the intercellular components. Commonly, surfactants are the preferred choice of solvents. They are amphiphilic molecules that have hydrophobic and hydrophilic parts and cause ruptures interacting with the cell membrane's phospholipids and, consequently, cell disruption. In addition to promoting cell disruption, they also have a role in flocculation. Another chemical method is osmotic shock. A highly concentrated solution causes a change in osmotic pressure, which ultimately leads to cell rupture. Osmotic shock treatments improve lipid extraction, especially the extraction of unsaturated fatty acids [48]. This method is non-toxic, but it is time-consuming when considered for industrial large-scale application. The use of various enzymes such as cellulases, proteases, and others is considered a highly selective method for cell disruption, often at commercial levels [49]. The application of enzymes is a time-consuming process that requires high capital investments for application on a large industrial scale.

Mechanical and physical cell disruption methods affect cell lysis by the action of force, electromagnetic waves, or the action of electrical and thermal energy. Unlike chemical methods, these methods are non-toxic and require less time, making them more suitable for industrial production [16]. Bead milling (BM), high speed or high-pressure homogenization (HPH) are methods that induce cell lysis through solid or liquid shear force. The BM is based on creating mechanical damage through the collision between the beads and the cells. The success of cell disrupting depends significantly on the choice of bead matter as well as on the density of biomass [50]. This process does not cause degradation of targeted biomolecules and the process itself is adaptable for large scale production. Still, the process consumes large amounts of energy, which makes its profitability questionable. The high-speed homogenization (HSH) method is based on the action of hydrodynamic cavitation caused by the high speed (10,000–20,000 rpm). The biggest drawback of this procedure is the high energy consumption as well as the denaturation of proteins. However, the duration of the procedure is short, making it interesting for scaling up.

The HPH method uses pressure of 20–120 MPa to create turbulence, friction, and liquid-shear stress that will consequently cause cell lysis. It is suitable for industrial application as well as HSH and BM. However, the high pressure causes an increase in temperature, which can adversely affect bioactive compounds. The morphological variability of the cells in terms of wall thickness can require the application of higher pressure, which can significantly increase the cost of the production [16]. Microwave irradiation is a simple and scalable physical method of cell disruption. The optimal operating frequency is 2,450 MHz and the cell walls disrupt under the electromagnetic effect interacting with dielectric and polar (e.g., water) molecules caused by microwave irradiation. This method is unsuitable when the target compounds are volatile but has been reported as successful for lipid extraction.

Cell disruption via ultrasound involves the use of waves of 20–100 kHz frequency. Ultrasonication is characterized by low operating costs and allows the optimization of parameters such as temperature, sound intensity, and time. The use of ultrasound alone is not enough to break the cells of some microalgae species, and additional chemical pretreatments are required. Pulsed electric field (PEF) is a novel and increasingly used method. The efficiency of PEF can be affected by several parameters such as the electric field strength, frequency, pulse (shape, width), operational time, physicochemical parameters (temperature, pH, and conductivity), and cell wall properties. This method is characterized by effective cell disruption and energy efficiency, making it suitable for large-scale production. Thermal treatments include standard procedures such as thermolysis, autoclaving, and steam explosion. The advantage of using thermal treatments is low maintenance costs. Although they are the simplest cell disruption methods, thermal treatments have several major drawbacks such as high energy consumption, low efficiency, generation of large amounts of undesirable cell debris, and applicability limited by the target molecule's thermal resistance.

5.3.2 MACROALGAE

Conventional and novel macroalgal pretreatment techniques (Figure 5.3) are used for macroalgal cell disruption. Conventional pretreatment techniques can be physical, chemical, and biological [37]. Physical pretreatment includes drying and milling of macroalgal material. Seaweeds are commonly air-dried, sun-dried, shade-dried, or oven-dried before any kind of extraction. However, it is evident that many parameters (temperature, light, time) influence the composition of bioactive compounds during drying [46, 52]. In recent years, researchers have used freeze-drying to preserve bioactive compounds and nutritional composition of seaweeds as high temperatures during conventional drying result in the loss of heat-sensitive compounds, such as PUFA, carotenoids, or polyphenols [37, 53]. Chemical pretreatment includes cell disruption using acids, salts, and surfactants, and biological pretreatment includes fermentation and the use of enzymes. Chemical and biological pretreatments are usually used as a first step in polysaccharides extraction [37].

Novel pretreatment techniques: mechanical disruption (BM, HPH, and hydrodynamic cavitation), compression puffing, and use of ultrasound, microwaves or PEF before drying are reported to enhance efficiency of extraction and are recently reviewed in Ummat et al. [37]. Pretreatments have significant impact on the quality and quantity of extracted compounds. After the pretreatment, dried macroalgal material is subjected to extraction using different methods, extraction conditions (temperature, time, pressure, pH), and choice of solvents, all of which affect bioactive compounds and nutraceuticals composition in seaweeds along with their biological activities. In some cases, a combination of the several extraction methods is used to enhance the recovery of the targeted compounds.

5.4 EXTRACTION METHODS

Generally, extraction is a mass transfer process that includes diffusion of the selected solvent into the dried or wet algal material, dissolution of bioactive compounds in the solvent, and separation of extracted compounds from the solvent. Besides cell disruption, the effectiveness of an extraction process depends on physico-chemical properties of the targeted compounds, the extraction conditions (type of solvent, solvent to dry mass ratio, extraction temperature and time) and their perspective use. The solvent also plays an important role in cell lysis because aggressive solvents can increase the extraction yield [54]. The choice of the solvent is based on the targeted compound and its polarity. The problem of the conventional methods of extraction is the use of toxic organic solvents. For example, in chlorophyll extraction firstly methanol was used. It was later replaced by acetone according to Edler's

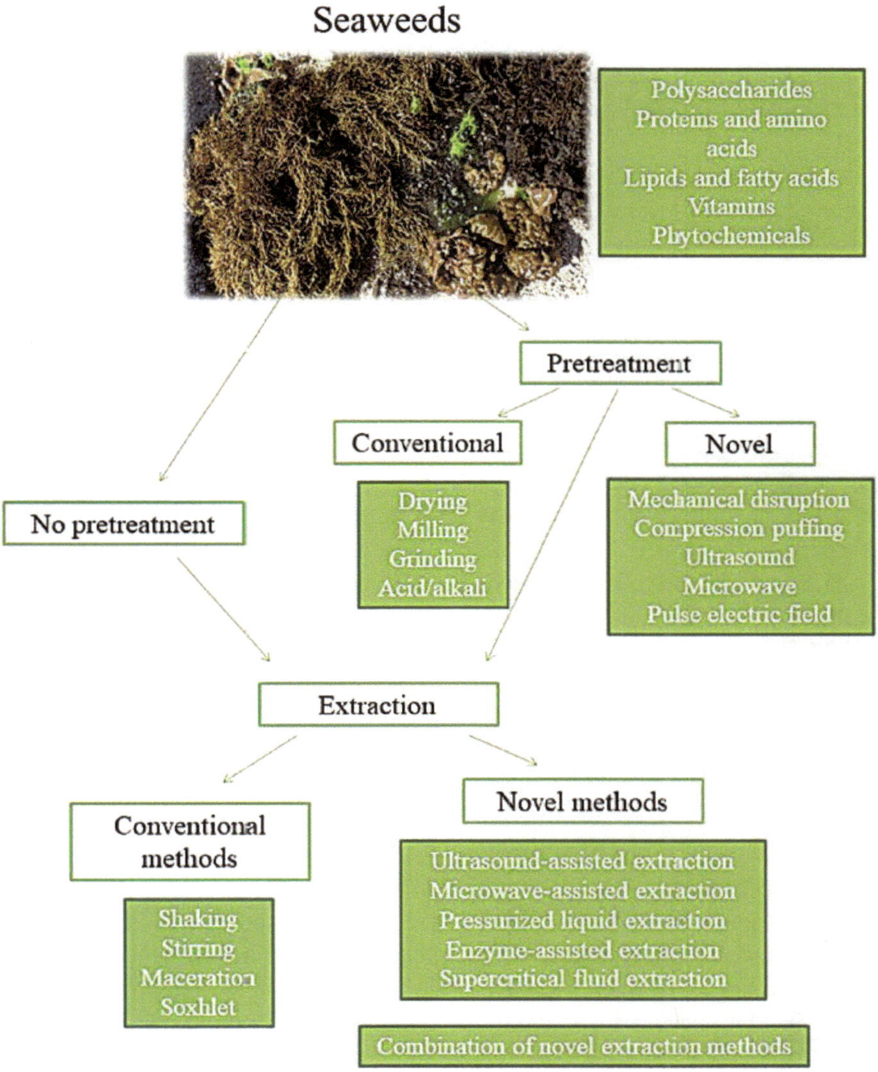

FIGURE 5.3 Pretreatment and extraction methods.

recommendations, and since 1995 ethanol has been used as a solvent for pigment extraction [54, 55]. Solvents such as chloroform, n-hexane, methanol, diethyl ether, and others are most commonly used for lipid extraction. To ensure safe and environmentally friendly extractions, their use tends to be reduced and gradually replaced with eco-friendly solvents such as ethanol, ethyl acetate, and acetone, or alternative solvents. Ionic liquids (ILs), deep eutectic solvents (DES), and liquid polymers are mentioned as promising options. IL are characterized by high thermal stability and have attracted interest due to 1,018 combinations of anions and cations in their synthesis [16].

This characteristic has a positive impact on the success of extraction and separation of different compounds. However, all ILs are not environmentally friendly, and extreme care should be taken in their selection. DES shares several similar chemical properties with IL. They are cheaper and less demanding in the synthesis process, which makes them a more favorable option. Similar to DES, natural deep eutectic solvents (NADES) include compounds that are products of primary metabolism, such as organic acids and sugars. Polyethylene glycol (PEG) is an example of a biodegradable liquid polymer and eco-friendly solvent [56, 57]. For example, in conventional lipid extraction methods, such as Bligh and Dyer (B&D) method [58], the Folch method [59], and the Soxhlet method [60] large amounts of toxic organic solvents are used [61, 62]. It is similar to other extraction methods. Therefore, the scientific community is looking for ways to develop novel, green extraction methods and optimize their parameters to increase the yield of targeted compounds, shorten the extraction time and reduce solvent consumption.

5.4.1 NOVEL EXTRACTION METHODS

Novel extraction methods include ultrasound-assisted extraction (UAE), microwave-assisted extraction (MAE), pressurized liquid extraction (PLE), enzyme-assisted extraction (EAE), and supercritical fluid extraction (SFE) [43, 64]. Researchers are reporting that seaweeds extracted with these novel methods have stronger biological activities (e.g., antioxidant, antimicrobial) in comparison to conventional methods [43, 46, 65, 66]. Regarding the extraction of microalgae, novel techniques recorded higher yields and quality of compounds such as lipids, pigments, etc. [67, 68].

5.4.1.1 ULTRASOUND-ASSISTED EXTRACTION (UAE)

UAE uses the power of ultrasound waves that can pass through liquid, solid, and gas media, causing regions of negative pressure. When that pressure exceeds the tensile strength of the liquid, cavitation happens, vapor bubbles are formed and they implode under strong ultrasound fields [69]. UAE significantly improves penetration of solvent into algal cells and extraction of targeted compounds contained within the cells, achieving higher extraction yield [70, 71]. It is a simple extraction method with lower equipment requirements than other novel methods. This makes it suitable for industrial scale [43, 69].

5.4.1.2 MICROWAVE-ASSISTED EXTRACTION (MAE)

Electromagnetic waves during MAE cause changes in the cell structure, increasing the heat and pressure inside the cell which ruptures the cell membranes, resulting in accelerated solvent penetration and releasing of compounds [71, 72]. One of the most important factors influencing the extraction process is a solvent, especially in MAE. Using a solvent with a higher dielectric constant and dielectric loss increases solvents' capacity to absorb microwave energy that leads to faster heating of the solvent. Besides, it is possible to combine different solvents to increase the yield of targeted compounds [73]. MAE requires less solvent and time, but it is more energy consuming compared to conventionally used methods [43].

5.4.1.3 OTHER NOVEL EXTRACTION METHODS

The EAE uses enzymes to hydrolyze complex cellular components, disrupting the integrity of cellular structure, making targeted compounds more accessible to the solvent [37, 43]. The enzymes are available as commercial products or extracted/purified from a different matrix (ea. Celluclast, Flavourzyme, Viscozyme, Alcalase, agarase, xylanase, Neutrase, Kojizyme, amyloglucosidase). They are commonly used for the extraction of proteins, polysaccharides or phenolics from seaweeds, based on the targeted compounds [43, 51, 74]. In the EAE of microalgae, the use of cellulose, snailase (a complex of more than 30 enzymes, including cellulose, hemicellulose, galactose, a proteolytic enzyme, 101 pectinase, β-glucuronidase, etc.), neutral protease, alkaline protease, papain, pectinase, hemicellulose, lipozyme, and trypsin is reported [75–77]. Compressed fluid extraction methods are supercritical water extraction (SWE), pressurized fluid extraction (PLE), and SFE. The SWE is performed with the solvent temperature above boiling point until the pressure is high enough to keep the solvent in the liquid state. On the other hand, SFE works on a different principle. The temperature and pressure during extraction are above the critical point of the solvent. The fluid's properties during this state are in between gas and liquid [78]. In SFE, CO_2 is the most commonly used due to its low critical conditions, availability, and non-toxicity. It returns to gas state at room temperature and obtained extract is solvent free [11, 79]. As the algal material absorbs the supercritical solvent, cell structures expand, and mass transfer between targeted compounds and the supercritical fluid is increased. Since supercritical CO_2 is a non-polar solvent, the addition of polar co-solvent improves the extraction of polar bioactive compounds [11, 37, 78]. The efficiency of the extraction and selectivity of the recovered compounds are optimized via different parameters (temperature, pressure, solvent flow rate, addition of co-solvents) [43]. The disadvantage of this method is the high cost of needed equipment and maintenance. PLE uses solvents under high pressure (up to 200 bar) and temperature (up to 200°C) that are below their critical points. The solvent's state under these conditions remains liquid [37]. This method requires small amounts of solvents. As the temperature during extraction increases, it affects the solubility of targeted compounds and increases the mass transfer. Also, high temperature decreases the viscosity and the surface tension of the solvents, increasing their penetrability into the algal material and extraction of targeted compounds [80]. The use of PLE at the industrial level is impeded by the high equipment cost and the difficulty of maintaining the extraction conditions of the fluids when applied on the large scale [79].

Various combinations of pretreatments and extraction methods are under research. There is a need for standard procedures for their industrial applications. Besides, researchers are combining novel extraction methods to utilize their synergy and improve extraction efficiency (ea. EAE with UAE or MAE, and UAE with MAE).

5.5 DIFFERENT ANALYTICAL METHODS IN THE ANALYSIS OF BIOACTIVE COMPOUNDS IN MICROALGAE AND MACROALGAE

Characterization and identification of compounds in microalgae, as well as in macroalgae, is a rather complex field as both are rich in nutraceuticals and bioactive compounds. The main analytical techniques used for the analysis of algal samples are UV/Vis spectrofotometrical (used for screening), liquid chromatography (LC), gas chromatography (GC), and nuclear magnetic resonance (NMR) coupled with different detectors, primarily mass spectrometry (MS) and used for more specific characterization [81].

5.5.1 SPECTROPHOTOMETRIC METHODS

The spectrophotometric methods offer economical, simple, fast, and technically easy performed method used for screening large quantity of samples to obtain a chemical index that indicates concentrations of total phenolics, flavonoids, and other compound [82]. The concentration of phenolic compounds in seaweed and microalgae extracts is typically determined spectrophotometrically as total phenolic content (TPC) using the Folin–Ciocalteu assay [45]. However, other co-extracted nonphenolic substances (amino acids, proteins, thiols, and vitamin derivatives) can interfere and overestimate the TPC [83, 84]. Folin–Ciocalteu assay should rather be measure of the total antioxidant capacity or total reducing substances than the TPC [83]. The TPC method is not specific; thus, chromatographic techniques are necessary to establish a structure-activity relation [82]. There is also a problem in comparison of results obtained by these methods. The researchers use different phenolic compounds as the standards for calibration (gallic acid, phloroglucinol) and report amounts from 0.02 to 277 mg phloroglucinol equivalents (PGE)/g extract, from 0.11 to 114 mg PGE/g dry algal weight, from 0.063 to 158.82 mg gallic acid equivalents (GAEs)/g extract, from 0.51 to 406.22 mg GAE/g dry algal weight, and from 0.001 to 4.83% dry algal weight in brown seaweeds [31].

5.5.2 CHROMATOGRAPHIC METHODS

For the identification of phenolic compounds from extracts' compounds mixtures, chromatographic methods should be used. They use specific characteristics of the molecules (adsorption (liquid-solid), partition (liquid-solid), affinity, and molecular weights) for their isolation and separation from the mixtures [16]. Isolation and quantification of phenolic compounds is rather difficult due to their structural similarity, reactivity with other compounds and large molecular weight (MW) [31]. The lack of commercially available standards (only phloroglucinol is available) is limiting the use of LC for the quantification of phlorotannin in seaweed extracts. This drawback makes MS a mostly qualitative tool for their analysis and identification. To observe the fragmentation pattern of the phenolic compounds and rapidly analyze these nonvolatile, polar, and thermolabile compounds, tandem mass spectrometry (MS/MS) has been used coupled to various LC systems [85, 86]. As algal extracts are mixtures of numerous compounds, different separation and purification methods are used for their easier identification. Methods like thin layer chromatography (TLC) and silica gel column chromatography are used for this purpose. Some of the disadvantages of these separation methods are long separation time, large solvent consumption and less sample loading [87]. Recently, high-performance countercurrent chromatography (HPCCC) and preparative high-performance liquid chromatography (Pre-HPLC) are being used alone or in combination to separate target compounds from complex mixtures, such as peptides or astaxanthins from microalgae, and nor isoprenoids or terpenoids from seaweeds [87–91]. The advantages of HPCCC are large loading capacity, simple operation and high product purity. The advantages of Pre-HPLC are high separation efficiency and accurate collection, making it suitable for industrial applications [87].

Identification and structural characterization of compounds isolated from microalgae and macroalgae was recently performed using LC coupled with different detectors (Tables 5.1 and 5.2). Overall, HPLC is an excellent technique as it achieves rapid and precise quantitative analysis, can be automated, and has high-sensitive detection. The high-value molecules from microalgae such as PUFA are of particular interest because of their many applications in cosmeceuticals, nutraceuticals, fine chemicals, pharmaceutics, and biodiesel [92]. In microalgae the average lipid content varies from 1 to 70% (w/w) depending on the species, temperature conditions, log phase and culture conditions (nutritive medium/environmental condition). In seaweeds the lipid content is usually less than 5%, consequently lipid yields are low. The efficiency of the further application of the algal lipid fraction is related to the ability of the extraction technique to obtain maximal lipid content without the damage to PUFA. Some microalgae species have a thick cell wall that makes the release of intracellular lipids difficult without large amount of solvents [93, 94]. Although to recover maximal algal lipids, an effective cell disruption method with suitable solvent mixtures has not been established yet, for large-scale extractions the most popular choice of organic solvent is hexane [95, 96]. Fractional distillation, SFE, ultra-sonication, Soxhlet, solvent extraction (SE), and microwave assisted extraction are currently been used [94], and fractional distillation has been shown as the best method for the separation of unsaturated fatty acids [94, 95], while MAE was identified as the most simple, easy, and effective [97].

Some of the analytical methods used in the analysis and characterization of lipophilic compounds (lipids, phytosterols, and carotenoids) in algae are gravimetric, spectroscopic, and chromatographic [98]. GC is a widely used analytical method that ensures high sensitivity and reproducibility for separation and detection of different compounds from microalgae and macroalgae [99, 100]. Through hydrolysis and/or derivatization, large components such as proteins, polysaccharides, and lipids are converted into smaller forms (amino acids, carbohydrates, and sterols), that GC is able to detect. Also, the derivatization process allows detection of various organic acids, free amino acids, free carbohydrates, free sterols and free fatty acids (FFAs) [100].

Monitoring of different polar compounds (organic acids, carbohydrates, amino acids, and several other intermediates of metabolic pathways) during cultivation (micro/macroalgae) or different seasons is very important for determining the nutritional quality of the algal proteins and polysaccharides. Algal lipids include polar compounds (phospholipids and glycolipids) and non-polar compounds (mono-glycerides (MAGs), diglycerides (DAGs), triglycerides (TAGs), and FFAs. Fatty acids are converted in the form of methyl esters (FAMEs) by transesterification and they are the most explored biomolecules by GC.

Peptides and polysaccharides are also purified and separated using chromatographic techniques before their identification. Fast and economic separation of peptides with the desired MW is achieved using UF membrane systems [101]. LC has also been used extensively for the purification of peptides, but it is not regularly used for the purification of polysaccharides. However, it can be used to analyze the monosaccharide composition of polysaccharides [101]. In ion-exchange chromatography (IEC), molecules are separated using the difference between their charge properties. Hence, biomolecules with ionizable groups (nucleic acids, proteins, peptides) can be separated with this method [16]. IEC is one of the classic methods for separation of polysaccharides from seaweeds (alginate, fucoidan, and carrageenan) [101].

Size exclusion chromatography (SEC), LC technique, is a widely used and accepted analytical method for the separation of peptides. It can give a rapid estimate of the peptide's MW. SEC has also been used for the determination of the MW distributions of multiple seaweed polysaccharides [101]. Besides chromatographic methods, NMR spectroscopy has been broadly used for polysaccharides' structure characterization and detection of the hydrophilic molecules of algae [101]. NMR enables optimal detection and quantification of unknown compounds from complex mixtures at low concentration, by combining information such as chemical shifts and coupled constants. This method benefits the fields of metabolomics and lipidomics as it enables detection of compounds less tractable to LC-MS and GC-MS methods [102]. It is a non-destructive, non-biased, and easily quantifiable method that requires no chemical derivatization or separation. The sensitivity of this method has been improved over the years, but it still requires a relatively large sample size, and the collection, processing, and interpretation of the data is quite complex [103].

TABLE 5.1 An Overview of Different Sample Processing Methods and Analytical Methods in Different Types of Microalgae

Microalgae	Target Compounds	Sample Processing	Analytical Method	Main Findings	References
Parachlorella kessleri	Lutein	*Sample*: 10 mg of dry biomass. *Solvent*: 5 ml in different proportions (n-heptane, ethanol, water). *Assisted by*: ME, UE, MPE. *Fractionation*: HPCCC.	HPLC-PDA	Since the applied method gave good results in terms of lutein production, it could be used for large-scale production in the future.	[89]
Thalassiosira sp. *Skeletonema* sp. *Chaetoceros* sp.	Fatty acids	*Sample*: 100 mg of dry biomass. *Solvent*: 0.2 mL chloroform, 0.2 methanol, 0.1 mL distilled water. *Assisted by*: UTE.	GC-FID	The results showed a high content of ω-3 fatty acids, especially EPA and DHA, which makes these species a competitive source of high-value biomolecules for use in various industries.	[23]
Scenedesmus sp.	Carotenoids Phenolic compounds	*Sample*: dry biomass. *Solvent*: SUPRAS (octaonic acid/ethanol/water). *Assisted by*: GBVE.	LC-MS/MS (APCI+) LC-MS/MS (ESI-)	SUPRAS provides environmentally friendly and cost-effective extraction with high yields of carotenoids and polyphenols. Since the extraction takes place under simple conditions, an additional advantage of this extraction is the simultaneous extraction of polar and non-polar molecules.	[104]
Haematococcus pluvialis	Astaxanthin	*Sample*: 4 g dry biomass. *Solvent*: 250 mL ethanol and ethyl acetate. *Assisted by*: MPE. *Fractionation*: HPCCC.	HPLC-PAD	HP-CCC in combination with HPLC is an effective chromatographic technique for obtaining astaxanthin monoester from *H. pluvialis* microalgae, which have been found to differ in their effect on cancer cells.	[88]
Haematococcus pluvialis	Astaxanthin	*Sample*: 1 mL of algal broth. *Solvent*: 5 mL butan-1-ol, n-heptane, MTBE, EtOAc, and DCM. *Assisted by*: HPCCC.	HPLC-PAD	The solvents used gave a high yield of astaxanthin from zoospores, so this extraction method has the potential for industrial application.	[105]

TABLE 5.1 *(Continued)*

Microalgae	Target Compounds	Sample Processing	Analytical Method	Main Findings	References
Porphyridium cruentum *Isochrysis galbana* *Phaeodactylum tricornutum* *Tetraselmis suecica* *Nannochloropsis gaditana*	Carotenoids	*Sample:* 10 mg of dry biomass. *Solvent:* 40 ml of acetone/methanol. *Assisted by:* VE, UE. *Saponification:* ethanolic potassium hydroxide (10% w/v), TBHQ dissolved in methanol (1% w/v).	HPLC-PAD	Higher levels of carotenoids compared to food were found, and carotenoid profiles were species-specific. B-carotene was the only carotenoid common to all species. Carotenoids are mainly present in free forms.	[106]
Spirulina platensis	Peptides	*Sample:* 50 g dry biomass. *Solvent:* 1,000 mL pure water. *Assisted by:* UFT. *Fractionation:* HPLC. *Sample:* 50 g dry biomass. *Solvent:* 1,000 mL pure water. *Assisted by:* USW. *Fractionation:* HPLC.	LC-MS/MS	Isolated and identified peptide from Spirulina platensis, showed antidiabetic activity. In the future, they could be used in the treatment of type 2 diabetes, but also as functional food supplements.	[90]
Chlorella vulgaris *Scenedesmus obliquus* *Aphanothece microscopic* Nägeli *Phormidium autumnale*	Polar compounds (amino acids, organic acids, sugars, and sugar alcohols) and non-polar compounds (sterols and fatty acids).	*Sample:* 100 mg dry biomass. *Solvent:* 3 mL of methanol, 1.5 mL of chloroform, and 1.2 mL of distilled water. *Assisted by:* UE, SE, VE, UTE.	GC-FID	The newly developed extraction method enabled the simultaneous extraction of polar and non-polar metabolites in a significantly reduced amount of organic solvents and algae mass. Higher concentrations of PUFA were obtained from *C. vulgaris* and Scenedesmus obliquus extracts.	[107]

TABLE 5.1 *(Continued)*

Microalgae	Target Compounds	Sample Processing	Analytical Method	Main Findings	References
Nannochloropsis gaditana *Chlorella* sp. *Haematococcus pluvialis* *Scenedesmus almeriensis* *Isochrysis galbana* *Tetraselmis suecica* *Karlodinium veneficu* *Protoceratium reticulatum*	Carotenoids	*Sample:* 5 g dry biomass/5 g of alumina. *Assisted by:* VE. *Saponification:* ethanol, hexane, water.	HPLC-PAD	Saponification conditions are closely species-specific. Intracellular pigment, xanthophylls, and carotenes are best collected in solvent mixtures of different polarities, but conditions must be defined since these are readily degradable compounds.	[108]
C. vulgaris *Scenedesmus regularis*	Carotenoids	*Sample:* 100 mg dry biomass. *Solvent:* 3 mL hexane-ethanol-acetone-toluene. *Assisted by*: SE. *Saponification:* 2 mL 40% methanolic KOH.	HPLC-PAD	The obtained results of carotenoid composition and total yield indicate the potential of *C. vulgaris* in lutein production and *S. regularis* in β-carotene production.	[109]

Abbreviation: ME: maceration extraction; UE: ultrasound extraction; MPE: Mortar and Pestle extraction; UTE: ultra-turrax extraction; GBVE: glass bead vortexing extraction; VE: vortex extraction; UFTE: ultrasound freeze-thaw extraction; USWE: ultrasound subcritical-water extraction; SE: shaker extraction; LC: liquid chromatography; HP: high performance; CCC: counter-current chromatography; MS: mass spectrometry; PUFA: polyunsaturated fatty acids; GC: gas chromatography; EPA: eicosapentaenoic acid; and DHA: docosahexaenoic acid.

TABLE 5.2 An overview of different extraction and analytical methods recently used for seaweeds.

Seaweed	Target Compounds	Sample Processing	Analytical Method	Main Findings	References
Chondrus crispus, *Mastocarpus stellatus*, and *Gigartina pistillata*	Phenolic compounds, pigments, organic acids, proteins, carbohydrates.	• Conventional heat assisted extraction (HAE): 30 g freeze-dried sample/L hydroalcoholic mixture, placed in water bath and stirred (19.5 to 120.5 min; 21.4 to 88.6°C; 0 to 100% ethanol). • High pressure assisted extraction (HPAE): 30 g freeze-dried sample/L hydroalcoholic mixture, placed in high hydrostatic pressure research device (10–110 min; 100–600 Mpa; 0–100% ethanol).	UFLC, HPLC-MS, HPLC, and AOAC methods for protein, fat, and carbohydrates.	HAE was the most efficient extraction technique (36.6 min, 88.6°C and 50.0% ethanol). All algae show potential for industrial application as a source of phenolic compounds and pigments.	[110]
Grateloupia turuturu	Polar lipids (esterified PUFAs)	Lipid extraction: 200 mg oven-dried sample with 3 mL of chloroform:methanol mixture (1:2, v/v). The organic phase was collected, and the remaining biomass residue was re-extracted twice. Total organic phase was dried under a stream of nitrogen gas.	HPLC-MS and GC-MS for polar lipids.	205 lipid species identified. 19 fatty acids were identified by GC-MS after the transmethylation. Palmitic acid (C16:0) was the most abundant fatty acid (22.03%) followed by EPA (C20:5n–3) with 20.86%. *G. turuturu* lipid extracts could be used for nutraceutical applications.	[111]
Padina pavonica	Phenolic compounds	• Shaking at room temperature for 24 h • Shaking in an incubator at 60°C for 1 and 2 h • UAE (40 kHz, 60°C for 1 an d 2 h) • MAE (200 W, 60°C for 5, 10, 15, and 30 min) • 1:10 freeze-dried/shade-dried/oven-dried algae to solvent ratio; water, ethanol 50%, and ethanol 70%.	Spectrophotometry: TPC, TTC, TFC	The best drying method for high TPC was freeze-drying. MAE was identified as the best extraction method and 50% ethanol as the best solvent.	[46]

TABLE 5.2 *(Continued)*

Seaweed	Target Compounds	Sample Processing	Analytical Method	Main Findings	References
Saccharina japonica and *Undaria pinnatifida*	Bioactive metabolites	UAE (1:10, methanol, 45 min)	HPTLC/bioautography/UHPLC-LTQ-MS^2/GC-MS.	*S. japonica* was is richer in pigments (carotenoids) than *U. pinnatifida.* Stearidonic acid, EPA, and arachidonic acid were identified as major bioactive compounds in seaweed extracts.	[112]
Asparagopsis taxiformis, Chondrus crispus, Delesseria sanguinea, Dumontia consortia, Furcellaria lumbricalis, Gracilaria vermiculophylla, Polyides rotundus, Ulva intestinalis, Ulva sp., *Alaria esculenta, Chorda filum, Dictyota dichotoma, Fucus evanescens, Fucus serratus, Fucus spiralis, Laminaria digitata, Saccharina latissima,* and *Sargassum muticum*	Halomethanes and phenolics	UAE (50 mg/1.25 mL methanol, 15 min).	GC-MS, UPLC-QTOF-MS-MS	Only *A. taxiformis* contained halomethanes (bromoform in high concentrations). All seaweeds contained high numbers of sulfated phenolic acids and unidentified sulfuric acid-containing compounds. The condensed tannin gallocatechin was tentatively identified in *Fucus* sp.	[113]

TABLE 5.2 *(Continued)*

Seaweed	Target Compounds	Sample Processing	Analytical Method	Main Findings	References
Palmaria palmata, L. digitata, S. latissima and *Ascophyllum nodosum*	Fatty acids and derivatives	UAE (30 g/100 mL hexane/dichloromethane:methanol/methanol:water, 10 min)	UHPLC-QTOF-HRMS	The six isolated compounds were tentatively identified as five fatty acids and one monoglyceride (stearidonic acid, EPA, alpha-linolenic acid, DHA, arachidonic acid, and monoacylglycerol). Selected isolated compounds showed synergic anti-parasitic activity *in vitro*.	[114]
Macrocystis pyrifera and *Chondracanthus chamissoi*	Amino acids, fatty acids, tocopherols, polyphenols.	• Extraction for amino acids: 1 g/5 mL HCl, mixed, and hydrolyzed (110 C, 4 h). • Extraction for fatty acids: 2 g/10 mL hexane, mixed, and stirred for 5 min, crude oil was methylated. • Conventional extraction: 0.25 g/10 mL methanol, centrifuged for 1 h at room temperature.	HPLC, GC, TPC	*M. pyrifera* had significantly higher content of PUFAs and tocopherols. Both seaweeds had significant TPC.	[115]
Hypnea musciformis and *Gelidium pusillum*	Phenolic compounds, carotenoids, proteins, lipids, carbohydrate.	Conventional extraction: 5 g/50 mL 80% acetone or 1 g/25 mL methanol, shaking for 90 min.	Spectrophotometry: TPC, TFC, TTC, TCC; AOAC methods for proteins, lipids, and carbohydrates.	*G. pusillum* had higher TPC, TCC, carbohydrates, and lipids contents while *H. musciformis* had higher TFC, TTC, and protein content.	[116]

TABLE 5.2 *(Continued)*

Seaweed	Target Compounds	Sample Processing	Analytical Method	Main Findings	References
G. turuturu and *Porphyra umbilicalis*	Phenolic compounds, amino acids.	• Hydroethanolic agitation (0.5 g/50 mL 80% ethanol, 1 h). • Water decoction (0.5 h/150 mL distilled water, heated to boiling, 30 min). • Water infusion (0.5 h/150 mL boiling distilled water, stirring, 10 min).	TPC, TFC, RP-HPLC-DAD, LC-DAD-ESI-MS	Samples extracted with water had higher TPC, TFC, and variety of compounds. Several mycosporine-like amino acids were identified, palythine, shinorine, and asterina-330 in all extracts of both algae, and porphyra-334 and aplysiapalythine A/palythinol in all extracts of *P. umbilicalis*.	[117]
Fucus virsoides	Volatile organic compounds, fatty acids, terpenoids, steroids, carotenoids.	• Headspace Solid-Phase Microextraction: 1 g sample was sealed in 5 mL glass vial, placed in a water bath at 60°C, 15 min for equilibration and 45 min for the extraction. • Hydrodistillation: 15 g of sample in Clavenger apparatus for 2 h with 1 mL of pentane: diethyl ether (1:2 v/v) as solvent trap. • UAE: 1:10, methanol:dichloromethane (1:1, v/v), 5 min 3 times.	GC-FID/MS, HPLC-ESI-HRMS	Oleic acid was dominant fatty acid, followed by arachidonic acid. High omega-6 and omega-3 content was found. Fucoxanthin and chlorophyll derivatives were identified.	[118]
Ulva sp., *Codium* sp., *Caulerpa* sp., *Dasya* sp., *Centroceras* sp., *Grateloupia* sp., *Ecklonia* sp., and *Sargassum* sp.	Phenolic compounds	Conventional extraction: 2 g/10 mL of 80% ethanol, homogenized, shaking in incubator at 4°C for 16 h.	TPC, TFC, TTC, LC-ESI-QTOF-MS/MS	54 phenolic compounds were tentatively identified in all seaweeds. *Centroceras* sp. had the highest number of phenolic compounds, followed by *Ecklonia* sp. and *Caulerpa* sp. *p*-hydroxybenzoic acid was the most abundant phenolic compound.	[119]

TABLE 5.2 *(Continued)*

Seaweed	Target Compounds	Sample Processing	Analytical Method	Main Findings	References
Halimeda cuneata, Pseudorhizoclonium africanum, Pseudocodium devriesii, Dictyota suhrii, Gracilaria salicornia, Hypnea rosea, Laurencia natalensis and *Jania adhaerens*	Volatile compounds	Conventional extraction: oven-dried seaweeds extracted with methanol overnight.	GC-MS	82 phytocompounds were identified. Phytol and Z-8-Methyl-9-tetradecenoic were present in all seaweeds.	[120]

Abbreviations: AOAC: Association of Official Agricultural Chemists; UFLC: ultra-fast liquid chromatography; LC: liquid chromatography; HP: high performance; U-Ultra/UP: ultra-performance; MS: mass spectrometry; HR: high resolution; RP: reversed-phase; PUFA: polyunsaturated fatty acids; GC: gas chromatography; EPA: eicosapentaenoic acid; DHA: docosahexaenoic acid; UAE: ultrasound-assisted extraction; MAE: microwave-assisted extraction; TPC: total phenolics content; TFC: total flavonoids content; TTC: total tannins content; TCC: total carotenoid content; TLC: thin layer chromatography; LTQ: linear trap quadrupole; QTOF: quadrupole time-of-flight; DAD: diode array detection; and ESI: electrospray ionization.

5.6 CONCLUSIONS

Microalgae and seaweed have been included in the human diet for thousands of years, but today, they are gradually becoming a new source of nutritive components and nutraceuticals. They are characterized by high content of proteins and high-quality lipids, numerous bioactive compounds, and most of all sustainability. They present a renewable source of nutrients and bioactive compounds that will ensure a new generation of products with enhanced nutritional and functional characteristics with multiple applications. The perspective of algae lies in the prevention of diseases, treatment, and improvement of chronic diseases symptoms, so they also serve as a source for novel drug development. Microalgae can be cultivated in large quantities as well as seaweed. However, technological barriers make it impossible to cover high market demand. The large capital investments, high energy consumption, but also non-standardized extraction methods present a challenge for the future. Therefore, future research should focus on solving technological problems related to cultivation techniques of microalgae and seaweed, issues related to the standardization of the extraction methods that would ensure maximum yield, economic efficiency and eco-friendly solutions for obtaining the target compounds. Besides the nutritional composition, the nutritive value of the algae should also be accessed through its digestibility and toxicity. Addressing these issues would enable the long-term sustainability and exploitation profitability of this marine resource.

KEYWORDS

- **chromatography methods**
- **exploitation profitability**
- **long-term sustainability**
- **macroalgae**
- **microalgae**
- **nutraceuticals**
- **polyunsaturated fatty acids**
- **seaweeds**

REFERENCES

1. Šimat, V., (2021b). Nutraceuticals and pharmaceuticals from marine fish and invertebrates. *Marine Drugs, 19*(7), 401. https://doi.org/10.3390/md19070401.
2. Šimat, V., Elabed, N., Kulawik, P., Ceylan, Z., Jamroz, E., Yazgan, H., Čagalj, M., et al., (2020). Recent advances in marine-based nutraceuticals and their health benefits. *Marine Drugs, 18*(12), 1–40. https://doi.org/10.3390/md18120627.
3. Lomartire, S., Cotas, J., Pacheco, D., Marques, J. C., Pereira, L., & Gonçalves, A. M. M., (2021). Environmental impact on seaweed phenolic production and activity: An important step for compound exploitation. *Marine Drugs, 19*(5), 1–20. https://doi.org/10.3390/md19050245.
4. Šimat, V., (2021a). Valorization of seafood processing by-products. In: Bhat, R., (ed.), *Valorization of Agri-Food Wastes and By-Products: Recent Trends, Innovations and Sustainability Challenges*. In press. Elsevier Inc.
5. Hamed, I., Özogul, F., & Regenstein, J. M., (2016). Industrial applications of crustacean by-products (chitin, chitosan, and chitooligosaccharides): A review. *Trends in Food Science and Technology, 48*, 40–50. https://doi.org/10.1016/j.tifs.2015.11.007.
6. Khawli, F. A., Pateiro, M., Domínguez, R., Lorenzo, J. M., Gullón, P., Kousoulaki, K., Ferrer, E., et al., (2019). Innovative green technologies of intensification for valorization of seafood and their by-products. *Marine Drugs, 17*(12), 1–20. https://doi.org/10.3390/md17120689.
7. Ozogul, F., Cagalj, M., Šimat, V., Ozogul, Y., Tkaczewska, J., Hassoun, A., Kaddour, A. A., et al., (2021). Recent developments in valorization of bioactive ingredients in discard/seafood processing by-products. *Trends in Food Science & Technology, 116*, 559–582. https://doi.org/10.1016/j.tifs.2021.08.007.
8. Šimat, V., Vlahović, J., Soldo, B., Generalić, M. I., Čagalj, M., Hamed, I., & Skroza, D., (2020). Production and characterization of crude oils from seafood processing by-products. *Food Bioscience, 33*, 100484. https://doi.org/10.1016/j.fbio.2019.100484.
9. Suresh, P. V., Kudre, T. G., & Johny, L. C., (2018). Sustainable valorization of seafood processing by-product/discard. In: *Waste to Wealth* (pp. 111–140). https://doi.org/10.1007/978-981-10-7431-8.
10. Christaki, E., Bonos, E., & Florou-Paneri, P., (2015). Innovative microalgae pigments as functional ingredients in nutrition. In: *Handbook of Marine Microalgae: Biotechnology Advances*. Elsevier Inc. https://doi.org/10.1016/B978-0-12-800776-1.00014-5.
11. Molino, A., Mehariya, S., Di Sanzo, G., Larocca, V., Martino, M., Leone, G. P., Marino, T., et al., (2020). Recent developments in supercritical fluid extraction of bioactive compounds from microalgae: Role of key parameters, technological achievements, and challenges. *Journal of CO_2 Utilization, 36*(2019), 196–209. https://doi.org/10.1016/j.jcou.2019.11.014.
12. Jayachandran, V., Panchanathan, M., & Kim S., (2015). In: Kim, S., (ed.), *Handbook of Marine Microalgae*.
13. Sumi, Y., (2009). *Microalgae Pioneering the Future — Application and Utilization*. NISTEP Science & Technology Foresight Center., Quarterly review No.34, 9–21. https://core.ac.uk/download/pdf/236667555.pdf (accessed on 30 September 2023).
14. Dudeja, P., & Gupta, R. K., (2017). Chapter 40 – nutraceuticals. In: *Food Safety in the 21st Century*. Elsevier Inc. https://doi.org/10.1016/B978-0-12-801773-9/00040-6.

15. Venugopal, V., (2020). Seaweed: Nutritional value, bioactive properties, and uses. In: *Marine Products for Healthcare*. https://doi.org/10.1201/9781420052640-13.
16. Corrêa, P. S., Morais, Jr. W. G., Martins, A. A., Caetano, N. S., & Mata, T. M., (2021). Microalgae biomolecules: Extraction, separation, and purification methods. *Processes, 9*(1), 1–40. https://doi.org/10.3390/pr9010010.
17. McMurry, J., (2015). *Organic Chemistry with Biological Applications*. Cengage Learning, 2014, Stamford, USA, pp. 877–905.
18. Thirumurugan, D., Cholarajan, A., Raja, S. S. S., & Vijayakumar, R., (2018). An introductory chapter: Secondary metabolites. *An Introduction to Natural Products Chemistry; Secondary Metabolites-Sources and Applications* (pp. 3–22). https://doi.org/10.5772/intechopen.79766.
19. Bérdy, J., (2005). Bioactive microbial metabolites a personal view. *The Journal of Antibiotics, 58*(1), 1–26.
20. Frassanito, R., Cantonati, M., Tardìo, M., Mancini, I., & Guella, G., (2005). On-line identification of secondary metabolites in freshwater microalgae and *Cyanobacteria* by combined liquid chromatography-photodiode array detection-mass spectrometric techniques. *Journal of Chromatography A, 1082*(1 SPEC. ISS.), 33–42. https://doi.org/10.1016/j.chroma.2005.02.066.
21. Patras, D., Moraru, C. V., & Socaciu, C., (2018). Screening of bioactive compounds synthesized by microalgae: A progress overview on extraction and chemical analysis. *Studia Universitatis Babes-Bolyai Chemia, 63*(1), 21–35. https://doi.org/10.24193/subbchem.2018.1.02.
22. Megawati, D. S., Fasya, A. G., Pratiwi, R. A., & Maghfiroh, N., (2020). Pharmacology potency of thin layer chromatography steroid isolates of *Chlorella* sp. chloroform fraction. *IOP Conference Series: Earth and Environmental Science, 456*(1). https://doi.org/10.1088/1755-1315/456/1/012012.
23. Bhattacharjya, R., Kiran, M. T., Tiwari, A., Saxena, A., Kumar, S. P., & Mishra, B., (2020). Bioprospecting of marine diatoms *Thalassiosira*, *Skeletonema* and *Chaetoceros* for lipids and other value-added products. *Bioresource Technology, 318*, 124073. https://doi.org/10.1016/j.biortech.2020.124073.
24. Singh, G., & Patidar, S. K., (2018). Microalgae harvesting techniques: A review. *Journal of Environmental Management, 217*, 499–508. https://doi.org/10.1016/j.jenvman.2018.04.010.
25. Zhu, L., Li, Z., & Hiltunen, E., (2018). Biotechnology for biofuels microalgae *Chlorella vulgaris* biomass harvesting by natural flocculant: Effects on biomass sedimentation, spent medium recycling and lipid extraction. *Biotechnology for Biofuels*, 1–10. https://doi.org/10.1186/s13068-018-1183-z.
26. Alhattab, M., Kermanshahi-pour, A., & Su-Ling, B. M., (2019). Dispersed air flotation of *Chlorella saccharophila* and subsequent extraction of lipids – effect of supercritical CO_2 extraction parameters and surfactant pretreatment. *Biomass and Bioenergy, 127*, 105297. https://doi.org/10.1016/j.biombioe.2019.105297.
27. Baierle, F., John, D. K., Souza, M. P., Bjerk, T. R., Moraes, M. S. A., Hoeltz, M., Rohlfes, A. L. B., et al., (2015). Biomass from microalgae separation by electroflotation with iron and aluminum spiral electrodes. *Chemical Engineering Journal, 267*, 274–281. https://doi.org/10.1016/j.cej.2015.01.031.
28. Niaghi, M., Mahdavi, M. A., & Gheshlaghi, R., (2015). Optimization of dissolved air flotation technique in harvesting microalgae from treated wastewater without

flocculants addition. *Journal of Renewable and Sustainable Energy*, *7*(1), 1–20. https://doi.org/10.1063/1.4909541.

29. Bejor, E. S., Mota, C. R., Ogarekpe, N. M., & Emerson, K. U., (2013). *Low-Cost Harvesting of Microalgae Biomass from Water*. 2(1), 1–11.
30. Molina, G. E., Belarbi, E. H., Acién, F. F. G., Robles, M. A., & Chisti, Y., (2003). Recovery of microalgal biomass and metabolites: Process options and economics. *Biotechnology Advances*, *20*(7, 8), 491–515. https://doi.org/10.1016/S0734-9750(02)00050-2.
31. Generalić, M. I., Skroza, D., Šimat, V., Hamed, I., Čagalj, M., & Perković, Z. P., (2019). Phenolic content of brown algae (*Pheophyceae*) species: Extraction, identification, and quantification. In: *Biomolecules* (Vol. 9, No. 6). MDPI AG. https://doi.org/10.3390/biom9060244.
32. Cai, J., Lovatelli, A., Aguilar-Manjarrez, J., Cornish, L., Dabbadie, L., Desrochers, A., Diffey, S., et al., (2021). Seaweeds and microalgae: An overview for unlocking their potential in global aquaculture development. In: *FAO Fisheries and Aquaculture Circular No. 1229*. FAO. https://doi.org/10.4060/cb5670en.
33. Cai, J., & Galli, G., (2021). *Top 10 Species Groups in Global Aquaculture 2019*. FAO Fisheries and Aquaculture Technical Paper.
34. Ferdouse, F., Løvstad, H. S., Smith, R., Murúa, P., & Yang, Z., (2018). *The Global Status of Seaweed Production, Trade and Utilization* (Vol. 124, p. 120). FAO Globefish Research Program.
35. Qin, Y., (2020). Health benefits of bioactive seaweed substances. In: *Handbook of Algal Science, Technology and Medicine*. Elsevier Inc. https://doi.org/10.1016/B978-0-12-818305-2.00029-2.
36. Ganesan, A. R., Tiwari, U., & Rajauria, G., (2019). Seaweed nutraceuticals and their therapeutic role in disease prevention. *Food Science and Human Wellness*, *8*(3), 252–263. https://doi.org/10.1016/j.fshw.2019.08.001.
37. Ummat, V., Sivagnanam, S. P., Rajauria, G., O'Donnell, C., & Tiwari, B. K., (2021). Advances in pre-treatment techniques and green extraction technologies for bioactives from seaweeds. *Trends in Food Science and Technology*, *110*, 90–106. https://doi.org/10.1016/j.tifs.2021.01.018.
38. Sharanagat, V. S., Singla, V., & Singh, L., (2020). Bioactive compounds from marine sources. In: Goyal, M. R., Rasul, S. H. A., & Kirubanandan, S., (eds.), *Technological Processes for Marine Foods-from Water to Fork: Bioactive Compounds, Industrial Applications and Genomics*. Apple Academic Press, Inc.
39. Suleria, H. A. R., Gobe, G., Masci, P., & Osborne, S. A., (2016). Marine bioactive compounds and health-promoting perspectives; innovation pathways for drug discovery. *Trends in Food Science and Technology*, *50*, 44–55. https://doi.org/10.1016/j.tifs.2016.01.019.
40. Fleurence, J., Morançais, M., & Dumay, J., (2018). Seaweed proteins. In: *Proteins in Food Processing* (2nd edn.). Elsevier Ltd. https://doi.org/10.1016/B978-0-08-100722-8.00010-3.
41. Mišurcová, L., Buňka, F., Vávra, A. J., Machů, L., Samek, D., & Kráčmar, S., (2014). Amino acid composition of algal products and its contribution to RDI. *Food Chemistry*, *151*, 120–125. https://doi.org/10.1016/j.foodchem.2013.11.040.
42. Miyashita, K., Mikami, N., & Hosokawa, M., (2013). Chemical and nutritional characteristics of brown seaweed lipids: A review. *Journal of Functional Foods*, *5*(4), 1507–1517. https://doi.org/10.1016/j.jff.2013.09.019.

43. Gullón, B., Gagaoua, M., Barba, F. J., Gullón, P., Zhang, W., & Lorenzo, J. M., (2020). Seaweeds as promising resource of bioactive compounds: Overview of novel extraction strategies and design of tailored meat products. *Trends in Food Science and Technology, 100*(2019), 1–18. https://doi.org/10.1016/j.tifs.2020.03.039.
44. Cotas, J., Leandro, A., Monteiro, P., Pacheco, D., Figueirinha, A., Gonçalves, A. M. M., Da Silva, G. J., & Pereira, L., (2020). Seaweed phenolics: From extraction to applications. *Marine Drugs, 18*(8). https://doi.org/10.3390/MD18080384.
45. Hermund, D. B., (2018). Antioxidant properties of seaweed-derived substances. In: *Bioactive Seaweeds for Food Applications*. Elsevier Inc. https://doi.org/10.1016/B978-0-12-813312-5.00010-8.
46. Čagalj, M., Skroza, D., Tabanelli, G., Özogul, F., & Šimat, V., (2021). Maximizing the antioxidant capacity of *Padina pavonica* by choosing the right drying and extraction methods. *Processes, 9*(4), 587. https://doi.org/10.3390/pr9040587.
47. Jacobsen, C., Sørensen, A. D. M., Holdt, S. L., Akoh, C. C., & Hermund, D. B., (2019). Source, extraction, characterization, and applications of novel antioxidants from seaweed. *Annual Review of Food Science and Technology, 10*, 541–568. https://doi.org/10.1146/annurev-food-032818-121401.
48. Rakesh, S., Dhar, D. W., Prasanna, R., Saxena, A. K., Saha, S., Shukla, M., & Sharma, K., (2015). Cell disruption methods for improving lipid extraction efficiency in unicellular microalgae. *Engineering in Life Sciences, 15*(4), 443–447. https://doi.org/10.1002/elsc.201400222.
49. Gomes, T. A., Zanette, C. M., & Spier, M. R., (2020). An overview of cell disruption methods for intracellular biomolecules recovery. *Preparative Biochemistry & Biotechnology, 50*(7), 635–654. https://doi.org/10.1080/10826068.2020.1728696.
50. Günerken, E., D'Hondt, E., Eppink, M. H. M., Garcia-Gonzalez, L., Elst, K., & Wijffels, R. H., (2015). Cell disruption for microalgae biorefineries. *Biotechnology Advances, 33*(2), 243–260. https://doi.org/10.1016/j.biotechadv.2015.01.008.
51. Ummat, V., Tiwari, B. K., Jaiswal, A. K., Condon, K., Garcia-Vaquero, M., O'Doherty, J., O'Donnell, C., & Rajauria, G., (2020). Optimization of ultrasound frequency, extraction time, and solvent for the recovery of polyphenols, phlorotannins, and associated antioxidant activity from brown seaweeds. *Marine Drugs, 18*(5). https://doi.org/10.3390/md18050250.
52. Amorim, A. M., Nardelli, A. E., & Chow, F., (2020). Effects of drying processes on antioxidant properties and chemical constituents of four tropical macroalgae suitable as functional bioproducts. *Journal of Applied Phycology, 32*(2), 1495–1509. https://doi.org/10.1007/s10811–020-02059-7.
53. Chan, J. C. C., Cheung, P. C. K., & Ang, P. O., (1997). Comparative studies on the effect of three drying methods on the nutritional composition of seaweed *Sargassum hemiphyllum* (Turn.) C. Ag. *Journal of Agricultural and Food Chemistry, 45*(8), 3056–3059. https://doi.org/10.1021/jf9701749.
54. Henriques, M., Silva, A., & Rocha, J., (2007). *Extraction and Quantification of Pigments from a Marine Microalga: A Simple and Reproducible Method* (pp. 586–593).
55. Edler, L., (1979). Recommendations of methods of marine biological studies in the Baltic sea: Phytoplankton and chlorophyll. *The Baltic Marine Biologists.*
56. Chew, K. W., Chia, S. R., Krishnamoorthy, R., Tao, Y., Chu, D. T., & Show, P. L., (2019). Liquid biphasic flotation for the purification of C-phycocyanin from

Spirulina platensis microalga. *Bioresource Technology*, 288. https://doi.org/10.1016/j.biortech.2019.121519.

57. Ruiz-ruiz, F., López-guajardo, E., Vázquez-villegas, P., Angel-chong, M. E., Nigam, K. D. P., Willson, R. C., & Rito-palomares, M., (2019). Chemical engineering & processing: Process intensification continuous aqueous two-phase extraction of microalgal c-phycocyanin using a coiled flow inverter. *Chemical Engineering & Processing: Process Intensification*, *142*, 107554. https://doi.org/10.1016/j.cep.2019.107554.
58. Bligh, E. G., & Dyer, W. J., (1959). A rapid method of total lipid extraction and purification. *Canadian Journal of Biochemistry and Physiology*, *37*(8), 911–917.
59. Folch, J., Lees, M., & Sloane, S. G. H., (1957). A simple method for the isolation and purification of total lipides from animal tissues. *The Journal of Biological Chemistry*, *226*(1), 497–509. https://doi.org/10.1016/s0021-9258(18)64849-5.
60. Kumar, R. R., Rao, P. H., & Arumugam, M., (2015). Lipid extraction methods from microalgae: A comprehensive review. *Frontiers in Energy Research*, *3*, 1–9. https://doi.org/10.3389/fenrg.2014.00061.
61. Ghasemi, N. F., González, G. L. M., Chan, W., & Schenk, P. M., (2016). Progress on lipid extraction from wet algal biomass for biodiesel production. *Microbial Biotechnology*, *9*(6), 718–726. https://doi.org/10.1111/1751-7915.12360.
62. Saini, R. K., & Keum, Y. S., (2018). Carotenoid extraction methods: A review of recent developments. *Food Chemistry*, *240*(2017), 90–103. https://doi.org/10.1016/j.foodchem.2017.07.099.
63. Zhang, R., Parniakov, O., Grimi, N., Lebovka, N., Marchal, L., & Vorobiev, E., (2019). Emerging techniques for cell disruption and extraction of valuable bio-molecules of microalgae *Nannochloropsis* sp. *Bioprocess and Biosystems Engineering*, *42*(2), 173–186. https://doi.org/10.1007/s00449-018-2038-5.
64. Šimat, V., Čagalj, M., Skroza, D., Gardini, F., Tabanelli, G., Montanari, C., Hassoun, A., & Özogul, F., (2021). Sustainable sources for antioxidant and antimicrobial compounds used in meat and seafood products. In: Toldrá, F., (ed.), *Advances in Food and Nutrition Research* (p. In Press, Corrected Proof). Academic Press. https://doi.org/10.1016/bs.afnr.2021.03.001.
65. Kadam, S., O'Donnell, C., Rai, D., Hossain, M., Burgess, C., Walsh, D., & Tiwari, B., (2015). Laminarin from Irish brown seaweeds *Ascophyllum nodosum* and *Laminaria hyperborea*: Ultrasound assisted extraction, characterization and bioactivity. *Marine Drugs*, *13*(7), 4270–4280. https://doi.org/10.3390/md13074270.
66. Yuan, Y., Zhang, J., Fan, J., Clark, J., Shen, P., Li, Y., & Zhang, C., (2018). Microwave assisted extraction of phenolic compounds from four economic brown macroalgae species and evaluation of their antioxidant activities and inhibitory effects on α-amylase, α-glucosidase, pancreatic lipase and tyrosinase. *Food Research International*, *113*, 288–297. https://doi.org/10.1016/j.foodres.2018.07.021.
67. Krishnan, S., Ghani, N. A., Aminuddin, N. F., Quraishi, K. S., Azman, N. S., Cravotto, G., & Leveque, J. M., (2020). Microwave-assisted lipid extraction from *Chlorella vulgaris* in water with 0.5%–2.5% of imidazolium based ionic liquid as additive. *Renewable Energy*, *149*, 244–252. https://doi.org/10.1016/j.renene.2019.12.063.
68. Zghaibi, N., Omar, R., Kamal, S. M. M., Biak, D. R. A., & Harun, R., (2019). Microwave-assisted brine extraction for enhancement of the quantity and quality of lipid production from microalgae *Nannochloropsis* sp. *Molecules*, *24*(19), 1–21. https://doi.org/10.3390/molecules24193581.

69. Kadam, S. U., Álvarez, C., Tiwari, B. K., & O'Donnell, C. P., (2015). Extraction of biomolecules from seaweeds. In: *Seaweed Sustainability: Food and Non-Food Applications*. Elsevier Inc. https://doi.org/10.1016/B978-0-12-418697-2.00009-X.
70. Mason, T. J., Paniwnyk, L., & Lorimer, J. P., (1996). The uses of ultrasound in food technology. *Ultrasonics Sonochemistry*, *3*(3), S253–S260. https://doi.org/10.1016/S1350-4177(96)00034-X.
71. Vernès, L., Vian, M., & Chemat, F., (2020). Ultrasound and microwave as green tools for solid-liquid extraction. In: *Liquid-Phase Extraction* (pp. 355–374). Elsevier. https://doi.org/10.1016/B978-0-12-816911-7.00012-8.
72. Veggi, P. C., Martinez, J., & Meireles, M. A. A., (2013). Fundamentals of microwave extraction. In: *Food Engineering Series* (pp. 15–52). Springer. https://doi.org/10.1007/978-1-4614-4830-3_2.
73. Routray, W., & Orsat, V., (2012). Microwave-assisted extraction of flavonoids: A review. *Food and Bioprocess Technology*, *5*(2), 409–424. https://doi.org/10.1007/s11947-011-0573-z.
74. Rodrigues, D., Sousa, S., Silva, A., Amorim, M., Pereira, L., Rocha-Santos, T. A. P., Gomes, A. M. P., et al., (2015). Impact of enzyme- and ultrasound-assisted extraction methods on biological properties of red, brown, and green seaweeds from the central west coast of Portugal. *Journal of Agricultural and Food Chemistry*, *63*(12), 3177–3188. https://doi.org/10.1021/jf504220e.
75. He, Y., Zhang, B., Guo, S., Guo, Z., Chen, B., & Wang, M., (2020). Sustainable biodiesel production from the green microalgae *Nannochloropsis*: Novel integrated processes from cultivation to enzyme-assisted extraction and ethanolysis of lipids. *Energy Conversion and Management*, *209*, 112618. https://doi.org/10.1016/j.enconman.2020.112618.
76. Huo, S., Wang, Z., Cui, F., Zou, B., Zhao, P., & Yuan, Z., (2015). Enzyme-assisted extraction of oil from wet microalgae *Scenedesmus* sp. G4. *Energies*, *8*(8), 8165–8174. https://doi.org/10.3390/en8088165.
77. Liang, K., Zhang, Q., & Cong, W., (2012). Enzyme-assisted aqueous extraction of lipid from microalgae. *Journal of Agricultural and Food Chemistry*, *60*(47), 11771–11776. https://doi.org/10.1021/jf302836v.
78. Chemat, F., Rombaut, N., Meullemiestre, A., Turk, M., Perino, S., Fabiano-Tixier, A. S., & Abert-Vian, M., (2017). Review of green food processing techniques. Preservation, transformation, and extraction. *Innovative Food Science & Emerging Technologies*, *41*, 357–377. https://doi.org/10.1016/j.ifset.2017.04.016.
79. Garcia-Vaquero, M., Sweeney, T., O'Doherty, J., & Rajauria, G., (2021). Recent advances in the use of greener extraction technologies for the recovery of valuable bioactive compounds from algae. *Recent Advances in Micro and Macroalgal Processing, Tiwari*, 96–122. https://doi.org/10.1002/9781119542650.ch4.
80. Ibañez, E., Herrero, M., Mendiola, J. A., & Castro-Puyana, M., (2012). Extraction and characterization of bioactive compounds with health benefits from marine resources: Macro and microalgae, cyanobacteria, and invertebrates. In: *Marine Bioactive Compounds* (pp. 55–98). Springer US. https://doi.org/10.1007/978-1-4614-1247-2_2.
81. Dunn, W. B., Broadhurst, D. I., Edison, A., Guillou, C., Viant, M. R., Bearden, D. W., & Beger, R. D., (2017). Quality assurance and quality control processes: Summary of a metabolomics community questionnaire. *Metabolomics*, *50*, 1–6. https://doi.org/10.1007/s11306-017-1188-9.

82. Matić, P., Sabljić, M., & Jakobek, L., (2017). Validation of spectrophotometric methods for the determination of total polyphenol and total flavonoid content. *Journal of AOAC International*, *100*(6), 1795–1803. https://doi.org/10.5740/jaoacint.17-0066.
83. Everette, J. D., Bryant, Q. M., Green, A. M., Abbey, Y. A., Wangila, G. W., & Walker, R. B., (2010). Thorough study of reactivity of various compound classes toward the Folin–ciocalteu reagent. *Journal of Agricultural and Food Chemistry*, *58*(14), 8139–8144. https://doi.org/10.1021/jf1005935.
84. Singleton, V. L., Orthofer, R., & Lamuela-Raventós, R. M., (1999). Analysis of total phenols and other oxidation substrates and antioxidants by means of Folin-ciocalteu reagent. *Methods in Enzymology*, *299*, 152–178.
85. Agregán, R., Munekata, P. E. S., Franco, D., Dominguez, R., Carballo, J., & Lorenzo, J. M., (2017). Phenolic compounds from three brown seaweed species using LC-DAD–ESI-MS/MS. *Food Research International*, *99*, 979–985. https://doi.org/10.1016/j.foodres.2017.03.043.
86. Ford, L., Theodoridou, K., Sheldrake, G. N., & Walsh, P. J., (2019). A critical review of analytical methods used for the chemical characterization and quantification of phlorotannin compounds in brown seaweeds. *Phytochemical Analysis*, *30*(6), 587–599. https://doi.org/10.1002/pca.2851.
87. Nie, J., Chen, D., Ye, J., Lu, Y., & Dai, Z., (2021). Preparative separation of three terpenoids from edible brown algae *Sargassum fusiforme* by high-speed countercurrent chromatography combined with preparative high-performance liquid chromatography. *Algal Research*, *59*, 102449. https://doi.org/10.1016/j.algal.2021.102449.
88. Fábryová, T., Tůmová, L., Da Silva, D. C., Pereira, D. M., Andrade, P. B., Valentão, P., Hrouzek, P., et al., (2020). Isolation of astaxanthin monoesters from the microalgae *Haematococcus pluvialis* by high performance countercurrent chromatography (HPCCC) combined with high performance liquid chromatography (HPLC). *Algal Research*, *49*, 101947. https://doi.org/10.1016/j.algal.2020.101947.
89. Fábryová, T., Kubáč, D., Kuzma, M., Hrouzek, P., Kopecký, J., Tůmová, L., & Cheel, J., (2021). High-performance countercurrent chromatography for lutein production from a chlorophyll-deficient strain of the microalgae *Parachlorella kessleri* HY1. *Journal of Applied Phycology*, *33*(4), 1999–2013. https://doi.org/10.1007/s10811-021-02434-y.
90. Hu, S., Fan, X., Qi, P., & Zhang, X., (2019). Identification of anti-diabetes peptides from spirulina platensis. *Journal of Functional Foods*, *56*, 333–341. https://doi.org/10.1016/j.jff.2019.03.024.
91. Kim, H. S., Fernando, I. P. S., Lee, S. H., Ko, S. C., Kang, M. C., Ahn, G., Je, J. G., et al., (2021). Isolation and characterization of anti-inflammatory compounds from *Sargassum horneri* via high-performance centrifugal partition chromatography and high-performance liquid chromatography. *Algal Research*, *54*, 102209. https://doi.org/10.1016/j.algal.2021.102209.
92. López, G., Yate, C., Ramos, F. A., Cala, M. P., Restrepo, S., & Baena, S., (2019). Production of polyunsaturated fatty acids and lipids from autotrophic, mixotrophic and heterotrophic cultivation of *Galdieria* sp. strain USBA-GBX-832. *Scientific Reports*, *9*(1), 1–13. https://doi.org/10.1038/s41598-019-46645-3.
93. Neto, A. M. P., Sotana De, S. R. A., Leon-Nino, A. D., Da Costa, J. D. A., Tiburcio, R. S., Nunes, T. A., Sellare De, M. T. C., et al., (2013). Improvement in microalgae lipid extraction using a sonication-assisted method. *Renewable Energy*, *55*, 525–531. https://doi.org/10.1016/j.renene.2013.01.019.

94. Kumar, B. R., Deviram, G., Mathimani, T., Duc, P. A., & Pugazhendhi, A., (2019). Microalgae as rich source of polyunsaturated fatty acids. *Biocatalysis and Agricultural Biotechnology, 17*(2018), 583–588. https://doi.org/10.1016/j.bcab.2019.01.017.
95. Adarme-Vega, T. C., Lim, D. K. Y., Timmins, M., Vernen, F., Li, Y., & Schenk, P. M., (2012). Microalgal biofactories: A promising approach towards sustainable omega-3 fatty acid production. *Microbial Cell Factories, 11*, 1–10. https://doi.org/10.1186/1475-2859-11-96.
96. Chauton, M. S., Reitan, K. I., Norsker, N. H., Tveterås, R., & Kleivdal, H. T., (2015). A techno-economic analysis of industrial production of marine microalgae as a source of EPA and DHA-rich raw material for aquafeed: Research challenges and possibilities. *Aquaculture, 436*, 95–103. https://doi.org/10.1016/j.aquaculture.2014.10.038.
97. Lee, J. Y., Yoo, C., Jun, S. Y., Ahn, C. Y., & Oh, H. M., (2010). Comparison of several methods for effective lipid extraction from microalgae. *Bioresource Technology, 101*, S75–S77. https://doi.org/10.1016/j.biortech.2009.03.058.
98. Pérez, J. P., Muñoz, A. A., Figueroa, C. P., & Agurto-Muñoz, C., (2021). Current analytical techniques for the characterization of lipophilic bioactive compounds from microalgae extracts. *Biomass and Bioenergy, 149*(2020). https://doi.org/10.1016/j.biombioe.2021.106078.
99. Du, X., Smirnov, A., & Jia, W., (2018). A preparatory study of how to construct consensus mass spectra of recurrent unknown metabolites from untargeted GC–MS metabolomics data. *International Journal of Mass Spectrometry, 427*, 73–78. https://doi.org/10.1016/j.ijms.2017.09.015.
100. Guidetti, V. R., Bittencourt, F. M., Jacob-Lopes, E., & Wagner, R., (2019). Analytical strategies for using gas chromatography to control and optimize microalgae bioprocessing. *Current Opinion in Food Science, 25*, 73–81. https://doi.org/10.1016/j.cofs.2019.02.008.
101. Lafarga, T., Acién-Fernández, F. G., & Garcia-Vaquero, M., (2020). Bioactive peptides and carbohydrates from seaweed for food applications: Natural occurrence, isolation, purification, and identification. *Algal Research, 48*. https://doi.org/10.1016/j.algal.2020.101909.
102. Bisht, B., Kumar, V., Gururani, P., Tomar, M. S., Nanda, M., Vlaskin, M. S., Kumar, S., & Kurbatova, A., (2021). The potential of nuclear magnetic resonance (NMR) in metabolomics and lipidomics of microalgae: A review. *Archives of Biochemistry and Biophysics, 710*, 108987. https://doi.org/10.1016/j.abb.2021.108987.
103. Wishart, D. S., (2008). Quantitative metabolomics using NMR. *TrAC – Trends in Analytical Chemistry, 27*(3), 228–237. https://doi.org/10.1016/j.trac.2007.12.001.
104. Keddar, M. N., Ballesteros-Gómez, A., Amiali, M., Siles, J. A., Zerrouki, D., Martín, M. A., & Rubio, S., (2020). Efficient extraction of hydrophilic and lipophilic antioxidants from microalgae with supramolecular solvents. *Separation and Purification Technology, 251*, 117327. https://doi.org/10.1016/j.seppur.2020.117327.
105. Bauer, A., & Minceva, M., (2019). Direct extraction of astaxanthin from the microalgae: *Haematococcus pluvialis* using liquid-liquid chromatography. *RSC Advances, 9*(40), 22779–22789. https://doi.org/10.1039/c9ra03263k.
106. Di Lena, G., Casini, I., Lucarini, M., & Lombardi-Boccia, G., (2019). Carotenoid profiling of five microalgae species from large-scale production. *Food Research International, 120*, 810–818. https://doi.org/10.1016/j.foodres.2018.11.043.

107. Vendruscolo, R. G., Facchi, M. M. X., Maroneze, M. M., Fagundes, M. B., Cichoski, A. J., Zepka, L. Q., Barin, J. S., et al., (2018). Polar and non-polar intracellular compounds from microalgae: Methods of simultaneous extraction, gas chromatography determination and comparative analysis. *Food Research International*, *109*, 204–212. https://doi.org/10.1016/j.foodres.2018.04.017.
108. Cerón-García, M. C., González-López, C. V., Camacho-Rodríguez, J., López-Rosales, L., García-Camacho, F., & Molina-Grima, E., (2018). Maximizing carotenoid extraction from microalgae used as food additives and determined by liquid chromatography (HPLC). *Food Chemistry*, *257*(2017), 316–324. https://doi.org/10.1016/j.foodchem.2018.02.154.
109. Aluç, Y., Başaran, K. G., & Tüzün, İ., (2018). Determination of carotenoids in two algae species from the saline water of Kapulukaya reservoir by HPLC. *Journal of Liquid Chromatography and Related Technologies*, *41*(2), 93–100. https://doi.org/10.1080/10826076.2017.1418376.
110. Carpena, M., Caleja, C., Pereira, E., Pereira, C., Ćirić, A., Soković, M., Soria-Lopez, A., et al., (2021). Red seaweeds as a source of nutrients and bioactive compounds: Optimization of the extraction. *Chemosensors*, *9*(6). https://doi.org/10.3390/chemosensors9060132.
111. Da Costa, E., Melo, T., Reis, M., Domingues, P., Calado, R., Abreu, M. H., & Domingues, M. R., (2021). Polar lipids composition, antioxidant, and anti-inflammatory activities of the Atlantic red seaweed *Grateloupia turuturu*. *Marine Drugs*, *19*(8), 1–24. https://doi.org/10.3390/md19080414.
112. Ristivojević, P., Jovanović, V., Opsenica, D. M., Park, J., Rollinger, J. M., & Velicković, T. Ć., (2021). Rapid analytical approach for bioprofiling compounds with radical scavenging and antimicrobial activities from seaweeds. *Food Chemistry*, *334*(2019). https://doi.org/10.1016/j.foodchem.2020.127562.
113. Nørskov, N. P., Bruhn, A., Cole, A., & Nielsen, M. O., (2021). Targeted and untargeted metabolic profiling to discover bioactive compounds in seaweeds and hemp using gas and liquid chromatography-mass spectrometry. *Metabolites*, *11*(5). https://doi.org/10.3390/metabo11050259.
114. Smith, B. C., Bornancin, L., Lu, Y., Simonsen, H. T., Martínez-Valladares, M., Peña-Espinoza, M., Mejer, H., et al., (2021). Bio-Guided fractionation and molecular networking reveal fatty acids to be principal anti-parasitic compounds in Nordic seaweeds. *Frontiers in Pharmacology*, *12*, 1–12. https://doi.org/10.3389/fphar.2021.674520.
115. Vilcanqui, Y., Mamani-Apaza, L. O., Flores, M., Ortiz-Viedma, J., Romero, N., Mariotti-Celis, M. S., & Huamán-Castilla, N. L., (2021). Chemical characterization of brown and red seaweed from Southern Peru, a sustainable source of bioactive and nutraceutical compounds. *Agronomy*, *11*(8), 1669. https://doi.org/10.3390/agronomy11081669.
116. Hossain, M. S., Sifat, S. A. D., Hossain, M. A., Salleh, S., Hossain, M., Akter, S., & Hossain, M. B., (2021). Comparative assessment of bioactive compounds, antioxidant capacity, and nutritional quality of red seaweeds and water spinach. *Regional Studies in Marine Science*, *46*, 101878. https://doi.org/10.1016/j.rsma.2021.101878.
117. Ferreira, J., Hartmann, A., Martins-Gomes, C., Nunes, F. M., Souto, E. B., Santos, D. L., Abreu, H., et al., (2021). Red seaweeds strengthening the nexus between nutrition and health: Phytochemical characterization and bioactive properties of *Grateloupia turuturu* and *Porphyra umbilicalis* extracts. *Journal of Applied Phycology*, *0123456789*. https://doi.org/10.1007/s10811-021-02529-6.

118. Jerković, I., Cikoš, A. M., Babić, S., Čižmek, L., Bojanić, K., Aladić, K., Ul'yanovskii, N. V., et al., (2021). Bioprospecting of less-polar constituents from endemic brown macroalga *Fucus virsoides J. Agardh* from the Adriatic Sea and targeted antioxidant effects in vitro and in vivo (zebrafish model). *Marine Drugs*, *19*(5), 235. https://doi.org/10.3390/md19050235.
119. Zhong, B., Robinson, N. A., Warner, R. D., Barrow, C. J., Dunshea, F. R., & Suleria, H. A. R., (2020). LC-ESI-QTOF-MS/MS characterization of seaweed phenolics and their antioxidant potential. *Marine Drugs*, *18*(6), 1–21. https://doi.org/10.3390/md18060331.
120. Dias, V., Bandeira, S., Chaúque, E., Lipassula, M., & Mussagy, A., (2020). Evaluation of phytocompounds and chemical elements present in selected species of seaweeds, to sustain future quantitative analysis for bioactive compounds. *Journal of Drug Delivery and Therapeutics*, *10*(5), 232–239. https://doi.org/10.22270/jddt.v10i5.4422.

CHAPTER 6

Technological and Analytical Aspects of Bioactive Compounds and Nutraceuticals from Seaweeds

FARHAN MOHIUDDIN BHAT,[1] YADWINDER SINGH DHALIWAL,[1] SARANA ROSE SOMMANO,[2] and CHARANJIT S. RIAR[3]

[1]*CSK Himachal Pradesh Agricultural University, Palampur. Himachal Pradesh, India*

[2]*Chiang Mai University, Bangkok. Thailand*

[3]*Sant Longowal Institute of Engineering and Technology, Sangrur, Punjab, India*

ABSTRACT

Seaweeds being primary producers are composed of unicellular and multicellular forms that provide habitat and food to all other fauna in the marine bodies during their life cycle of growth, multiplication, and disintegration. About 90% of the marine seaweeds comprise of algae and it contributes nearly 50% of the photosynthesis in the whole world. Seaweeds are exported throughout the world in dried, pickled, semi-processed, and processed forms for consumption purposes. These are also marketed in extracted forms like agar, carrageen, alginates, algin, and other phycocolloidal forms that are finding application in dairy industries, pharmaceuticals, paints, surgical items, cosmetics, and many more as a stabilizing and thickening agent. Besides, these are also rich sources of iron, calcium, magnesium, and iodine that are essential for the performance of vital body functions, including bone formation, thyroid functioning, muscle building, and oxygen transportation. Seaweeds are being used as food, medicines, cattle feed, and fertilizers.

Bioactive Compounds and Nutraceuticals from Dairy, Marine, and Nonconventional Sources: Extraction Technology, Analytical Techniques, and Potential Health Prospects.
Charanjit Singh Riar, Parmjit S. Panesar (Eds.)

Seaweeds are the reservoirs of numerous bioactive components not available in terrestrial plants that possess health benefits superior to land-grown plants

6.1 INTRODUCTION

"Seaweed" is the generic name given to numerous species of marine plants and algae grown naturally in different water bodies like oceans, rivers, lakes, springs, and other sources. These are present as microscopic and giant macroscopic kelp-like forests that provide the vase for the existence of extensive marine food webs. Most of the seaweeds exist as medium-sized marine plankton possessing diverse shades of colors like black, red, green, brown, and yellow, providing habitat and food to marine creatures and inhabitants residing around the shoreline areas [1]. Some of the marine plankton, like planktonic algae, drifting plants, and photosynthetic bacteria, ensure the natural balance of atmospheric constituents by generating about 50% of global oxygen and utilizing atmospheric carbon dioxide, thereby resulting in a reduction of nutrient pollution and acidification of localized water bodies [2]. Thus, the global farming of seaweed has an enhanced potential for mitigating climate change due to their carbon negative tendency.

The cultivation and harvesting of seaweeds for human uses has been practiced from prehistoric times and has now opened avenues for global agricultural operations for manufacturing food materials, natural chemicals, biofuels, fertilizers, and livestock feeds. China, Japan, Indonesia, Malaysia, the Philippines, North Korea, South Korea, and Tanzania are the leading seaweed-producing countries that have boosted the economies of these nations along with mitigating pressure on overexploiting fishing [3]. Oceans, comprising about 71% of the earth's surface and 97% of the water, have become a suitable alternative to arable land for meeting the ever-increasing feeding demands of the growing population mainly in Asia and Africa. Seaweeds preserve the diversity of localized fishes and invertebrates by providing an additional niche and restoring coral reefs [4]. Due to their impact on maintaining the stability of atmospheric gases, seaweeds are playing a crucial part in the maintenance of the global ecology. The physiological activities of seaweeds are confirmed to be the source of roughly 90% of the oxygen and 80% of the organic compounds on the earth's surface. The major seaweed derivatives Carrageenan, alginates, and Agar are finding applications in various industrial food recipes, pharmaceuticals, and cosmetics [5]. Seaweeds are rich sources of naturally producing iodine that meets the daily requirement of iodine, especially in countries consuming seaweeds for food. Seaweed-extracted organic iodine is being used as a nutritional supplement in common salt to prevent iodine deficiency in the general populace and to promote healthy functioning of the thyroid gland.

The pigmented components in seaweeds, including chlorophylls, carotenoids, and phycobiliproteins, are associated with antioxidant properties [6]. Seaweeds are globally exported in various forms such as dried, pickled, semi-processed, and processed for consumption purposes. Additionally, they are also marketed in extracted forms like agar, carrageen, alginates, algin, and other phycocolloidal forms. These extracted forms are used as stabilizing and thickening agents in industries such as dairy, pharmaceuticals, paints, surgical items, cosmetics, and more. Seaweeds are not only valuable for their functional properties, but they are also rich sources of essential nutrients including iron, calcium, magnesium, and iodine. These nutrients play crucial roles in vital body functions such as bone formation, thyroid functioning, muscle building, and oxygen transportation. Moreover, seaweeds are utilized as food, medicines, cattle feed, and fertilizers in various Asian and European maritime countries, including Ireland, Wales, and Tanzania [7]. Algal foods offer one of the few vegetarian alternatives for cobalamin (vitamin B_{12}) in the diet. As vitamin B_{12} (Cobalamin) is not synthesized by terrestrial plants due to which fruits and vegetables are not good sources of vitamin B_{12}. Seaweeds mainly algal planktons are good sources of cobalamin and thus serve as a best alternative for vegetarians to eradicate the deficiency of vitamin B_{12} [8]. Seaweed aquaculture has been found to be an effective tool in reducing eutrophication problems and maintaining a healthy and stable ecosystem in marine environments [9]. This system of integrated farming, along with seaweed aquaculture, leads to a reduction of wastewater generation by absorbing inorganic wastes like ammonia and nitrates from marine bodies.

6.1.1 BIOACTIVE SEAWEEDS APPLICATIONS IN FUNCTIONAL FOOD FORMULATIONS

Seaweeds are the reservoirs of numerous bioactive components not available in terrestrial plants that possess health benefits superior to land-grown plants. Due to changing lifestyles, consumers are focusing mainly on naturally grown functional food alternatives, so the application of seaweed-extracted bioactive components in formulation of functional foods and nutraceuticals are gaining immense popularity in modern era beyond the basic nutrition.

The seaweed derived bioactives have been found to rectify different types of nutritional deficiencies, promote optimal health, and reduce the risk of chronic diseases. The bioactive components derived from seaweeds have been found to influence the biological functions of living beings by modifying their physical, chemical, and biological mechanisms [10]. Seaweeds are rich sources of calcium, iron, magnesium, potassium, zinc, folate, copper, and other minerals and vitamins, that are having nutritional benefits and are essential for healthy functioning of the body. The bioactive compounds in seaweeds have been found effective in killing of carcinogenic cells [11].

The extraction of proteins, amino acids, Polysaccharides, fatty acids, vitamins, minerals, phenols, terpenoids, and pigmented components from various seaweeds have been found to possess excellent antioxidant, antiviral, antibacterial, anticancer, anticoagulant, antidiabetic, antiallergic, anti-inflammatory, antihypertensive, and radioprotective properties [12]. The presence of Sulfated polysaccharides in seaweeds have been found to possess anticoagulant, antiviral, and immunoinflammatory properties that finds in application in formulation of functional foods and nutraceuticals [13]. Functional oligosaccharides derived from seaweeds have been recognized to proliferate normal flora of gastrointestinal tract (GIT) along with suppression of pathogenic organisms. It has also been found effective in preventing dental caries, boosting of immunity, enhancing mineral absorption and retains the ability of antioxidant, antibiotic, anti-diabetic by reducing hike in blood glucose level [14]. The presence of xanthophyll, fucoxanthin in seaweed has been associated with enhancing metabolism by burning of fat cells in adipose tissue, reducing of oxidative stresses, killing of cancerous cells by arresting cell cycle, aids in antiproliferation, inducing apoptosis, and suppressing angiogenesis and thus retains potential applications in functional food formulation. The extraction of phloroglucinol-based phlorotannin compounds from various seaweeds (brown and red algae) are excellent ingredients in function food formations due to its unique health benefits like antioxidant properties, anti-HIV activity, antidiabetic, antiproliferative, anti-inflammatory, radioprotective, antihypertensive, anti-Alzheimer disease, antimicrobial, and anti-matrix metalloproteinase ability [15].

Seaweeds contain carotenoids, flavonoids, vitamins A, C, and E, that are potent antioxidants and protects the body from cell damage due to free radicals. Some seaweed is also a rich source of omega-3 fatty acids and vitamin B_{12}. The presence of natural fiber alginate in seaweeds has been found to restrict the absorption of fat from the human intestinal gut and thus play an important role in the reduction of weight in obese persons. Thus, the application of alginates as thickening agents is suited for the formulation of food products for overweighed or obesity patients. The main bioactive component in seaweed, carrageenan, has the ability to form a gel that the body cannot digest and thus functions like a soluble fiber. This property of carrageenan gives a fuller feeling in the stomach for a longer time and ensures a reduction in body weight and cholesterol by lowering fat absorption, accelerating the breakdown of stored fat, and enhancing fat burning of fatty tissues. It also acts as a prebiotic for microbial organisms residing in the gut lining of the human body and decreases the population of harmful species such as *Clostridium septicum* and *Streptococcus pneumonia* [16]. Seaweeds are rich in essential fatty acids mainly eicosapentaenoic acid and docosahexaenoic acid, that are having an important role performing various physiological functions in body like cellular signaling, immunomodulation, brain development, regulation of transcription factors, prevention of some cancer, neurodegenerative, cardiovascular, and autoimmune diseases [17].

Seaweeds are excellent sources of health-endorsing molecules as compared to terrestrial plants in respect of ω-3 fatty acids, amino acids, vitamins A, B, C, and E all of which are considered essential bases for cosmeceuticals. The heavily sulfated green seaweed polysaccharides ulvans extracted from species of the Ulvales retains bioactive properties due to their hydrophilic polyanionic features and have been reported to possess antioxidant, antimicrobial, anticoagulant, antiviral, anticancerous, anti-hyperlipidemic, and immunoregulatory effects [18]. Astaxanthin is an important carotenoid produced by the freshwater green algae *Haematococcus pluvialis* that has been found to be useful in nutraceutical formation due to its highly antioxidant properties, and protective properties against metabolic syndrome, cancer, cardiac heart diseases, inflammatory diseases, diabetes, and neurodegenerative diseases [19]. Seaweeds contain brightly colored water-soluble Phycobiliproteins proteins that consist of four major types as per their color and absorption ability, with allophycocyanin having bluish green color, red pigmented consists of phycoerythrin, phycocyanin is blue pigmented and phycoerythrocyanin retains orange pigmentation [20]. These phycobiliproteins have been regarded to possess pharmacological properties, including antioxidant, enhancing immune system and reduction of cardiovascular diseases (CVDs), cancers, and age-related disorders like multiple sclerosis and cataract [21]. Seaweeds are promising natural resources possessing bioactive components recognized as polysaccharides, fatty acids, carotenoids, glycolipids, steroids, phenolics, flavonoids, and terpenoids. These components are associated with extensive biological functions, including antioxidant, anti-viral, phytotoxic, antimalarial, insecticidal, antibacterial, and cytotoxic properties. The carotenoids found in seaweeds consist of mainly β-carotene, fucoxanthin, and tocopherol, all of which exhibit highly antioxidative properties by quenching of singlet oxygen species and scavenging of free radical species [6].

Seaweeds are consumed as substitutes of many food products that are labeled as fat-free, gluten-free, mineral rich, vitamin-rich, low sugar and low caloric foods. The seaweeds are used as salt replacers, reducing the risk of chronic ailment caused by NaCl consumption. The substitution of 1% seaweeds of *Undaria pinnatifida, Porphyra umbilicalis, Palmaria palmate* has been found to reduce the salt in frankfurters and cooked hams up to 50% without affecting the texture and other sensory attributes of the structured food products [22]. Seaweeds, being rich sources of proteins, can be utilized in the formulation of high protein-rich food items. The seaweed *Porphyra tenera and Porphyra yezoensis* have been validated to possess higher protein content than legumes, and these proteins have also found application as coloring agents in food recipes, such as phycoerythrin, from red seaweeds. These proteins can be used as alternative to animal proteins for formulation of athletic diets [23] (Figures 6.1 and 6.2; Tables 6.1 and 6.2).

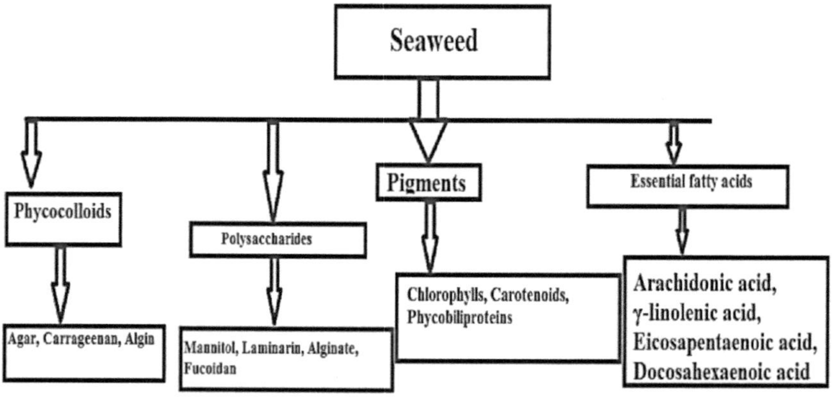

FIGURE 6.1 Bioactive components extracted from seaweeds.

6.2 TECHNIQUES FOR EXTRACTION OF BIOACTIVE COMPOUNDS FROM SEAWEEDS

6.2.1 SUPERCRITICAL FLUID EXTRACTION (SFE)

Supercritical fluid extraction (SFE) is based on the principle of extraction with carbon dioxide and water as supercritical fluids that are raised above

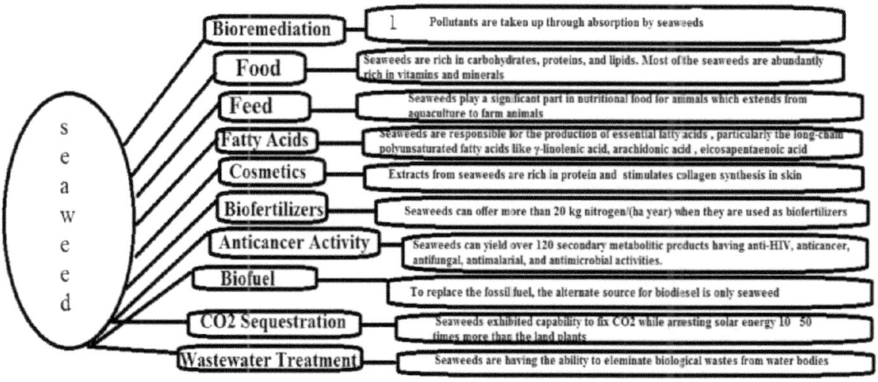

FIGURE 6.2 Biological importance of seaweed.

their critical pressure and temperature to penetrate deeply into the seaweed cells and enhance extraction. The temperature utilized in this method usually ranged from 40–90°C with pressure conditioned between 100 and 400 bar depending on the type of fluid and morphological characteristics of the solid matrix. It is an advanced technique to separate bioactive compounds from seaweed utilizing enhanced solvating power of fluids above their critical point, resulting in higher purity of the extracting compounds. SFE has been found to be effective in extracting low polar seaweed phytochemicals including polyphenols, fatty acids, tocopherols, carotenoids, and fucoxanthin that are finding application in the different sectors of food and pharmaceutical industries [38].

6.2.2 ULTRASOUND-ASSISTED EXTRACTION (UAE)

Ultrasound-assisted extraction (UAE) is an extraction technique using ultrasound waves that propagate as compression and rarefaction waves at high frequencies greater than 20 KHz up to 100 kHz. The passage of ultrasound waves through the extracting matrix induces the creation of bubbles at high and low pressure, resulting in acoustic cavitation. This process of cavitation creates an increasing contact surface area between the solvent and seaweed, resulting in better penetration of solvent and faster diffusion of solute. The implosion of bubbles during cavitation creates micro-turbulence inducing mechanical energy that causes lyses of cell walls and efficient extraction of bioactive compounds from biological matrix. This technique of extraction

TABLE 6.1 Bioactive Compounds from Seaweeds and Their Functions

Seaweed	Bioactive Components	Class	Functional Uses	References
Brown seaweed, sea urchins, sea cucumbers	Fucoidan	Polysaccharide	Anticoagulant and antithrombotic, antivirus, antitumor, and immunomodulatory, anti-inflammatory, blood lipids reducing, antioxidant, and anticomplementary properties, activity against hepatopathy, uropathy, and renalpathy and gastric protective effects.	[24]
Brown macroalgae, *Fucus* spp., golden-brown, unicellular microalgae, the diatoms.	Fucoxanthin	Xanthophyll	Induce G1 cell-cycle arrest and apoptosis in various cancer cell lines and tumor growth. Fucoxanthin also reduces weight and improves blood lipid profiles.	[25]
Ascophyllum nodosum, *Laminaria digitata*, L. hyperborea and *Fucus serratus*	Betadine	Trimethylglycine	Helps in the metabolism of a chemical called homocysteine. Homocysteine is involved in the normal function of many different parts of the body, including blood, bones, eyes, heart, muscles, nerves, and the brain.	[26]
Brown algae	Alginate	Polysaccharide	Alginate has several applications in the pharmaceutical industry such as formation of gels, stabilizing agents, and localized drug delivery.	[27]
Brown seaweeds	Phlorotannins	Polyphenolic compounds	Used as functional food ingredients with good health effects, including anti-HIV, antiproliferative, anti-inflammatory, radioprotective, antidiabetic, anti-Alzheimer disease, antimicrobial, antihypertensive, and anti-matrix metalloproteinase activities.	[28]
Cyanobacteria, rhodophytes, cryptomonads, glaucocystophytes	Phycobiliproteins	Proteins	Anticancer, antioxidant, antiproliferative, antiallergic applications in the food, pharmaceutical, and cosmetic industries	[29]
Wakame	Folates	Vitamins	Folate helps to form DNA and RNA and is involved in protein metabolism. It plays a key role in breaking down homocysteine, an amino acid that can exert harmful effects in the body if it is present in high amounts. Folate is also needed to produce healthy red blood cells.	[30]

TABLE 6.2 Some of the Common Genera of Seaweeds and Their Uses

Genus	Pictographic Description	Phylum	Uses	References
Caulerpa		*Chlorophyta*	*Caulerpa contains a high amount of iron, magnesium, and* calcium. The main pigments of Caulerpa are chlorophyll a and b. It has *a high diversity of chemical compounds which have pharmaceutical potential.*	[31]
Fucus		*Ochrophyta*	*Fucus can attack and kill the methicillin-resistant Staphylococcus aureus super bacterium. It has an alkali content of about 2.5%–5% that is mainly sodium carbonate, used in soap making, glassmaking, and other industries.*	[32]
Gracilaria		*Rhodophyta*	***Gracilaria* sp. polysaccharides have been shown to be an effective prophylactic agent during *in vitro* and *in vivo*** *experiments against Japanese encephalitis viral infection. The sulfated oligo-saccharides from Gracilaria sp. seem to be promising candidates for further development as antiviral agents.*	[33]

TABLE 6.2 *(Continued)*

Genus	Pictographic Description	Phylum	Uses	References
Laminaria		*Ochrophyta*	*Laminaria is a source of the relatively rare element iodine, which is commonly used to promote thyroid health. The soluble form of Laminarin has high antitumor activity, can be used in the treatment of antibiotic-resistant nosocomial infections.*	[34]
Macrocystis		*Ochrophyta*	*The ash of such seaweeds was an important source of potash and iodine. Many kelps produce algin, a complex carbohydrate (polysaccharide) useful in various industrial processes, including tire manufacture.*	[35]
Monostroma	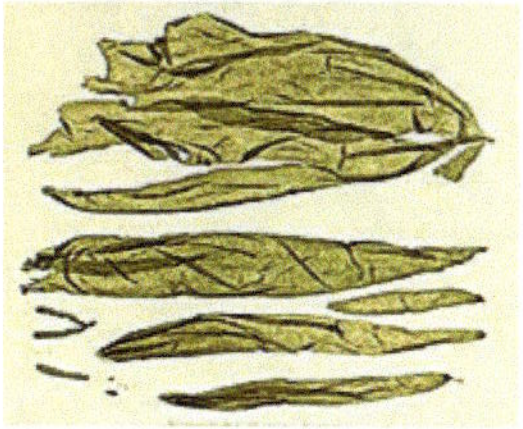	*Chlorophyta*	This is an edible green alga extensively used as nori in sushi.	[7]

TABLE 6.2 *(Continued)*

Genus	Pictographic Description	Phylum	Uses	References
Porphyra	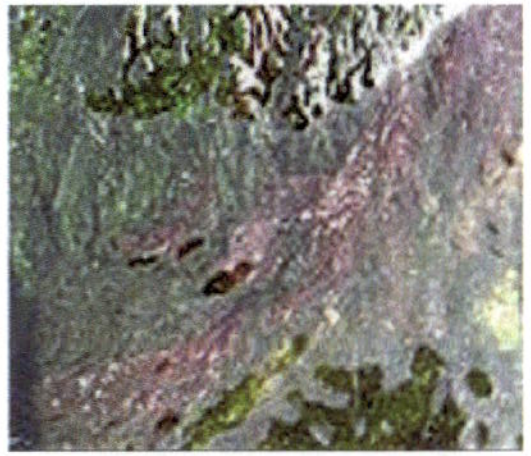	*Rhodophyta*	*The marine red alga Porphyra has been cultivated extensively in many Asian countries as an edible seaweed used to wrap the rice and fish that compose the Japanese food sushi and the Korean food gimbap.*	[36]
Sargassum		*Ochrophyta*	*Sargassum is rich in iodine, bromine, mineral salts, and vitamins, and that also contains properties that benefit the absorption of fats. The polysaccharides found in sargassum kelp helps to control blood sugar and blood pressure. Sargassum has high antibacterial and antioxidant properties.*	[37]

operates at lower temperature that in turn ensures lesser damage of thermo labile bioactive compounds [39].

6.2.3 *PRESSURIZED SOLVENT EXTRACTION (SE)*

Pressurized solvent extraction (SE) or accelerated SE is an extraction technique employing elevated temperature and pressure of 50–300°C and 35–200 bar for solvents to achieve higher extraction efficiency. This process requires shorter extraction time and is non-hazardous. The elevated temperature increases the extraction rate by enhancing solvent penetration through decreasing density, surface tensions, and viscosity, while the higher pressure maintains the solvent in liquid form at a higher temperature, resulting in improved extraction efficiency by increasing mass transfer in the seaweed cells [40].

6.3 CONCLUSION

Seaweeds ensure the natural balance of atmospheric constituents by generating about 50% of global oxygen and utilizing of atmospheric carbon dioxide, thereby resulting in reduction of nutrient pollution and acidification of localized water bodies. Seaweeds are rich in essential fatty acids, such as eicosapentaenoic acid and docosahexaenoic acid, which play important roles in various physiological functions in the body. These functions include cellular signaling, immunomodulation, brain development, regulation of transcription factors, and prevention of certain cancers, neurodegenerative diseases, cardiovascular diseases, and autoimmune diseases. Seaweeds are also excellent sources of health-promoting molecules compared to terrestrial plants, providing ω-3 fatty acids, amino acids, and vitamins A, B, C, and E, which are all essential for cosmeceuticals. Additionally, seaweeds can be consumed as substitutes for various food products, offering benefits such as being fat-free, gluten-free, mineral-rich, vitamin-rich, low in sugar, and low in calories. They can also be used as salt replacers, reducing the risk of chronic ailments caused by excessive sodium chloride consumption. Seaweeds contain carotenoids, flavonoids, vitamins A, C, and E, that are potent antioxidants and protects the body from cell damage due to free radicals.

KEYWORDS

- **acidification**
- **eicosapentaenoic acid**
- **free radicals**
- **functional food**
- **seaweeds**
- **solvent extraction**
- **ω-3 fatty acids**

REFERENCES

1. James, William, D., Berger, & Timothy, G., (2006). *Andrews' Diseases of the Skin: In Clinical Dermatology* (pp. 25–28). Saunders Elsevier.
2. Duarte, Carlos, M., Wu, J., Xiao, X., Bruhn, A., & Krause-Jensen, D., (2017). Can seaweed farming play a role in climate change mitigation and adaptation? *Frontiers in Marine Science, 4.*
3. Ask, E. I., (1990). *Cottonii and Spinosum Cultivation Handbook* (p. 52). Philippines: FMC Biopolymer Corporation.
4. Zertruche-Gonzalez, & Jose, A., (1997). *Coral Reefs: Challenges and Opportunities for Sustainable Management.* The World Bank.8213-4235-5.
5. Siahaan, E. A., Pangestuti, R., & Se-Kwon, K., (2018). Seaweeds: Valuable ingredients for the pharmaceutical industries. In: Rampelotto, P. H., & Trincone, A., (eds.), *Grand Challenges in Marine Biotechnology* (pp. 49–95). Springer International Publishing.
6. Holdt, S. L., & Kraan, S., (2011). Bioactive compounds in seaweed: Functional food applications and legislation. *Journal of Applied Phycology, 23,* 543597.
7. Shields, R. J., & Lupatsch, I., (2012). *Algae for Aquaculture and Animal Feeds* (Vol. 21, p. 2337). Technikfol-Genabscha¨tzung Theor. Praxis.
8. Waldmann, A., Koschizke, J. W., Leitzmann, C., & Hahn, A., (2004). Homocysteine and cobalamin status in German vegans. *Public Health Nutrition, 7*(3), 467–472.
9. Chung, I. K., Kang, Y. H., Yarish, C., Kraemer, G. P., & Lee, J. A., (2002). Application of seaweed cultivation to the bioremediation of nutrient-rich effluent. *Algae, 17,* 187–194.
10. Gupta, V., Kumar, M., Brahmbhatt, H., Reddy, C., Seth, A., & Jha, B., (2011). Simultaneous determination of different endogenetic plant growth regulators in common green seaweeds using dispersive liquid–liquid microextraction method. *Plant Physiology and Biochemistry. 49*(11), 1259–1263.
11. Shanab, S. M. M., Mostafa, S. S. M., Shalaby, E. A., & Mahmoud, G. I., (2012). Aqueous extracts of microalgae exhibit antioxidant and anticancer activities. *Asian Pacific Journal of Tropical Biomedicine, 2*(8), 608–615.

12. Kim, S., (2013). In: Kim, S., (ed.), *Marine Nutraceuticals: Prospects and Perspectives* (p. 464). CRC press, Switzerland.
13. Zhang, L., Gionfriddo, E., Acquaro, V., & Pawliszyn, J., (2018). Direct immersion solid-phase microextraction analysis of multi-class contaminants in edible seaweeds by gas chromatography-mass spectrometry. *Analytica Chimica Acta, 1031*, 83–97.
14. Steevensz, A. J., Mackinnon, S. L., Hankinson, R., Craft, C., Connan, S., Stengel, D. B., & Melanson, J. E., (2012). Profiling phlorotannins in brown macroalgae by liquid 366 sustainable seaweed technologies chromatography-high resolution mass spectrometry. *Phytochemical Analysis, 23*(5), 547–553.
15. Terme, N., Boulho, R., Kucma, J. P., Bourgougnon, N., & Bedoux, G., (2018). Radical scavenging activity of lipids from seaweeds isolated by solid-liquid extraction and supercritical fluids. *Oilseeds and Fats, Crops and Lipids, 25*(5).
16. Inoue, N., Yamano, N., Sakata, K., Nagao, K., Hama, Y., & Yanagita, T., (2009). The sulfated polysaccharide porphyran reduces apolipoprotein B100 secretion and lipid synthesis in HepG2 cells. *Bioscience Biotechnology and Biochemistry, 73*, 447–449.
17. Gonçalves, A. M. M., Marques, J. C., & Gonçalves, F., (2017). Fatty acids' profiles of aquatic organisms: Revealing the impacts of environmental and anthropogenic stressors. In: Catala, A., (ed.), *Fatty Acids* (p. 68544). Intech Open; London, UK.
18. Mao, W. J., Fang, F., Li, H. Y., Qi, X. H., Sun, H. H., Chen, Y., & Guo, S. D., (2008). Heparinoid-active two-sulfated polysaccharides isolated from marine green algae *Monostroma nitidum. Carbohydrate Polymer, 74*, 834–839.
19. Cysewski, G. R., & Lorenz, R. T., (2004). Industrial production of microalgal cell-mass and secondary products-species of high potential: *Haematococcus*. In: Richmond, A., (ed.), *Handbook of Microalgal Culture, Biotechnology and Applied Phycology* (p. 281288). Blackwell Science, Oxford, UK.
20. Lee, S. H., Kang, S. M., Ko, S. C., Lee, D. H., & Jeon, Y. J., (2012). Octaphlorethol A, a novel phenolic compound isolated from a brown alga, *Ishige* foliacea, increases glucose transporter 4-mediated glucose uptake in skeletal muscle cells. *Biochemical and Biophysical Research Communications, 420*(3), 576–581.
21. Pape, M. A. L., Grua-Priol, J., Prost, C., & Demaimay, M., (2004). Optimization of dynamic headspace extraction of the edible red algae *Palmaria palmata* and identification of the volatile components. *Journal of Agriculture and Food Chemistry, 52*(3), 550–556.
22. Vilar, E. G., Ouyang, H., Sullivan, M. G., Kerry, J. P., Hamill, R. M., Grady, M. N., Mohammed, H. O., & Kilcawley, K. N., (2020). Effect of salt reduction and inclusion of 1% edible seaweeds on the chemical, sensory and volatile component profile of reformulated frankfurters. *Meat Science, 161*, 108001.
23. Fleurence, J., Morançais, M., & Dumay, J., (2018). *Seaweed Proteins* (2nd edn.) Elsevier Ltd.: Amsterdam, the Netherlands. ISBN 9780081007297.
24. Maeda, H., (2015). "Nutraceutical effects of fucoxanthin for obesity and diabetes therapy: A review. *Journal of Oleo Science, 64*(2), 125–132.
25. Kanazawa, K., Ozaki, Y., Hashimoto, T., Das, S. K., Matsushita, S., Hirano, M., & Nakatsuka, M., (2008). Commercial-scale preparation of biofunctional fucoxanthin from waste parts of brown sea algae *Laminalia japonica. Food Science and Technology Research, 14*(6), 573–582.
26. Calvo, P., Nelson, L., & Kloepper, J. W., (2014). Agricultural uses of plant biostimulants. *Plant Soil, 383*(1, 2), 3–41.

27. Venugopal, V., (2011). *Marine Polysaccharides: Food Applications* (p. 396). CRC Press Taylor and Francis Group, Boca Raton, FL, USA.
28. Shin, T., Ahn, M., Hyun, J. W., Kim, S. H., & Moon, C., (2014). Antioxidant marine algae phlorotannins and radioprotection: A review of experimental evidence. *Acta Histochemistry, 116*, 669–674.
29. Liu, H. T., Li, Y. F., Luan, T. G., Lan, C. Y., & Shu, W. S., (2007). Simultaneous determination of phytohormones in plant extracts using SPME and HPLC. *Chromatographia, 66*(7, 8), 515–520.
30. Thomas, P. M., Flanagan, V. P., & Pawlosky, R. J., (2003). Determination of 5-methyltetrahydrofolic acid and folic acid in citrus juices using stable isotope dilution-mass spectrometry. *Journal of Agriculture and Food Chemistry, 51*(5), 1293–1296.
31. Perryman, S. E., Lapong, I., Mustafa, A., Sabang, R., & Rimmer, M. A., (2017). Potential of metal contamination to affect the food safety of seaweed (*Caulerpa* spp.) cultured in coastal ponds in Sulawesi, Indonesia. *Aquaculture Reports, 5*, 27–33.
32. Clow, A., & Clow, N. L., (1952). *Chemical Revolution* (pp. 65–90). Freeport, N.Y.: Books for Libraries Press.
33. Kazłowski, B., Chiu, Y. H., Kazłowska, K., Pan, C. L., & Wu, C. J., (2012). Prevention of Japanese encephalitis virus infections by low-degree-polymerization sulfated saccharides from *Gracilaria* sp. and *Monostroma nitidum*. *Food Chemistry, 133*(3), 866–874.
34. Crepineau, F., Roscoe, T., Kaas, R., Kloareg, B., & Boyen, C., (2000). Characterization of complementary DNAs from the expressed sequence tag analysis of life cycle stages of *Laminaria digitata* (*Phaeophyceae*). *Plant Molecular Biology, 43*(4), 503–513.
35. Britannica, (2020). The editors of encyclopaedia. *Kelp*. Encyclopedia Britannica. https://www.britannica.com/science/kelp (accessed on 14 September 2023).
36. Aoki, Y., & Kamei, Y., (2006). Preparation of recombinant polysaccharide-degrading enzymes from the marine bacterium *Pseudomonas* sp. ND137 for the production of protoplasts of *Porphyra yezoensis*. *European Journal of Phycology, 41*, 321–328.
37. Wang, M., & Hu, C., (2017). Predicting sargassum blooms in the Caribbean Sea from MODIS observations: Sargassum bloom prediction. *Geophysical Research Letters, 44*(7), 3265–3273.
38. Bordoloi, A., & Goosen, N., (2020). *Green and Integrated Processing Approaches for the Recovery of High-Value Compounds from Brown Seaweeds* (p. 95). Elsevier Ltd.: Amsterdam, The Netherlands. ISBN 9780081027103.
39. Khalil, H. P. S. A., Lai, T. K., Tye, Y. Y., Rizal, S., Chong, E. W. N., Yap, S. W., Hamzah, A. A., et al., (2018). A review of extractions of seaweed hydrocolloids: Properties and applications. *Express Polymer. Letters, 12*, 296–317.
40. Grosso, C., Valentão, P., Ferreres, F., & Andrade, P. B., (2015). Alternative and efficient extraction methods for marine-derived compounds. *Marine Drugs, 13*, 3182–3230.

PART III

Technology of Bioactive Compounds and Nutraceuticals from Nonconventional Sources

CHAPTER 7

Technological and Analytical Aspects of Bioactive Compounds and Nutraceuticals from Spices and Condiments Sources

JOSÉ DE JESÚS LIRA-RICÁRDEZ and LUCÍA ORTEGA CABELLO

Biological Systems Department, Metropolitan Autonomous University-Xochimilco Campus, Mexico City, Mexico

ABSTRACT

The worldwide use of spices is related not only to their use as flavor and odor adjuvants but also to their biological properties such as antioxidant, anti-inflammatory, cytotoxic, and antiviral activities, among others, due to the presence of bioactive compounds. These activities have made spices interesting sources of nutraceuticals. The extraction of bioactive compounds from spices must start with the selection of the appropriate drying process, which is important also in the preservation of the spices by removing water and minimizing possible chemical degradation. Afterwards, an extraction technique is carefully chosen depending on the interested compounds. These techniques can be either classical, such as maceration or Soxhlet apparatus, or modern approaches involving eco-friendly procedures such as microwave-assisted, enzyme-assisted, and supercritical fluids. Later on, structural elucidation or quantification by using analytical techniques such as gas and high-performance liquid chromatography (HPLC), mass spectrometry (MS), infrared (IR) spectroscopy, and nuclear magnetic resonance (NMR) spectroscopy must be carried out based on their chemical characteristics. These techniques have also been useful for the authentication and chemotaxonomical analysis of closely related species that have been used as adulterants in commercial preparations. Current trends aim to develop

Bioactive Compounds and Nutraceuticals from Dairy, Marine, and Nonconventional Sources: Extraction Technology, Analytical Techniques, and Potential Health Prospects.
Charanjit Singh Riar, Parmjit S. Panesar (Eds.)

novel extraction techniques to enhance the production of nutraceuticals with the use of statistical tools such as chemometric measurements and response surface methodologies, thus allowing to explore metabolomics for a better knowledge of nutraceutical compounds found in spices.

7.1 INTRODUCTION

Botanical resources, such as legumes, vegetables, and medicinal plants are inexhaustible sources of naturally occurring compounds that play important roles in well-being maintenance, symptomatology prevention, and/or disease treatment, related to their vast structural complexity [1]. In recent years, these compounds have been clinically analyzed to bear out their pharmacological efficacy against the worldwide population's most important pathologies, such as cancer, CVDs, infectious diseases, neuropsychiatric disorders, obesity, and many more; to rely on novel and alternative options for treatment; thus awakening the interest of the pharmaceutical and nutritional guilds to offer mankind a well-delivered administration form and dosage for their massive use in the population.

Food-related bioactive compounds have been aggregated into the description of "nutraceuticals," a 1980s-born term that includes natural raw materials or microbial living species (modified/unmodified whole food, plant extracts, phytopharmaceutical preparations, and purified or semi-purified phytochemicals, and human microbiome-related species) with checked activity as health enhancer or clinically demonstrated pharmacological efficacy against health disorders [2, 3]. Nutraceuticals are found in conventional foods, functional foods (enriched or fortified foods), and dietary supplements (multivitamins, amino acids, minerals, fiber, and prebiotics), that are used for certain population groups, such as the elderly, oncological patients, athletes, and weight loss patients; representing them an interesting marketing population for the nutraceutical market, which their total retail sales in the USA for 2017 surpassed US $8 billion, and by the end of 2014, the overall sales in China exceeded US $120 billion [4].

As part of nutraceutical raw material, spices and condiments have been of historical relevance as health enhancers related to their distinctive organoleptic properties (flavor, odor, pungency, and color) widely used in cooking and food decorating to stimulate the appetite or potentiate the taste of the meal. Moreover, their diverse bioactive compounds content (including terpenes, phenolics, alkaloids, among others) make them an important source of naturally therapeutic options for prevention and/or treatment [2]. One of the most important worldwide spice-related nutraceuticals is curcumin, isolated from turmeric rhizome (Curcuma longa L.) which exerts a "panacea statu quo" for its anti-inflammatory, anti-cancer, anti-diabetic, antioxidant, hepatoprotective potential, and neuro-psychiatric enhancer on depression, anxiety, and Alzheimer's disease; and recently has been studied for anti-viral potential against HIV and SARS-CoV-2 [5, 6]. Curcumin-related nutraceuticals are one of the most worldwide lucrative markets, with an estimation of US $500 million and a projected compound annual growth rate of 13% throughout 2018–2025 [7]. Along with turmeric, some European, Indo-Pacific, and Latin American spices and condiments are known since ancestral times for their therapeutic properties and recently, their phytochemicals have been clinically tested and marketed as nutraceuticals, including cinnamon [8], chili pepper [9], garlic [10], ginger [11], onion [12], and many more; expanding the fan of alternative therapeutic options for health maintenance [13].

The development of nutraceutical products involves the application of novel techniques for bioactive compound extraction, such as microwave-assisted, ultrasound-assisted, supercritical fluid, and enzyme-assisted extraction (EAE) [14]; and their optimization for other sources, including food waste reusing, and biotechnological methods [15]; along with the novel pharmaceutical delivery systems to perform adequate absorption and systemic distribution for phytochemicals with low water solubility and poor permeability, such as nanotechnological biopolymers, nanoparticles, nanospheres, and metallic-complexed delivery systems [16]. Likewise, analytical procedures for identification, characterization, and quantification of phytochemicals and other compounds (as heavy metals, decomposition products, and preservatives) allow assuring quality control of known nutraceutical products and corroborating the properties and activity of novel bioactive compounds with therapeutic potential, aided by instrumental techniques such as high-performance liquid chromatography (HPLC), gas chromatography (GC), NMR, capillary electrophoresis (CE), and their hyphenated combinations (GC-MS, HPLC-MS, HPLC-NMR, CE-MS) [17]. These technological improvements are applied to enhance the optimization of formulation development, adequate effectiveness, and full safety commercialization of spices and condiments-related nutraceuticals around the world.

7.2 SPICES AND CONDIMENTS

7.2.1 DEFINITION AND CLASSIFICATION

To define the term "spice," there is a need to explore the concept according to different organizations because this concept can carry some controversies as will be seen in the following paragraphs, although they gather similar characteristics that may be joined in one. The International Organization for Standardization (ISO) defines a spice and a condiment as "vegetable products or mixtures thereof, free from extraneous matter, used for flavoring, seasoning, and imparting aroma to foods." Nevertheless, Peter and Shylaja (2012) agreed with a need to differentiate the term spice from herbs because they could be considered as synonyms; the authors defined herbs as the dried leaves of aromatic plants, while spices are dried parts from all other organs of plants.

Another concept of spice to be included is the one by the Food and Agriculture Organization (FAO) of the United Nations as "vegetable products such as leaves, flowers, seeds, and roots that are rich in essential oils and aromatic principles. They are used mainly as condiments." In this concept, FAO considers the following as such: pepper, pimento, vanilla, cinnamon, cloves, nutmeg, anise, ginger, and spices (a mixture of other spices); with the characteristic that spices are dried. In a similar manner as Peter and Shylaja (2012), FAO also refers to the term herb as plants with aromatic leaves; even though in FAO's concept dried leaves are also considered as spices, while Peter and Shylaja do not, making the first contradiction here [18]. According to the Cambridge Dictionary, a spice is "a substance made from a plant, used to give a special flavor to food," while the American Spice Trade Association (ASTA) defines a spice as "any aromatic vegetable substance in the whole, broken, or ground form, except for those substances which have been traditionally regarded as foods, such as onions, garlic, and celery; whose significant function in food is seasoning rather than nutritional; that is true to name; from which no portion of any volatile oil or other flavoring principle has been removed." In this regard, the CFR-Code of Federal Regulations displays a list of spices and natural seasoning and flavorings.

Analyzing the given concepts of spice, it can be concluded that all of them agree with the fact that they are dried ingredients of vegetable origin that can come from different plant organs, except those from leaves, which are called herbs according to Peter and Shylaja (2012) and FAO, although there would be a need to deepen the exploration of the inclusion of leaves in the concept of spice according to FAO as previously noticed, considering the ASTA's concept of spice there could be perhaps an explanation if the leaf is not only dried but also grounded or broken. Spices also contain aromatic compounds, rich in essential oils (EOs), which make spices useful for seasoning and flavoring.

However, there could be a disagreement in the concept of spice according to ASTA for two reasons: first, spices are only considered for flavoring rather than nutritional value; this can be debated because of the chemical constitution of some spices to be seen later in this chapter where such compounds have demonstrated medicinal activities and there is also literature that defends the importance of spices in health; and second, onion, garlic, and celery according to FAO as well as literature [19] are considered also as spices. Based on what was researched in the concept, spice can be defined as a vegetable product rich in aromatic compounds, which can come from any plant organ previously dried and can be used whole, broken, or grounded. There are ways to classify the spices and condiments, according to their taxonomy, to the plant organs and the flavoring or aromatic properties [19], as can be seen in Figure 7.1.

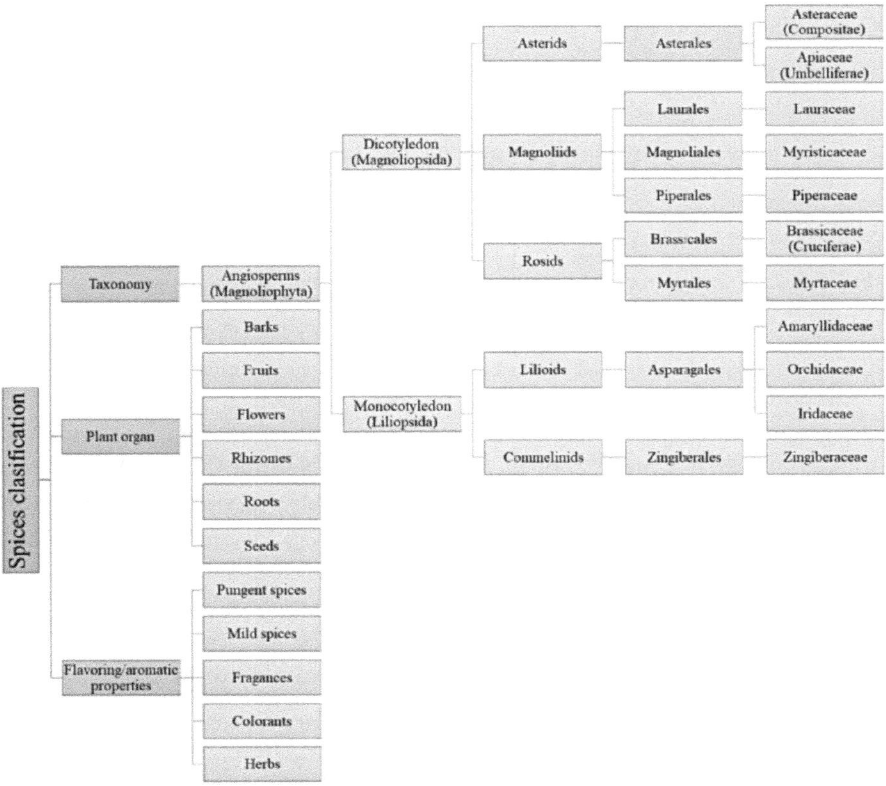

FIGURE 7.1 Classification of spices according to different criteria.

Source: Adapted from Ref. [19].

It can be seen that these classification criteria are interconnected because of their chemical composition. For example, ginger and turmeric; both belong to the *Zingiberaceae* family and the rhizome is the part used, it is known that depending on the part of the plant used, specific metabolites can be found (Figure 7.2). In this regard, the extracts of the rhizomes contain an important amount of flavonoids and phenolic acids [20] such as gingerol and Shogaol in the case of ginger [21], and curcumin (a diarylheptanoid) regarding turmeric [22], these metabolites are the compounds responsible for their pungent flavor [23].

FIGURE 7.2 Location of natural products according to plant organ.

Source: Adapted from Ref. [24].

Another classification that has not been considered is according to their geographical origin where allspice, cardamom, chili pepper, cinnamon, clove, ginger, nutmeg, pepper, turmeric, and vanilla came from tropical regions, and coriander, cumin, oregano, mustard, peppermint, sesame seed, sage, and thyme came from template regions [18], as can be seen in Figures 7.3–7.6. In America (Figure 7.3) it can be seen spices such as *Bixa orellana* (achiote or annatto), *Capsicum annuum* (chili pepper, Cayenne pepper, and paprika), *Pimenta dioica* (pimienta gorda or allspice), *Schinus molle* (pirul or pink peppercorn), and *Vanilla platifolia* (vanilla) and are mainly present in Latin America.

Regarding Europe it is mainly focused on the Mediterranean region such as *Apium graveolens* (celery), *Coriandrum sativum* (coriander), *Foeniculum vulgare* (fennel), *Glycyrrhiza glabra* (licorice), *Origanum majorana* (marjoram), and *Origanum vulgare* (oregano); except for *Artemisia dracunculus* (tarragon) and *Armoracia rusticana* (horseradish) which are found in Russia.

FIGURE 7.3 Spices from America.

FIGURE 7.4 Spices from Europe.

In Africa and the Middle East region are found many spices such as *Allium cepa* (garlic), *Allium sativum* (onion), *Anethum graveolens* (dill), *Brassica nigra* (black mustard), *Carum carvi* (caraway), *Coriandrum sativum* (coriander), *Crocus sativus* (saffron), *Ocimum basilicum* (basil) *Thymus vulgaris* (thyme), *Trigonella foenum-graecum* (fenugreek), and *Sinapis alba* (white mustard).

FIGURE 7.5 Spices from Africa and the Middle-East region.

Another area rich in spices is the Indo-Pacific region, among the spices that can be found on this site are *Cinnamomum verum* (cinnamon), *Citrus* ssp. (lemon and orange zest), *Curcuma longa* (turmeric), *Elettaria cardamomum* (cardamom), *Illicium verum* (anise star), *Myristica fragrans* (nutmeg), *Piper nigrum* (black pepper), *Sesamum indicum* (sesame), *Syzygium aromaticum* (clove), *Tasmannia lanceolata* (Tasmanian pepper), *Wasabia japonica* (wasabi), *Zingiber officinale* (ginger), and *Zanthoxylum piperitum* (Sichuan pepper).

7.2.2 HEALTH BENEFITS OF SPICES

Since ancient times, spices and condiments have been used by ethnic populations around the world as potential medicines to reduce health risks and treat several diseases, such as body ache, digestive disorders (bloating, constipation, diarrhea), fever, inflammation, rheumatism, etc.; which have been included as part of the complementary and alternative medicine (CAM) options [25]. In literature, it has been demonstrated the perception and use

FIGURE 7.6 Spices from the Indo-Pacific region.

of spices for health benefits in the Midwestern United States local population in a cross-sectional study design. In this study, ginger presented the highest percentage of perceived efficacy with 64% of the studied population, followed by garlic (58%) and cinnamon (56%), other spices considered were chili pepper, turmeric, cilantro, cloves, black pepper, curry, and fenugreek, in which their intake is associated with an increased life expectancy and risks reduction from suffering degenerative diseases [26] (Figure 7.7).

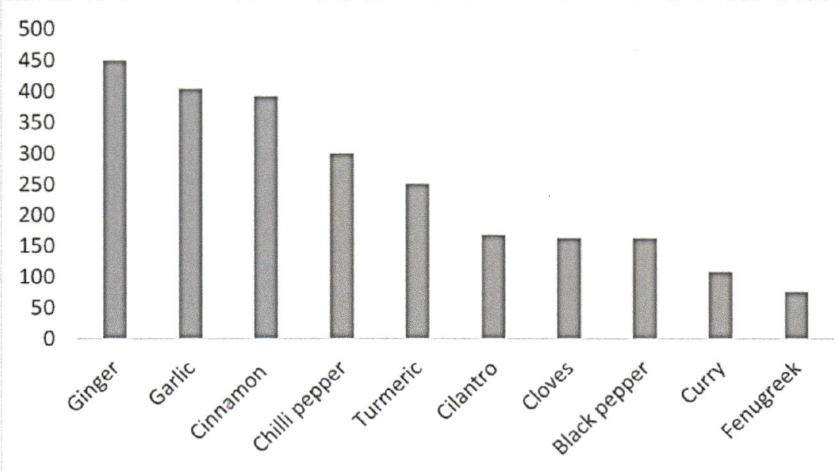

FIGURE 7.7 Perceived efficacy of spices for an n = 703 subject.

Source: Adapted from Ref. [27].

In the last decade, pharmacological preclinical and clinical assays of spices and their related chemical compounds have demonstrated their potential as antioxidant, anti-inflammatory, antimicrobial, antiviral, cytotoxic, hypoglycemic, hypolipidemic, neuroprotective, wound healing, and food preservatives activities [27]; and recently, several secondary metabolites (SMs) from spices and condiments have been studied for their potential on treatment for SARS-CoV-2 infection as antivirals or immunity boosters [29, 30]. These shreds of evidence have awakened a pharmaceutical interest in the development of potent natural-derived therapeutic options which contain their bioactive compounds known as nutraceuticals (Table 7.1), such as alkaloids, phenylpropanoids, polyphenols (tannins and flavonoids), and terpenes [31].

Several nutraceuticals have been studied extensively for their biological activities, including their molecular targets to associate their participation on cellular and physiological processes, such as downregulation of proinflammatory cytokines [13], neuronal enzymatic activation, and/or inhibition [25], induction of programmed cell death (apoptosis) [32], and regulation of microRNA expression [33].

TABLE 7.1 Some Biological Activities of Nutraceuticals Present in Spices

Activity	Natural Product	Spices	References
Antioxidant	Flavonoids, phenolic acids	*Curcuma longa, Murraya koenigii*	[34]
	Phenolic acids, flavonoids, tannins, stilbenes, anthocyanins, lignans.	*Allium sativum, Allium cepa, Zingiber officinale*	[35]
	Phenolic acids, flavonoids	*Brassica nigra, Nigella sativa, Trachysper mum ammi, Sesamum indicum*	[36]
	Phenolic compounds	*Apium graveolens, Thymus vulgaris, Salvia Rosmarinus, Origanum vulgare, Mentha piperita, Ocimum basilicum*	[37]
Anti-inflammatory	Carotenoids, capsaicinoids, saponins, alkaloids, terpenes, phenolic acids, phenylpropanoids, lignans, steroids, flavonoids	*Ocimum basilicum, Capsicum annuum, Capsicum frutescens, Nigella sativa*	[38]
	Terpenes, phenylpropanoids	*Myristica fragrans*	[39]
	Terpenoids	*Origanum vulgare*	[40]
	Curcuminoids	*Curcuma longa*	[41]

TABLE 7.1 *(Continued)*

Activity	Natural Product	Spices	References
Cytotoxic	Monoterpenes and polyphenols	*Thymus vulgaris*	[42]
	Alkaloids	*Piper nigrum*	[43]
	Carotenoids	*Crocus sativus*	[44]
	Polyphenols, anthraquinones, sulfur compounds	*Capsicum annuum, Allium sativum, Allium cepa, Zingiber officinale, Murraya koenigii*	[45]
	Glucosinolates, terpenes, phenolic compounds, alkaloids, saponins	*Cinnamomum verum, Curcuma longa, Brassica napus, Ocimum basilicum, Nigella sativa, Punica granatum*	[46]
Antimicrobial	Allylbencenes, essential oils, terpenoids, aldehydes, acetates, alcohols, sulfonates	*Origanum vulgare, Thymus vulgaris, Zingiber officinale, Cinnamomum verum, Syzygium aromaticum, Allium sativum, Murraya koenigii*	[47]
	Sesquiterpenes, curcuminoids, capsaicinoids, essential oils	*Zingiber officinale, Allium sativum, Curcuma longa, Allium cepa, Capsicum frutescens*	[48]
	Essential oils	*Cinnamomum verum, Syzygium aromaticum, Cuminum cyminum, Pimpinella anisum*	[49]
	Essential oils, alkaloids, saponins	*Piper nigrum, Syzugium aromaticum, Curcuma longa, Pimpinella anisum, Nigella sativa, Allium sativum, Zingiber officinale, Cinnamomum verum, Myristica fragrans*	[50]

7.2.3 CHEMICAL COMPOSITION OF SPICES

Specific extraction strategies, as well as analytical techniques, can be performed based on the plant organ that contains the bioactive compounds, and their chemical properties as well. In the case of the aril, which is a fleshy structure found in seeds rich in glycosides and flavor/aroma-related compounds as well as oils [51]. An example of spice obtained from the aril is the nutmeg [19], known to be used in traditional medicine as antimicrobial, antioxidant, antidiabetic, anticancer, and central nervous system stimulant activity, associated with their phenolic content; mainly phenylpropanoids, lignans, and neolignans. According to the reports by Ref. [52] on a nutmeg variety cultivated the total phenolics content from the dried fruit of a commercially cultivated variety in 2001 (764 mg) of phenolic compounds, was composed of 132 mg of neolignans (licarin A and B), 352 mg of phenolic aldehydes and 280 mg of phenylpropanoids (mainly myristicin and elemicin). The non-polar total extract also includes steroids (stigmasterol and β-sitosterol), triterpenoids (cycloartenol), and triglycerides (TAGs) (trimyristin) (Figure 7.8).

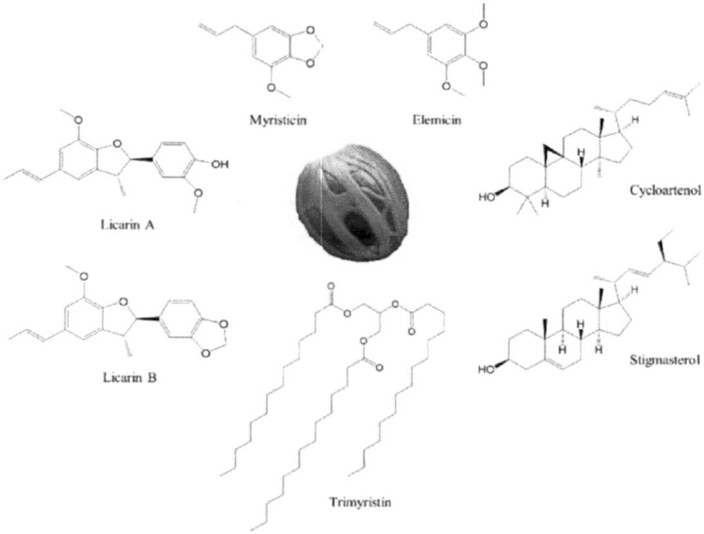

FIGURE 7.8 Some natural products extracted from the aril of nutmeg.

Another plant organ used in the nutmeg is the kernel [19] which is the softer part of the fruit, this part is commonly seen in corn and in apricots in which EOs are commonly extracted from this plant organ [53]. In the case of nutmeg contains monoterpenes (α-pinene, 4-terpineol, α-terpineol, γ-terpinene, and β-pinene) and phenylpropanoids (myristicin, elemicin, methyl eugenol, safrole, eugenol, and methyl isoeugenol) as well as lignans and neolignans (tetrahydrofuroguaiacin, saucernetindiol, verrucosin, nectandrins A and B, fragransin C_1 and epoxylignan among others) which are the main metabolites present in the kernel [54] (Figure 7.9).

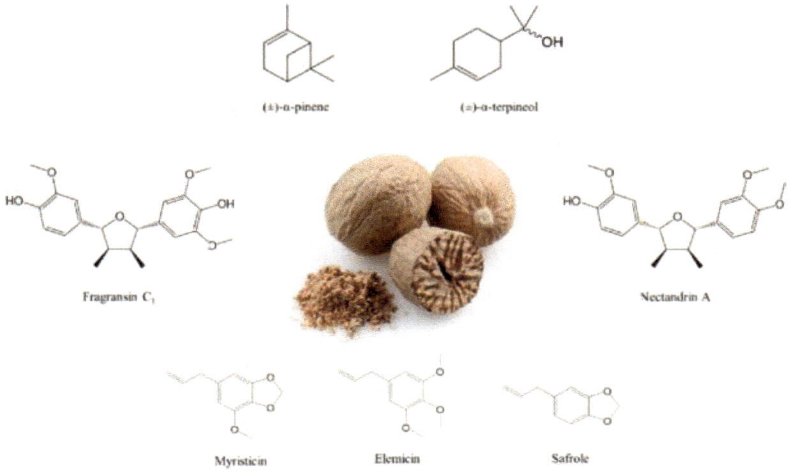

FIGURE 7.9 Some natural products extracted from the nutmeg kernel.

Tree bark, such as cinnamon, is another plant organ used as a spice [19], which is a spice widely used because of its aroma constituted mainly by *trans*-cinnamaldehyde (from 65 to 80%) in its EO possessing antityrosinase activity, another natural products found in cinnamon are condensed tannins (A- and B-type procyanidins) and flavonoids (catechin) with antioxidant activity. Because of these products cinnamon has been used as anti-inflammatory, nematicidal, larvicidal, antimycotic, and anticancer [55, 56]. Other compounds found on cinnamon are α-linalool (a monoterpene), cinnamyl acetate, eugenol (phenylpropanoids), and benzyl benzoate [57] (Figure 7.10).

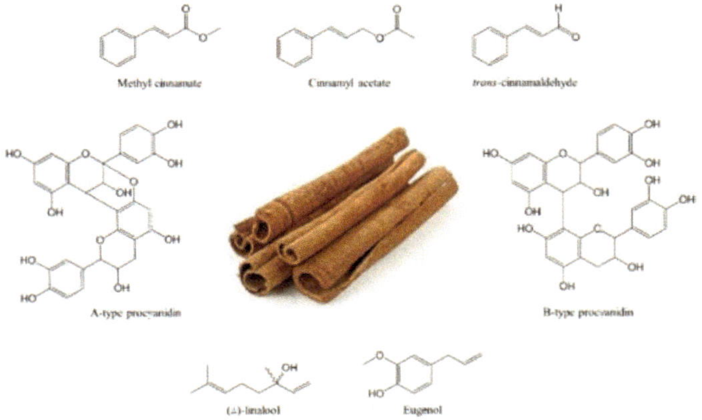

FIGURE 7.10 Some natural products are present in cinnamon bark.

Fruits are very important in natural product research due to the variety of compounds that can be found (Figure 7.2), as it is in the case of the black pepper [19], which has presented gastrointestinal, anti-inflammatory, antiepileptic, and antipyretic activities as well as its use in respiratory diseases. This spice contains a high amount of alkaloids, such as piperine, and other bioactive components like the sesquiterpenes nerolidol, copaene, and β-caryophyllene, monoterpenes such as limonene, α- and β-pinene, δ-3-carene, β-phellandrene, and among others [58] (Figure 7.11).

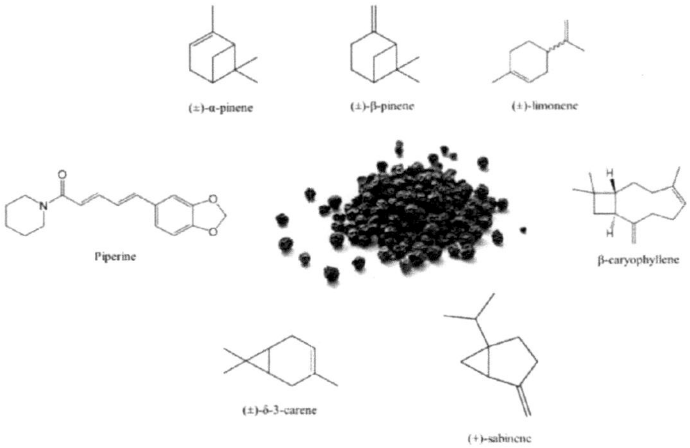

FIGURE 7.11 Some natural products of black peppercorn.

Another plant organ used is the bud as it is with clove [59], in which has been found previously mentioned metabolites such as β-caryophyllene, eugenol, and pinene; other metabolites found on clove are phenolic acids (gallic and caffeic acids), flavonoids (kaempferol, rhamnetin, myricetin, quercetin), chromones (eugenin), among other metabolites. These metabolites are associated with several pharmacological activities like antibacterial, anticancer, antifungal, anti-inflammatory, antioxidant, acaricidal, among others [60] (Figure 7.12).

The specialized stems, such as garlic and onion bulbs, are also used as spices and traditional herbal remedies. In the case of garlic, it has been found that alkaloids such as β-carbolines are the most abundant phytochemicals (7.2%), followed by tannins, saponins (sativoside and voghieroside), sapogenins (agapanthagenin, β-chlorogenin, diosgenin, gitogenin, among others), and flavonoids (epicatechin and apigenin) (4.8, 4.3, and 2.18% respectively)

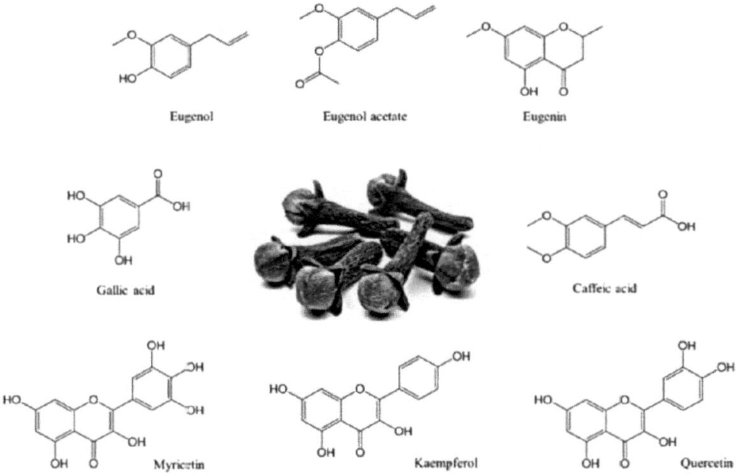

FIGURE 7.12 Some natural products of clove bud.

as well as steroids (0.5%) and terpenoids (0.4%); other compounds that are characteristic of garlic are sulfur compounds such as ajcene, allicin, and allitridin among others. Due to its composition, garlic possesses antiviral, antibacterial, antifungal, antioxidant, antiatherosclerotic, and anticancer properties [61–65] (Figure 7.13).

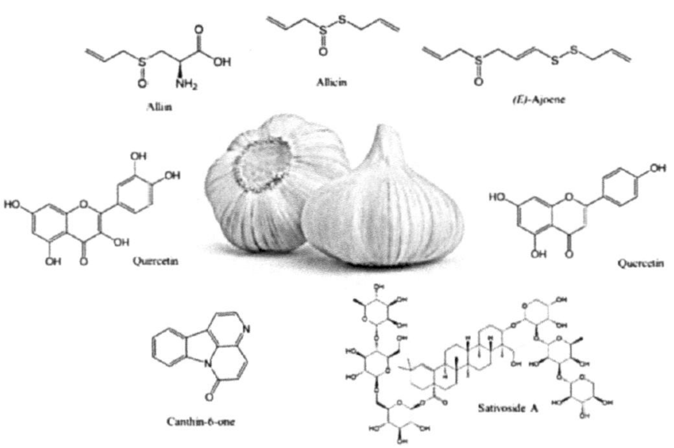

FIGURE 7.13 Some natural products of garlic bulbs.

A worldwide expensive spice like saffron, obtained from the pistil of *Crocus sativus* [59], is known for its therapeutical potential as an analgesic, antidepressant, and anti-inflammatory as well as antioxidant, associated with the content of carotenoids (crocin and crocetin), apocarotenoids (safranal), anthocyanins (petunidin and delphinidin glycosides), as well as kaempferol glycosides (flavonoid derivatives) [66, 67] (Figure 7.14).

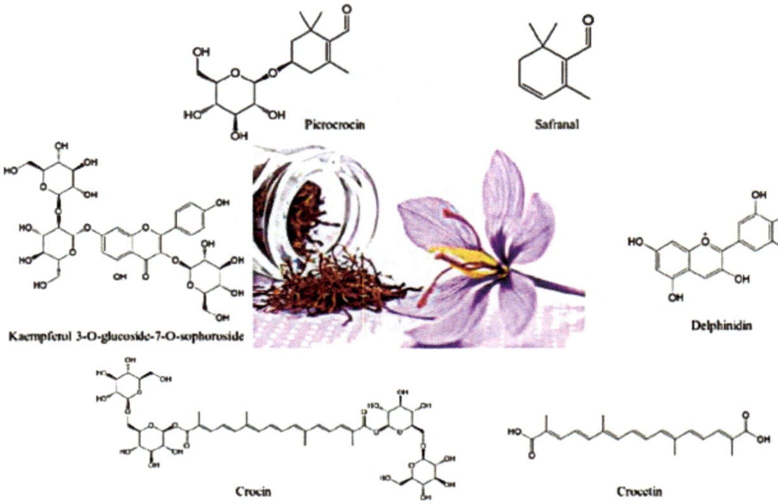

FIGURE 7.14 Some natural products from the saffron pistil.

The leaf is a plant organ rich in aromatic compounds, many spices from this plant organ are known such as basil, bay leaf, mint among others [59]; although this contradicts the definition of spices which could also be seen as herbs, it will be depending on the presentation of the spice, which could be either fresh or dried. In the case of the basil leaf is rich in terpenes (α- and β-pinene, 1,8-cineol, linalool, ocimene, borneol, geraniol), phenolic acids (caffeic, vanillic, chlorogenic, and *p*-hydroxybenzoic acids), phenylpropanoids (eugenol, chavicol, estragole) and flavonoids (quercetin, rutin, and apigenin) [68]. The composition of basil leaves confers certain activities such as antimicrobial, antioxidant well as skin protection, and treatment of dermatological disorders [69] (Figure 7.15).

Oleoresins or latex are also used as spices, as is the case of asafetida extracted from the rhizomes of herbs from the *Ferula* genus, it has been part of folk medicine because of its neuroprotective, antioxidant, antispasmodic, antimicrobial, anticarcinogenic, and antihelmintic activities, among others. Compounds that can be found in this spice are coumarins (umbelliprenin, tadshiferin, galbanic acid, conferol, gummosin, *epi*-samarcandin, franesiferol

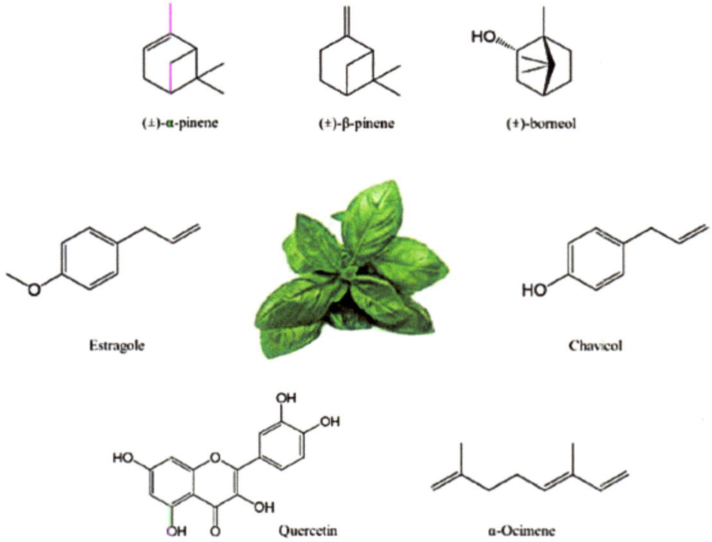

FIGURE 7.15 Some natural products from basil leaves.

A, B, and C), sulfur-containing compounds (asadisulfide, foetisulfide A and C), terpenoids (picealactone C, taraxin, fetidone A and B) and phenolic compounds such as vanillin [70] (Figure 7.16).

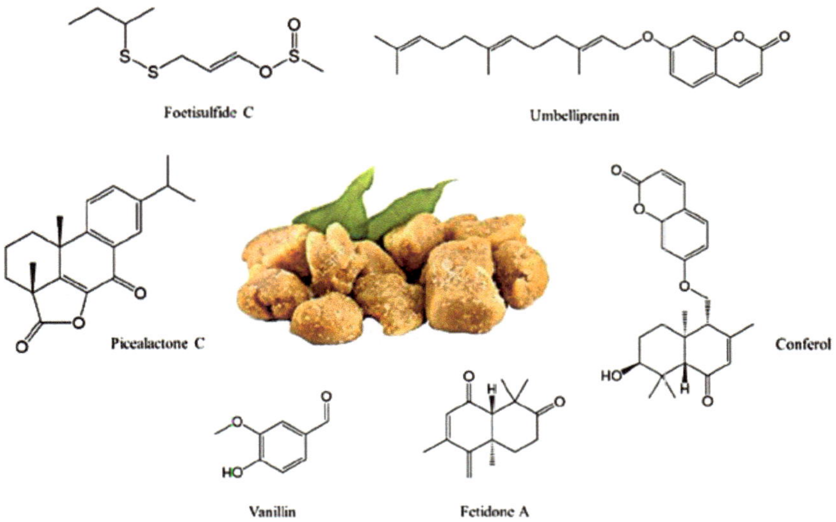

FIGURE 7.16 Some natural products from asafetida latex.

The roots of horseradish is another spice used, where the main compounds present are isothiocyanates, which are hydrolyzed products from glucosinolates (sinigrin, gluconasturtiin, glucoalyssine, glucohesperine, glucoibarine) among the identified isothiocyanates are allylisothiocyanate, pentyl isothiocyanate, and benzyl isothiocyanate [71]. Other compounds found on horseradish root are kanamycin (an aminoglycoside), coumarins (aesculetin and scopoletin), phenolic acids (caffeic acid), and flavonoids (kaempferol derivatives) among other compounds [72, 73]. Isothiocyanates are known to possess antibacterial and chemopreventive activities [74] (Figure 7.17).

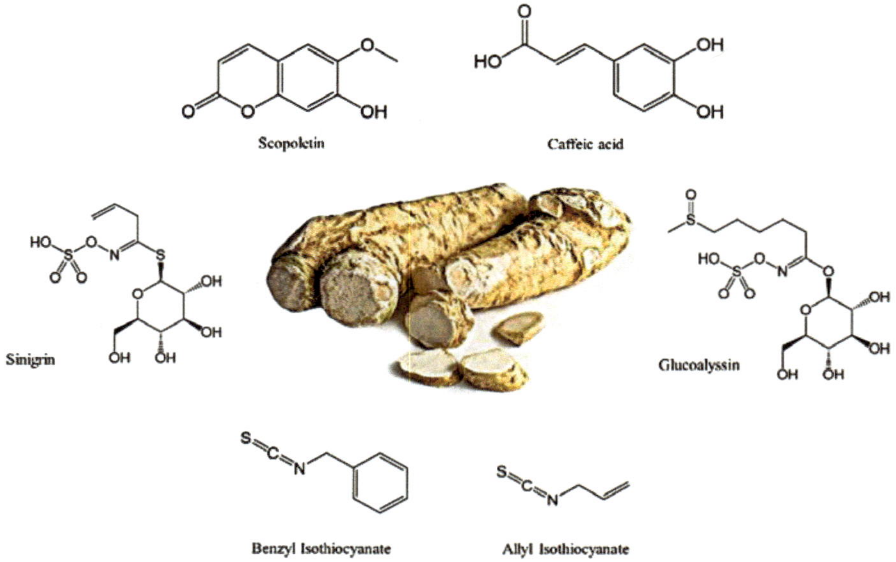

FIGURE 7.17 Some natural products from horseradish roots.

The seed is the last, but not least important plant organ to be examined is the seed such as fenugreek seed with a characteristic bittersweet flavor, as well as medicinal activities such as hypocholesterolemic, antibacterial, antidiabetic, hepatoprotective, and anticancer, among others. This seed contains the saponin fenugrin B, alkaloids (trigonelline, gentianine, carpaine, neurine, betaine, trimethylamine), steroids (diosgenin, trigogenin, gitongenin) as well as polyphenolic compounds (rhaponticin and isovitexin), lactones (sotolon), carboxylic acids (acetic acid, butanoic acid, isovaleric acid, caproic acid) [75] (Figure 7.18).

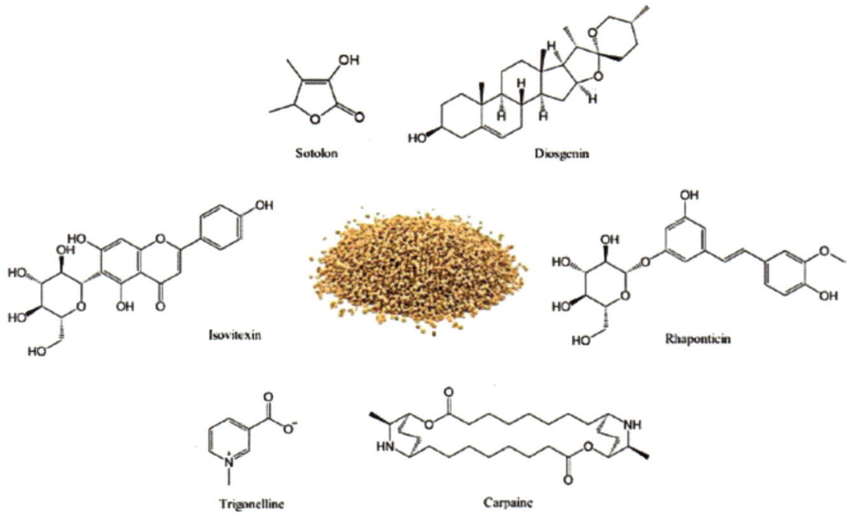

FIGURE 7.18 Some natural products from fenugreek seed.

7.3 EXTRACTION TECHNIQUES OF BIOACTIVE COMPOUNDS FROM SPICES

The chemical properties of bioactive compounds are crucial for extraction methodology. The first step in the extraction procedure is drying of the plant material, representing 30 to 50% of the total cost of production. Maximum drying temperature will depend mainly on the chemical compound to be extracted; for glycosidic compounds, 100°C is recommended, mucilages at 65°C, and EOs from 35 to 45°C [76]. Freeze-drying, sunlight drying, shade drying, oven drying, and microwave drying are the most used drying techniques for plant pretreatment. A careful selection of the drying method must be done because it will affect the quality of the extract as can be seen later in Table 7.2. The choice of extraction technique, which can be either classical such as maceration, Soxhlet, hydrodistillation, or modern and eco-friendly like supercritical fluids, microwave, ultrasound among other techniques, will depend on the molecular characteristics of the compound, type of solvents, and their affinity [77].

7.3.1 DRYING OF SPICES

To extract the bioactive compounds of spices, they must have proper pretreatment to avoid microbial contamination and enhance the preservation

TABLE 7.2 Drying Techniques are Applied on Spices Depending on the Plant Organ

Plant Organ	Plant Material	Drying Methods	Observations	References
Aril	*Myristica fragrans*	Microwave-assisted fluidized bed, sun drying, and convective drying.	60–70% drying time reduction and the highest yield in oil, oleoresin, and myristicin with microwave-fluidized bed.	[83]
Rhizome	*Zingiber officinale*	Microwave, freeze-drying, oven drying, and air drying.	Freeze-drying presented the highest phenolic and carotenoid content while air drying presented the lowest.	[84]
		Hybrid solar dryer at 50, 60, and 70°C.	Sliced samples presented the highest drying rate.	[76]
	Curcuma longa and *Curcuma xanthorrhiza.*	Freeze-drying and oven drying.	Both techniques yielded a similar amount of curcuminoids due to thermostability.	[85]
Bark	*Cinnamomum zeylanicum*	Sun-drying, shade-drying, oven drying, and freeze-drying.	Oven drying presented the highest total phenolic and flavonoid content.	[86]
		Art cave	Dryness is achieved between 50 and 60°C within 2.5 hours.	[87]
		Hot air-drying	Trans-cinnamaldehyde content increased with temperature with loses on sesquiterpenes and oxygenated terpenes	[88]
Fruit or berry	*Piper nigrum*	Horizontal solar dryer and open sun-drying	Horizontal solar dryers were more efficient, and the carbohydrate and protein content was higher and presented a better antioxidant activity.	[89]
		Sun-drying, mechanical drying	Mechanical drying decreases drying time with the lowest moisture content	[90]
		Fluidized bed dryer and sun drying.	A fluidized bed was more efficient.	[91]
	Capsicum annuum	Cold plasma pretreatment with hot air drying.	Retention of red pigment incremented as well as antioxidant activity.	[92]

TABLE 7.2 *(Continued)*

Plant Organ	Plant Material	Drying Methods	Observations	References
	Capsicum chinense	Convective-drying	The retention of active compounds was achieved at 70°C.	[93]
Leaf	*Laurus nobilis*	Air-drying, oven-drying, freezing, and freeze-drying.	Oven-drying and air-drying conserved the volatiles more than the other drying methods.	[94]
	Coriandrum sativum	Convective drying, vacuum microwave drying, combined drying.	Combined drying decreased the aroma quality.	[95]
	Origanum vulgare and *Origanum onites*	Sun-drying, shade drying, and oven drying.	Oil yield was higher with shade drying and presented the highest antioxidant activity.	[96]
	Ocimum basilicum	Sun-drying, freeze-drying, convective drying, and microwave-drying.	Microwave drying was the more efficient method.	[97]
Bud	*Syzygium aromaticum*	Oven drying	The relative amount of eugenol increased.	[98]
		Oven drying, sun drying.	Eugenol content decreases with sun drying.	[99]
		Microwave-drying, oven-drying, and air-drying.	Microwave-drying was the most effective to keep the antibacterial and antioxidant properties.	[100]
Root	*Armoracia rusticana*	Hot-air drying, infrared-assisted hot air-drying, freeze-drying, pulse-spouted infrared freeze-drying.	Pulse-spouted infrared freeze-drying kept the highest phenolic and volatile compounds content as well as antioxidant activity.	[101]
Seed	*Sinapsis alba*	Thin layer dryer at different temperatures.	Sterols are lost at high temperatures (120–140°C).	[102]
	Pimpinella anisum	Sun, oven, and shade drying.	Shade drying was the best technique to obtain essential oil and antioxidant activity.	[103]
	Illicium verum	Infrared radiation-hot air at different temperatures.	At 70°C the highest retention of essential oil was achieved.	[104]
Pistil	*Crocus sativus*	Reflectance window and oven drying.	The reflectance window allows saffron to keep its functionality as a spice.	[105]

TABLE 7.2 *(Continued)*

Plant Organ	Plant Material	Drying Methods	Observations	References
Bulbs	*Allium cepa* varieties violet of Galmi, Goudami, and White of Galmi	Solar drying and air drying.	Both methods yielded similar composition.	[106]
	Allium cepa	Oven drying, shade drying, sun drying.	Oven drying was the most cost-effective procedure for keeping the nutritional value.	[107]
		Industrial oven, electric dryer, and sun-drying.	The industrial oven was the preferred drying method.	[108]
	Allium sativum	Near-infrared fluidized bed drying, atmospheric freeze-drying, near-infrared vacuum drying.	Near-infrared vacuum drying presented the highest drying rate.	[109]
		Infrared drying at different temperatures and power.	Drying time was faster with increasing temperature, below 80°C no degradation of thiosulfinates was significant.	[110]

of foods, thus preventing the generation of toxins that may be hazardous to health. For this purpose, drying is an essential operation that decreases the moisture content based on mass and heat transfer processes. However, it must be carefully carried out to avoid the loss of nutritional and bioactive compounds. Among the drying techniques that have been used on spices are solar-drying, hot air-drying, freeze-drying, and microwave drying along with their combinations that provide a synergistic effect resulting in a better quality of dried spices [78], Figure 7.19 shows the classification of some drying techniques used on spices.

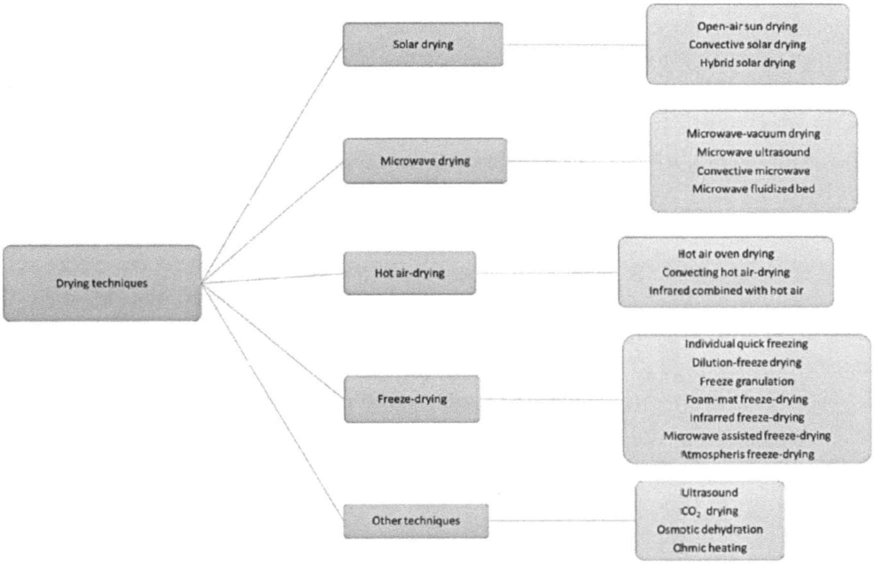

FIGURE 7.19 Classification of drying techniques.

Source: Adapted from Refs. [78, 79].

Even though the drying process is an essential step for the treatment of spices, there must be a consideration that it will be a loss of valuable components due to the thermolability of some of these compounds; in which the implementation of fluidized bed technologies has been studied to extract a higher amount of bioactive compounds because of an efficient heat transfer as it is in the case of the fruit of *Piper nigrum* and the bulb of *Allium sativum* in which the drying process was more efficient because it accelerated the drying process, as well as with the aril of *Myristica fragrans* where a fluidized bed was combined with microwave, achieving a higher amount of oleoresin and myristicin (Table 7.2). Sun or shade drying can offer low costs with the disadvantage of prolonged times and loss of valuable components because of oxidation reactions and low moisture removal. Freeze-drying, although it reduces time and improves the quality of the spice by increasing the number of bioactive compounds such as phenolic compounds and carotenoids from the rhizome of *Zingiber officinale*, its cost prevents this technique from being widely used [80, 81].

IR radiation is increasing its use due to the cost reduction and better product quality in combination with other techniques such as hot air and freeze-drying to either keep a higher amount of bioactive compounds or to increase EO extraction efficiency as was observed with the roots of *Armoracia rusticana* and the seeds of *Illicium verum* to mention a few examples; while vacuum microwave is also attractive because it shortens the heat exposure as was observed with the leaf of *Ocimum basilicum* and the bud of *Syzygium aromaticum* [80]. Novel techniques like ohmic heating require an electric current passing through the sample and evoking an electropermeabilization of the cell membrane, followed by decreasing drying temperature at 50°C, causing cellular damage, and accelerating the loss of water, which has been probed on the aril of the fruit of *Momordica cochinchinensis* [82], which could be useful for extraction of bioactive compounds from arils used as spices as it is in the case of nutmeg. Table 7.2 summarizes the above-mentioned examples as well as other spices, where classical and novel drying techniques have been used on spices depending on the plant organ.

7.3.2 EXTRACTION OF BIOACTIVE COMPOUNDS

Among the extractions methods to consider are classical methods such as maceration and Soxhlet extraction, in which either mass transfer between immiscible phases, diffusion, or distribution processes take place (Figure 7.20), where solvent characteristics, temperature, and time play important roles in the extraction efficiency of the desired compounds, along with their solvent affinity and physical properties (photostability and thermostability) [111].

However, because of the prolonged extraction times, as well as the generation of hazardous residues, the generation of eco-friendly extraction methods have emerged such as microwave-assisted extraction (MAE), supercritical fluid extraction (SFE), and enzyme-assisted extraction (EAE), among others; in which not only the used solvent and temperature are the

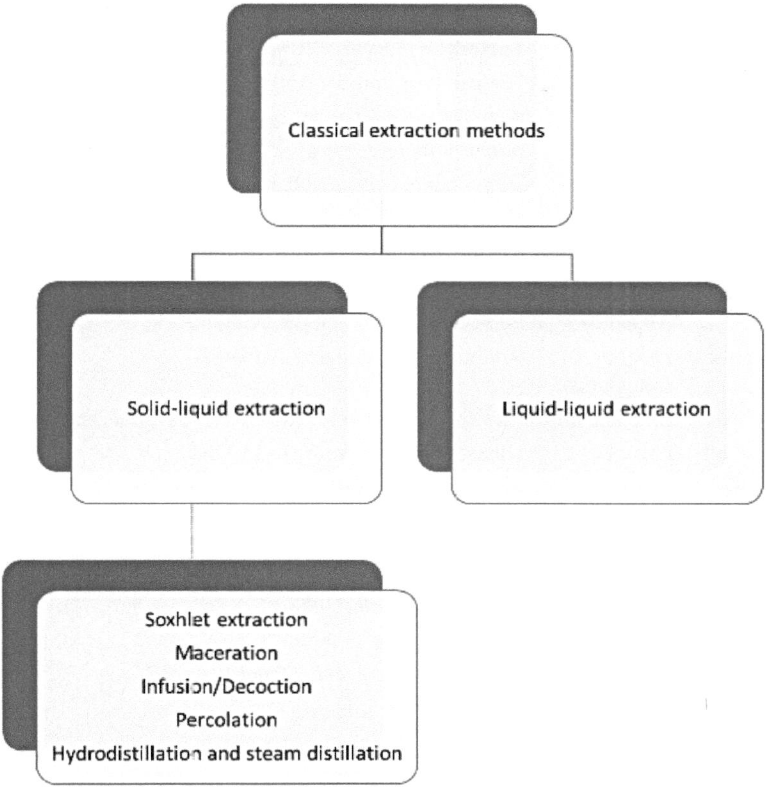

FIGURE 7.20 Classical extraction methods.

main variables but also pressure, enzyme concentration, pH, and electrical potency of the microwave apparatus, among others are also important. Table 7.3 will be reviewed the type of compounds that can be extracted from these modern techniques.

As can be seen, many factors can be considered to improve the yield of certain metabolites, in which SE of *Curcuma aeruginosa* yielded different compounds depending on the solvent due to their polarity or in the case of supercritical fluids in which temperature, pressure, and residence time allowed the obtention of different compounds depending on the spice. Table 7.4 demonstrates these and other examples depending on the extraction method as well as conditions that lead to the obtention of enriched extracts.

TABLE 7.3 Modern Extraction Techniques and Metabolites Extracted [112, 113]

Techniques	Compounds	Factors
Pressurized liquid extraction	Phenolic compounds	Temperature from 50–100°C.
		Solvents (methanol, ethanol, and mixtures with water at low proportion).
	Essential oils	Temperature between 60 and 100°C.
		Time from 5–30 minutes.
		Mixtures of ethanol:water as well as hexane, acetonitrile, ethyl acetate, and dichloromethane.
	Alkaloids	Solvent with different temperatures.
Supercritical fluid extraction	Phenolic compounds	Addition of ethanol and methanol.
	Essential oils	Pressure, temperature, solvent flow, particle size, co-solvent.
Ultrasound-assisted extraction	Phenolic compounds	Ultrasonic power from 80–150 W.
		Solvent methanol or ethanol up to 80%
		Temperature 40–80°C.
		Sonication time (10–60 minutes).
	Essential oil	Sonication power 30 W.
		Sonication time 30 minutes
Microwave-assisted extraction.	Phenolic compounds.	Solvent methanol and ethanol.
		Solvent ratio 5–30 mL/g.
		Temperature from 50–80°C.
		Microwave power 180–600 W.
	Essential oils	Solvent-free (3 KW, 30 minutes, 100°C) with hydrodistillation
	Alkaloids	Solvent (acidified solutions).
	Saponins	Type of saponins.
		Steroid saponins (300 W, 20 mL/g of ethanol:water 70:30, 50°C, 15 minutes).
		Triterpene saponins (500 W, 80°C, 3 minutes, 20 mL/g of ethanol:water 70:30.
		The vapor pressure of the solvent (use of a vacuum system).
Enzyme-assisted extraction	Essential oils and oleoresins	Cell wall components (cellulases, hemicellulases, and pectinases or mixtures).
		pH.
		Reaction time.
		Temperature.
		Enzyme:spice ratio.

TABLE 7.4 Different Extraction Methods Performed on Spices

Extraction Method	Spice and Extraction Conditions	Chemical Nature of Compounds	References
Solvent extraction	*Curcuma aeruginosa* with methanol followed by addition of chloroform.	Methanolic phase: sugars, organic acids, and alkanes. Chloroform phase: monoterpenoids and sesquiterpenoids.	[114]
	Oleoresin from *Curcuma longa* with hexane and agitation.	Curcuminoids	[115]
	Xylopia aethiopica, Ricinodendron heudelotii, Piper capense, and *Aframomum citratum* with methanol with stirring for 24 hours.	Phenolic compounds, tannins, and flavonoids.	[116]
Hydro distillation or steam distillation.	*Myristica fragrans*	Terpenoids	[117]
	Myristica fragrans at 100°C for 7 hours.	Essential oil	[118]
	Cinnamomum verum, Cuminum cyminum, Elettaria cardamomum, Wisteria, and *Coriandrum sativum* for 8 hours.	*Cinnamomum verum*: aldehydes and terpenoids. *Cuminum cyminum*: terpenoids. *Elettaria cardamomum*: terpenoids. *Wisteria*: Terpenoids. *Coriandrum sativum*: aldehydes and terpenoids.	[119]
	Coriandrum sativum, Cuminum cyminum, Foeniculum vulgare, and *Brassica juncea.*	Terpenoids, aldehydes, and isothiocyanates.	[120]
	Syzygium aromaticum, Ocimum basilicum, and *Pimpinella anisum* in ethanol for 7 days.	*Pimpinella anisum*: phenylpropanoids, aldehydes, and ketones. *Syzygium aromaticum*: terpenoids and phenolic compounds. *Ocimum basilicum*: terpenoids.	[121]
	Ocimum basilicum, thymus serpyllum, Rosmarinus officinalis, Sativa officinalis, Artemisia dracunculus, and *Coriandrum sativum* with methanol at room temperature for 5 months.	Phenolic compounds	[122]

TABLE 7.4 *(Continued)*

Extraction Method	Spice and Extraction Conditions	Chemical Nature of Compounds	References
Soxhlet apparatus	*Zingiber zerumbet* sequential extraction with petroleum ether, benzene, isopropanol, ethyl acetate, and ethanol at 50°C for 8 hours.	Isopropanol: sesquiterpenes and flavonoids. Benzene: phenolic compounds and curcuminoids. Ethyl acetate: highest content of flavonoids.	[123]
	Curcuma longa with acetone.	Curcuminoids	[115]
Ultrasound-assisted extraction	*Curcuma aeruginosa* with MTBE.	Monoterpenoids and sesquiterpenoids.	[114]
Microwave-assisted extraction	*Zingiber zerumbet* with ethanol 40% v/v.	Zerumbone	[77]
	Myristica fragrans	Essential oil	[118]
	Myristica fragrans complemented with hydrodistillation.	Essential oil	[124]
Subcritical water extraction.	*Curcuma longa* at 10 bars and 140°C	Curcuminoids	[125]
CO_2 supercritical fluids extraction.	*Curcuma longa* at 25 MPa and 333 K.	Essential oil	[126]
	Syzygium aromaticum, Ocimum basilicum and *Pimpinella anisum* chamber 1 at 65°C, 180 bars and chamber 2 at 50°C for 24 hours with aerogels.	*Pimpinella anisum*: phenylpropanoids, aldehydes, and ketones. *Syzygium aromaticum*: terpenoids and phenolic compounds. *Ocimum basilicum*: terpenoids.	[121]
	Curcuma longa (350 bar, 65°C, 150 min) and *Curcuma amada* (300 bar, 40°C, 30 min) with ethanol as cosolvent.	Phenolic acids, volatiles, and curcuminoids.	[127]
	Piper nigrum 250 bar, 50°C, 60 min with ethanol 30% v/w.	Oleoresin and piperine.	[128]
	Coriandrum sativum L. 3,000 psi.	Linalool and quercetin glycoside.	[129]
Enzyme-assisted extraction	*Crocus sativus* with Celluclast BG® and Xylanase AN® in acidified medium pH 4, 50°C.	Polyphenolic compounds and anthocyanins.	[130]
	Zingiber officinale with Accellerase® 1,500 with ultrasound 4 hours.	Increased yield of gingerols and shojails compared to Soxhlet extraction.	[131]
	Ocimum canum Sims with Viscozyme® at different concentrations, different times at 50°C.	Terpenoids	[132]
	Cinnamomum camphora with cellulase at 51.47°C, pH 6.15 and 5.39 hours.	Terpenoids	[133]

7.4 ANALYSIS OF SPICES AND BIOACTIVE COMPOUNDS

Instrumental techniques are usually used for the analysis of extracted bioactive compounds from either plants or microorganisms, among the techniques used for such characterization are infrared spectroscopy (FT-IR), ultraviolet-visible spectrometry (UV-Vis), nuclear magnetic resonance (NMR), MS with different ionization modes coupled with either GC or high-performance liquid chromatography (HPLC). In which some of the compounds found on spices such as flavonoids, carotenoids, alkaloids, and among others have a specific UV-Vis pattern, NMR chemical shift, and MS fragmentation pattern that allow different researchers to identify certain structural characteristics similar to phytochemical screening in which a chemical reacts with a structural moiety of the compounds, thus allowing to infer qualitatively the chemical nature of compounds present within the extract, as it is in the case of the study by Gad & Bouzabata [134], with *Curcuma longa* where characteristic signals of curcuminoids were identified by reporting specific IR frequencies associated with specific functional groups as well as the maximum wavelength in UV-Vis spectra and their final identification with the use of HPLC of specific curcuminoids, or in the case of *Armoracia rusticana* studied by Hertz et al. [73] identified different compounds by using different chromatographic techniques. In Table 7.5, it can be seen some analytical techniques of bioactive compounds found on spices depending on the chemical nature.

Other uses of these analytical techniques are to either identify taxonomical variations between spices or to authenticate spices and determine quality standards because of food alterations with the purpose to add more weight to the spice with sand, dirt, artificial color, starch, animal waste among other adulterants in which depending on their livestock sometimes can be easily identified and others cannot where analytical techniques such as scanning electron microscopy, thin layer chromatography (TLC), GC-MS, and HPLC-MS. These techniques present the disadvantage of time-consuming, destructive, special capacitation required as well as lack of accuracy, these difficulties have made difficult the authentication of spices [157].

To solve this problem, other techniques such as FT-IR spectroscopy have been used, because little sample preparation is necessary, fast, and can be coupled to chemometric analysis to enhance the analysis. This analysis has been carried out in black pepper, curry, turmeric, paprika, ginger, onion, anise

TABLE 7.5 Analytical Techniques were Performed on Some Spices

Spice	Analytical Technique and Conditions	Identified Compounds	References
Crocus sativus	HPLC on Ascentis® C18 column, 37°C, photodiode array detector (DAD) at 440 nm, acetonitrile: 1% formic acid, gradient elution.	Crocetin and crocins	[136]
	HPLC-MS on Ultimate XB C18, 30°C, methanol: 0.4% formic acid, gradient elution, DAD at 260, 310, and 440 nm, 0.8 mL/min, 10 μL injection coupled to electron spray ionization (ESI), mode positive and negative ion, m/z 50–1,200, dry gas temperature 350°C, gas flow 12 L/min, 35 psi, capillary voltage 4.5 kV, collision voltage 1 V, mobile phase flow 0.4 mL/min.	Apocarotenoids	[137]
	HPLC with Nucleosil RP C18 column, methanol:water, gradient elution, 1 mL/min, 20 μL injection, UV-Vis detector at 440 nm.	Crocetin esters	[138]
	GC-MS with HP-5 capillary column, He as carrying gas, at 1 mL/min, injector temperature 230°C, split ratio 1:15, detector temperature at 270°C, gradient heating, 70 eV, m/z 50–500 amu.	Safranal	
	GC-MS with Innowax FSC column, He as carrying gas, 0.8 mL/min, gradient heating, injector temperature 250°C, split ratio 40:1, electronic ionization impact (EI), 70 eV, m/z 35–450.	Safranal, α and β-isophorone,	[139]
	HPLC-MS/MS Supelco Discovery C18 column, 0.1% formic acid in methanol:water with 15%acetonitrile and 0.1% formic acid, gradient elution, ESI with triple quadrupole/linear ion trap, negative ion mode, m/z 100–1,200, N as collision gas, collision energy 30, declustering potential 20, gas source 50, temperature 550°C, ion spray voltage 4,500.	Picrococin, crocin A, B, and C, crocetin and its esters	
Curcuma longa	HPLC EC-C8 column, acetonitrile: 0.2% acetic acid (55:45) at 0.5 mL/min, UV-Vis detector at 425 nm, 20 μL injection.	Curcuminoids	[140]
	HPLC Phenomenex Luna C18(2) column, acetonitrile: 0.1% formic acid (50:50), 0.8 mL/min, 40°C, 425 nm, injection volume 10 μL.	Curcuminoids	[141]
	HPLC waters symmetry C18 column, methanol:0.2% acetic acid:acetonitrile, gradient elution, 1 mL/min.	Bisdemethoxycurcumin, desmethoxycurcumin, and curcumin.	[134]

TABLE 7.5 *(Continued)*

Spice	Analytical Technique and Conditions	Identified Compounds	References
	UV-Vis extract solubilized in methanol 200–800 nm.	Different absorbance	
	FT-IR with KBr from 4,000–400 cm^{-1}, resolution of 4 cm^{-1} and 16 scan per sample.	3,400 cm^{-1} (hydroxy group), 3,000–2,800 cm^{-1} (methyl and methylene), 1,740–1,680 cm^{-1} (carbonyl group), 1,510 cm^{-1} (aromatic strectching).	
Armoracia rusticana	GC-flame ionization detector (FID)/MS HP5 column, injector 200C, MS electronic impact 70 eV, He as carrying gas at 1 mL/min, oven temperature from 60 to 280°C at 3°C/min, detector temperature at 300°C for FID and 250°C for MS.	Glucosinolates, isothiocyanates, thiocyanates.	[142]
	HPLC-MS/MS with Agilent Zorbax SB-C18, 30°C, 0.1% formic acid:methanol, gradient elution, injection volume 5 μL, ESI with triple quadrupole detector, negative ion mode, nitrogen temperature 350°C, nebulizer pressure 45 psi, drying gas flow 9 L/min, capillary voltage 3,500 V, fragment voltage 100 V, collision energy 25–30 eV.	Glucosinolates, isothiocyanates, thiocyanates	[143]
	UHPLC-MS-MS/MS with Agilent Zorbax Entend-C18 rapid resolution column, 30°C, 0.01% aqueous formic acid: 0.01% formic acid in acetonitrile, gradient elution, 0.4 mL/min, ESI with a time of flight detector, positive, and negative ionization mode, m/z 100–1,700 for MS and 50–900 for MS/MS, capillary voltage 3.5 kV, source temperature 300°C, nebulizer gas flow 8 mL/min at 35 psi, skimmer 65 V, fragment voltage 175 V.	Glucosinolates, amino acids, phenolic acid, fatty acids, flavonoid glucosides.	[73]
	GC-MS with BP5MS column, initial oven temperature, gradient heating, He as carrying gas at 1.2 mL/min, m/z 50–300, 70 eV, EI.	Isothiocyanates, nitriles, aldehydes, phenolic alcohols, ketones.	
Piper nigrum	UHPLC-DAD with Accucore Vanquish C18 column, water:methanol 80:20, 0.5 mL/min, injection volume 10 μL, 340 nm.	Piperine	[145]

TABLE 7.5 *(Continued)*

Spice	Analytical Technique and Conditions	Identified Compounds	References
	FT-IR with KBr pellet, 400–4,000 cm^{-1}	3,009 cm^{-1} (C-H), 2,937 and 2,841 cm^{-1} (methylene), 1,636 and 1,609 cm^{-1} (diene C=C), 1,611, 1,583, and 1,491 cm^{-1} (phenyl C=C), 1,633 cm^{-1} (amide carbonyl), 1,448 cm^{-1} (methylene), 995 cm^{-1} (C-H of phenyl), 847, 831, and 804 cm^{-1} (out of plane C-H of trisubstituted phenyl group) with differences on peak intensities.	
	NMR with deuterated DMSO, 300 MHz for proton and 75 MHz for carbon, 295 K, 16, and 256 scans for proton and carbon, respectively.	7 ppm, 6 ppm, 3–4 ppm, 2.5 ppm, 0.8–1.8 ppm with differences in their signal intensities and an additional signal at 2 ppm for Vietnam variety.	
	HPLC-DAD with Agilent Eclipse XDB C18, acetonitrile:water (45:55), 1.1 mL/min, 20°C, injection volume 20 μL, 346 nm.	Piperine	[146]
	HPLC-MS with JASCO J-Pak Symphonia C18, 0.1% aqueous formic acid: 0.1% formic acid in acetonitrile (30:70), 0.4 mL/min, injection volume 5 μL, ESI in positive and negative ion mode, ion trap detector, gas flow 60, spray voltage 4.5 kV, m/z 100–1,000 and 115–1,000 for positive and negative modes, respectively.	Monoterpenes and sesquiterpenes	[147]
	GC-MS with an SHRCI-5MS column, injector at 280°C, He as carrying gas 1 mL/min, gradient heating, m/z 33–400.	Terpenes, piperamides	[148]
	HPLC-MS with a Zorbax Eclipse XDB C18 column, 25°C, 0.5% formic acid:acetonitrile at 0.5 mL/min, gradient elution, positive ion mode, sheath gas at 35, auxiliary, and sweep gas at 15, spray voltage 10 V, tube lens 55 V, m/z 50–1,000.	Piperamides	
	Thermal desorption direct analysis in real-time mass spectrometry m/z 50–1,000, positive ion mode, heater at 500°C, gradient heating 30°C–600°C.	Piperamides	

TABLE 7.5 *(Continued)*

Spice	Analytical Technique and Conditions	Identified Compounds	References
Vanilla planifolia	UHPLC-MS with Agilent Zorbax Eclipse Plus C18, ESI time of flight detector, positive ion mode, m/z 50–1,000, water: methanol, gradient elution, N as drying gas (8 L/min, 330°C), nebulizer pressure 60 psi, capillary voltage 3,500 V.	Flavonoids, phenolic acids, lignans, stilbenes.	[149]
	GC-vacuum ultraviolet, injector temperature 250°C, He as carrying gas 30 cm/s, data acquisition rate 10 Hz, Rtx-5 and Stabilwax, gradient heating.	Vanillin, vanillic acid, 4-hydroxybenzaldehyde, guaiacol on natural extracts.	[150]
		Piperonal, ethyl vanillin, vanillin, 4-hydroxybenzaldehyde, caffeine, and vanillic acid on modified extracts.	
	Supercritical fluid chromatography with a Shim-pack UC-X RP column, CO_2:methanol, gradient elution, 2 mL/min, 50°C, DAD at 300, 217, and 254 nm.	Vanillin, p-hydroxybenzaldehyde, vanillic acid, and p-hydroxybenzoic acid.	[151]
	GC-FID/MS with DB1 fused-silica nonpolar column, gradient heating, injector, and detector temperatures at 250 y 300°C, respectively, N as carrying gas 1 mL/min, quadrupole detector, He as carrying gas 1 mL/min, 70 eV, 230°C, detector temperature 250°C.	Monoterpenes, ketones, esters, phenols, alcohols, sesquiterpenes, aldehydes.	[152]
Cinnamomum verum	GC-MS with an RTx-MS column, He as carrying gas, 50 mL/min, gradient heating, 100 kPa, 1.69 mL/min for column flow, injector temperature 300°C, ion source 200°C, and interface 250°C.	Cinnamaldehyde, terpenes, sterols	[153]
	HPLC with a Zorbax Eclipse Plus C18 column, injection volume 10 μL, 0.1% orthophosphoric acid:acetonitrile, 1 mL/min, gradient elution, DAD at 290, 265, 285, and 280 nm.	Cinnamaldehyde, cinnamyl alcohol, coumarin, and coumaric acid.	[154]
	HPLC with a Shimadzu C18 column, UV-Vis detector at 280 nm 0.1% acetic acid:acetonitrile (80:20), 1 mL/min.	Cinnamic acid, methyl eugenol, and cinnamaldehyde.	[155]
	HPLC with Waters XBridge column, 0.5% acetic acid:acetonitrile, gradient elution, 1 mL/min, DAD detector from 210 to 500 nm.	Cinnamic acid and derivatives.	[156]

among other spices with an accuracy above 90% [157]. TLC using silica gel as stationary phase is another technique that has been useful on spices for authentication and adulterant detection in raw material without a purification step such as allspice for eugenol detection; chili pepper for carotenoids and capsaicinoids, as well as illegal dyes; cardamom for phenolic compounds; cinnamon for coumarins, cinnamic acid (CiA), and cinnamaldehyde; among other spices for specific compounds. Regarding the mobile phase the main solvent mixtures that have been observed are hexane:ethyl acetate, ethyl acetate:methanol:water, toluene:ethyl acetate, benzene:isopropyl ether, petroleum ether:acetone:diethylamine, ethyl acetate:acetonitrile, butanol:acetic acid:water, hexane:acetone:formic acid, hexane:chloroform, hexane:chloroform: triethylamine among others depending on the type of compound to be analyzed [158].

Another technique that has been used for spice authentication is NMR and its variants based on the main metabolites of the spice as reported with picrocrocin and crocetin glycosides from Iranian saffron, methyl eugenol from basil, anethole from fennel seed, and piperine from black pepper, as well as on thyme, chili, coriander, cinnamon, onion, turmeric, garlic, and mint. NMR, as well as FT-IR, must be accompanied by multivariate analysis such as discriminant analysis, however, concentrations above 20% are required to detect the presence of adulterants [159–162]. UV-Vis spectrometry is another technique that can be used but with limited application due to the complexity in the electronic interactions with molecules, where the presence of multiple bonds as well as heteroatoms are the main characteristics to be searched in this kind of analysis. Despite its difficulty, it has been used to distinguish methanolic extracts of different varieties of thyme, oregano, saffron [161].

As can be seen, there is a need to develop non-targeted methods in which a profound use of analytical techniques such as the ones mentioned above along with a profound knowledge of the main metabolites present on spices is useful to facilitate spectral interpretation as well as to obtain validated methods and chemometric software [161]. In this regard metabolomics is an emerging field that allows the integration of analytical as well as chemometric and statistical techniques for a complete analysis of the metabolome of samples by using techniques such as LC or GC, UV-Vis, FT-IR, MS, and NMR for the qualitative and quantitative determination of metabolites, thus facilitating the chemometric analysis [163].

7.5 CONCLUSIONS AND FUTURE PROSPECTS

Current techniques for spices and condiments preparation, conditioning, and extraction of bioactive compounds become one of the most important issues to highlight for the mass production of worldwide nutraceuticals, along with the development of novel pharmaceutical delivery systems for phytochemicals. This step has been enhanced with pharmaceutical procedures already applied to conventional and functional foods, providing a less-or-non inalterability of bioactive compounds with the use of different physicochemical and biotechnological methodologies, such as SFE, ultrasound-assisted extraction (UAE), or enzyme-assisted extraction. Currently, a short-scale pressurized SE like hard cap expresso extraction has been studied as a fast, efficient, and low-cost method for the extraction of compounds of interest that are compatible with polar solvents, showing high recovery percentages compared with physical extraction techniques [164, 165]. Likewise, modern optimization of extraction of phytochemicals is performed along with the novel use of statistical tools such as response surface methodology, involving relationships between extracting operation conditions and expected phytochemicals concentrations on these extracts, representing an option for long-scale extraction of bioactive compounds for the production of nutraceuticals [144, 135, 166, 165].

Subsequently, modern analytical techniques are an important tool for the identification and quantification of obtained bioactive compounds from spices and condiments extracts, corroborating nutraceuticals dosage content, inalterability, and less-or-non toxin amount. That can be possible mainly aided for chromatographic analysis such as HPLC, UHPLC, and GC-MS. Recently, the optimization of these instrumental methodologies has been the raising of new procedures, such as multidimensional chromatography, that can be released with a two-column system with the same or different instrumentation between them (LC × LC, GC × GC, LC × GC) for the correct separation of phytochemicals with poor resolution and co-elution problems, representing a real-time quantitative analysis for compounds of interest like volatile EOs, decomposition products and chiral compounds [28]. Integrating all these techniques will allow a better production of spices and condiments related nutraceuticals for use of alternative medicine to prevent and treat the most common worldwide diseases, such as SARS-CoV-2 infection, cancer, neuropsychiatric disorders, and metabolic disorders.

KEYWORDS

- **cancer**
- **chiral compounds**
- **chromatography**
- **metabolic disorders**
- **neuropsychiatric disorders**
- **phytochemicals**
- **SARS-CoV-2 infection**

REFERENCES

1. Wolfender, J. L., Eugster, P. J., Bohni, N., & Cuendet, M., (2011). Advanced methods for natural product drug discovery in the field of nutraceuticals. *Chimia, 65*(6), 400–406.
2. Das, L., Bhaumik, E., Raychaudhuri, U., & Chakraborty, R., (2011). Role of nutraceuticals in human health. *Journal of Food Science and Technology, 49*(2), 173–183. Available from: https://link.springer.com/article/10.1007/s13197-011-0269-4 (accessed on 14 September 2023).
3. Siddiqui, R. A., & Moghadasian, M. H., (2020). Nutraceuticals and nutrition supplements: Challenges and opportunities. *Nutrients, 12*(6), 1593. Available from: https://www.mdpi.com/2072-6643/12/6/1593/htm (accessed on 14 September 2023).
4. Williamson, E. M., Liu, X., & Izzo, A. A., (2020). Trends in use, pharmacology, and clinical applications of emerging herbal nutraceuticals. *British Journal of Pharmacology [Internet], 177*(6), 1227–1240. Available from: https://bpspubs.onlinelibrary.wiley.com/doi/full/10.1111/bph.14943 (accessed on 14 September 2023).
5. Salehi, B., Stojanović-Radić, Z., Matejić, J., Sharifi-Rad, M., Anil, K. N. V., Martins, N., et al., (2019). The therapeutic potential of curcumin: A review of clinical trials. *European Journal of Medicinal Chemistry, 163*, 527–545.
6. Rattis, B. A. C., Ramos, S. G., & Celes, M. R. N., (2021). Curcumin as a potential treatment for COVID-19. *Frontiers in Pharmacology, 12*, 1068.
7. Kotha, R. R., & Luthria, D. L., (2019). Curcumin: Biological, pharmaceutical, nutraceutical, and analytical aspects. *Molecules, 24*(16), 2930. Available from: https://www.mdpi.com/1420-3049/24/16/2930/htm (accessed on 14 September 2023).
8. Selvaraj, K., Katare, D. P., Chand, S., & Chaudhary, N., (2021). *Trachyspermum ammi* and *Cinnamomum verum* as nutraceuticals: Spices rich in therapeutically significant protein tyrosine phosphatases. *Journal of Food Biochemistry [Internet], 45*(6), e13750. Available from: https://onlinelibrary.wiley.com/doi/full/10.1111/jfbc.13750 (accessed on 14 September 2023).

9. Villa-Rivera, M. G., & Ochoa-Alejo, N., (2020). Chili pepper carotenoids: Nutraceutical properties and mechanisms of action. *Molecules, 25*(23), 5573. Available from: https://www.mdpi.com/1420-3049/25/23/5573/htm (accessed on 14 September 2023).
10. Ansary, J., Forbes-Hernández, T. Y., Gil, E., Cianciosi, D., Zhang, J., Elexpuru-Zabaleta, M., et al., (2020). Potential health benefit of garlic based on human intervention studies: A brief overview. *Antioxidants, 9*(7), 619. Available from: https://www.mdpi.com/2076-3921/9/7/619/htm (accessed on 14 September 2023).
11. Srinivasan, K., (2017). Ginger rhizomes (*Zingiber officinale*): A spice with multiple health beneficial potentials. *Pharma Nutrition, 5*(1), 18–28.
12. Arshad, M. S., Sohaib, M., Nadeem, M., Saeed, F., Imran, A., Javed, A., et al., (2017). *Status and Trends of Nutraceuticals from Onion and Onion by-Products: A Critical Review [Internet], 3*(1), 1280254. Available from: http://www.editorialmanager.com/cogentagri; https://www.tandfonline.com/doi/abs/10.1080/23311932.2017.1280254 (accessed on 14 September 2023).
13. Kunnumakkara, A. B., Sailo, B. L., Banik, K., Harsha, C., Prasad, S., Gupta, S. C., et al., (2018). Chronic diseases, inflammation, and spices: How are they linked? *Journal of Translational Medicine, 16*(1), 1–25. Available from: https://translational-medicine.biomedcentral.com/articles/10.1186/s12967-018-1381-2 (accessed on 14 September 2023).
14. Gil-Chávez, G. J., Villa, J. A., Ayala-Zavala, J. F., Heredia, J. B., Sepulveda, D., Yahia, E. M., et al., (2013). Technologies for extraction and production of bioactive compounds to be used as nutraceuticals and food ingredients: An overview. *Comprehensive Reviews in Food Science and Food Safety [Internet], 12*(1), 5–23. Available from: https://onlinelibrary.wiley.com/doi/full/10.1111/1541-4337.12005 (accessed on 14 September 2023).
15. Kumar, A., Mosa, K. A., Ji, L., Kage, U., Dhokane, D., Karre, S., et al., (2017). *Metabolomics-Assisted Biotechnological Interventions for Developing Plant-Based Functional Foods and Nutraceuticals, 58*(11), 1791–807 https://doi.org/101080/1040839820171285752 [Internet]. Available from: https://www.tandfonline.com/doi/abs/10.1080/10408398.2017.1285752 (accessed on 14 September 2023).
16. Helal, N. A., Eassa, H. A., Amer, A. M., Eltokhy, M. A., Edafiogho, I., & Nounou, M. I., (2019). Nutraceuticals' novel formulations: The good, the bad, the unknown and patents involved. *Recent Patents on Drug Delivery & Formulation, 13*(2), 105–156.
17. Bernal, J., Mendiola, J. A., Ibáñez, E., & Cifuentes, A., (2011). Advanced analysis of nutraceuticals. *Journal of Pharmaceutical and Biomedical Analysis, 55*(4), 758–774.
18. Douglas, M., Heyes, J., & Smallfield, B., (2005). *Herbs, Spices and Essential Oils: Post-Harvest Operations in Developing Countries* (pp. 1–5). UNIDO and FAO.
19. Peter, K. V., & Shylaja, M. R., (2012). Introduction to herbs and spices: Definitions, trade and applications. *Handbook of Herbs and Spices* (2nd edn., Vol. 1, pp. 1–24).
20. Schmeda-Hirschmann, G., Jordan, M., Gerth, A., & Wilken, D., (2005). Secondary metabolite content in rhizomes, callus cultures and in vitro regenerated plantlets of solidago chilensis. *Zeitschrift für Naturforschung C [Internet], 60*(1, 2), 5–10. Available from: https://www.degruyter.com/document/doi/10.1515/znc-2005-1-202/html
21. Salgado, F., (2011). El jengibre (*Zingiber officinale*). *Revista Internacional de Acupuntura, 5*(4), 167–173.
22. Hewlings, S. J., & Kalman, D. S., (2017). Curcumin: A review of its effects on human health. *Foods, 6*(10), 92. Available from: https://www.mdpi.com/2304-8158/6/10/92/htm (accessed on 14 September 2023).

23. Prasad, S., & Tyagi, A. K., (2015). Ginger and its constituents: Role in prevention and treatment of gastrointestinal cancer. *Gastroenterology Research and Practice, 2015.*
24. De-la-Cruz, C. I., Riley-Saldaña, C. A., & González-Esquinca, A. R., (2012). Secondary metabolites during early development in plants. *Phytochemistry Reviews, 12*(1), 47–64. Available from: https://link.springer.com/article/10.1007/s11101-012-9250-8 (accessed on 14 September 2023).
25. Kannappan, R., Gupta, S. C., Kim, J. H., Reuter, S., & Aggarwal, B. B., (2011). Neuroprotection by spice-derived nutraceuticals: You are what you eat! *Molecular Neurobiology, 44*(2), 142–159. Available from: https://link.springer.com/article/10.1007/s12035-011-8168-2 (accessed on 14 September 2023).
26. Isbill, J., Kandiah, J., & Khubchandani, J., (2017). Use of ethnic spices by adults in the United States: An exploratory study. *Health Promotion Perspectives [Internet], 8*(1), 33–40. Available from: https://hpp.tbzmed.ac.ir/FullHtml/hpp-19193 (accessed on 14 September 2023).
27. Jiang, T. A., (2019). Health benefits of culinary herbs and spices. *Journal of AOAC International [Internet], 102*(2), 395–411. Available from: https://academic.oup.com/jaoac/article/102/2/395/5658185 (accessed on 14 September 2023).
28. Rasheed, D. M., Serag, A., Abdel, S. Z. T., & Farag, M., (2021). Novel trends and applications of multidimensional chromatography in the analysis of food, cosmetics and medicine bearing essential oils. *Talanta, 223,* 121710.
29. Ibrahim, M. A. A., Abdelrahman, A. H. M., Hussien, T. A., Badr, E. A. A., Mohamed, T. A., El-Seedi, H. R., et al., (2020). In silico drug discovery of major metabolites from spices as SARS-CoV-2 main protease inhibitors. *Computers in Biology and Medicine, 126*, 104046.
30. Gupta, S., Singh, V., Varadwaj, P. K., Chakravartty, N., Katta, A. V. S. K. M., Lekkala, S. P., et al., (2020). *Secondary Metabolites from Spice and Herbs as Potential Multitarget Inhibitors of SARS-CoV-2 Proteins.* https://doi.org/101080/0739110220201837679 [Internet]. Available from: https://www.tandfonline.com/doi/abs/10.1080/07391102.2020.1837679 (accessed on 14 September 2023).
31. Bhattacharya, E., Pal, U., Dutta, R., Bhowmik, P. C., & Mandal, B. S., (2021). Antioxidant, antimicrobial and DNA damage protecting potential of hot taste spices: A comparative approach to validate their utilization as functional foods. *Journal of Food Science and Technology*, 1–12. Available from: https://link.springer.com/article/10.1007/s13197-021-05122-4 (accessed on 14 September 2023).
32. Lu, M., Cao, Y., Xiao, J., Song, M., & Ho, C. T., (2018). Molecular mechanisms of the anti-obesity effect of bioactive ingredients in common spices: A review. *Food & Function [Internet], 9*(9), 4569–4581. Available from: https://pubs.rsc.org/en/content/articlehtml/2018/fo/c8fo01349g (accessed on 14 September 2023).
33. Desjarlais, M., Wirth, M., Lahaie, I., Ruknudin, P., Hardy, P., Rivard, A., et al., (2020). Nutraceutical targeting of inflammation-modulating microRNAs in severe forms of COVID-19: A novel approach to prevent the cytokine storm. *Frontiers in Pharmacology, 11*, 2108.
34. Sepahpour, S., Selamat, J., Manap, M. Y. A., Khatib, A., & Razis, A. F. A., (2018). Comparative analysis of chemical composition, antioxidant activity and quantitative characterization of some phenolic compounds in selected herbs and spices in different solvent extraction systems. *Molecules, 23*(2), 402. Available from: https://www.mdpi.com/1420-3049/23/2/402/htm (accessed on 14 September 2023).

35. Yang, D., Dunshea, F. R., & Suleria, H. A. R., (2020). LC-ESI-QTOF/MS characterization of Australian herb and spices (garlic, ginger, and onion) and potential antioxidant activity. *Journal of Food Processing and Preservation [Internet], 44*(7), e14497. Available from: https://ifst.onlinelibrary.wiley.com/doi/full/10.1111/jfpp.14497 (accessed on 14 September 2023).
36. Singhal, P., Singla, N., Sakhare, D., & Sharma, A., (2016). A comparative evaluation of in-vitro antioxidant activity of some commonly used spices of Northern India. *The Natural Products Journal, 7*(2), 131–136.
37. Nikolic, J. S., Mitic, V. D., Stankov, J. V. P., Dimitrijevic, M. V., & Stojanovic, G. S., (2019). Chemometric characterization of twenty-three culinary herbs and spices according to antioxidant activity. *Journal of Food Measurement and Characterization [Internet], 13*(3), 2167–2176. Available from: https://link.springer.com/article/10.1007/s11694-019-00137-0 (accessed on 14 September 2023).
38. Dzoyem, J. P., McGaw, L. J., Kuete, V., & Bakowsky, U., (2017). Anti-inflammatory and anti-nociceptive activities of African medicinal spices and vegetables. *Medicinal Spices and Vegetables from Africa: Therapeutic Potential Against Metabolic, Inflammatory, Infectious and Systemic Diseases*, 239–270.
39. Matulyte, I., Jekabsone, A., Jankauskaite, L., Zavistanaviciute, P., Sakiene, V., Bartkiene, E., et al., (2020). The essential oil and hydrolats from Myristica fragrans seeds with magnesium aluminometasilicate as excipient: Antioxidant, antibacterial, and anti-inflammatory activity. *Foods, 9*(1), 37. Available from: https://www.mdpi.com/2304-8158/9/1/37/htm (accessed on 14 September 2023).
40. Pina, L. T. S., Ferro, J. N. S., Rabelo, T. K., Oliveira, M. A., Scotti, L., Scotti, M. T., et al., (2018). *Alcoholic Monoterpenes Found in Essential Oil of Aromatic Spices Reduce Allergic Inflammation by the Modulation of Inflammatory Cytokines, 33*(12), 1773–1777. https://doi.org/101080/1478641920181434634 [Internet]. Available from: https://www.tandfonline.com/doi/abs/10.1080/14786419.2018.1434634 (accessed on 14 September 2023).
41. Edwards, R. L., Luis, P. B., Varuzza, P. V., Joseph, A. I., Presley, S. H., Chaturvedi, R., et al., (2017). The anti-inflammatory activity of curcumin is mediated by its oxidative metabolites. *Journal of Biological Chemistry [Internet], 292*(52), 21243–21252. Available from: http://www.jbc.org/article/S0021925820326910/fulltext (accessed on 14 September 2023).
42. Khafaji, S. S. O., (2018). Subject review: Pharmacological application of thyme. *Advances in Animal and Veterinary Sciences* (Vol. 6, pp. 366–371). Nexus Academic Publishers.
43. Joshi, D. R., Schrestha, A. C., & Adhikari, N. (2018). A review on diversified use of the king of spices: Piper nigrum (black pepper) [Internet]. *International Journal of Pharmaceutical Sciences and Research* (pp. 4089–4101). Available from: https://ijpsr.com/bft-article/a-review-on-diversified-use-of-the-king-of-species-piper-nigrum-black-pepper/ (accessed on 14 September 2023).
44. Bakshi, H. A., Faruck, H. L., Yadav, S. A., & Tambuwala, M. M., (2020). The remarkable pharmacological efficacy of saffron spice via antioxidant, immunomodulatory, and antitumor activities. *Saffron: The Age-Old Panacea in a New Light*, 245–262.
45. Muhammad, A., Ibrahim, M. A., Erukainure, O. L., Malami, I., & Adamu, A., (2018). Spices with breast cancer chemopreventive and therapeutic potentials: A functional foods based-review. *Anti-Cancer Agents in Medicinal Chemistry, 18*(2), 182–194.

46. Khanal, A., Devkota, H. P., Kaundinnyayana, S., Gyawali, P., Ananda, R., & Adhikari, R., (2021). Culinary herbs and spices in Nepal: A review of their traditional uses, chemical constituents, and pharmacological activities. *Ethnobotany Research and Applications [Internet], 21*, 1–18. Available from: https://ethnobotanyjournal.org/index.php/era/article/view/2753 (accessed on 14 September 2023).
47. Liu, Q., Meng, X., Li, Y., Zhao, C. N., Tang, G. Y., & Li, H. B., (2017). Antibacterial and antifungal activities of spices. *International Journal of Molecular Sciences, 18*(6), 1283. Available from: https://www.mdpi.com/1422-0067/18/6/1283/htm (accessed on 14 September 2023).
48. Maharjan, R., Thapa, S., & Acharya, A., (2019). Evaluation of antimicrobial activity and synergistic effect of spices against few selected pathogens. *Tribhuvan University Journal of Microbiology [Internet], 6*, 10–18. Available from: https://www.nepjol.info/index.php/tujm/article/view/26573 (accessed on 14 September 2023).
49. Condò, C., Anacarso, I., Sabia, C., Iseppi, R., Anfelli, I., Forti, L., et al., (2018). *Antimicrobial Activity of Spices Essential Oils and its Effectiveness on Mature Biofilms of Human Pathogens, 34*(4), 567–574. https://doi.org/101080/1478641920181490904 [Internet], Available from: https://www.tandfonline.com/doi/abs/10.1080/14786419.2018.1490904 (accessed on 14 September 2023).
50. D'Souza, S. P., Chavannavar, S. V., Kanchanashri, B., & Niveditha, S. B., (2017). *Pharmaceutical Perspectives of Spices and Condiments as Alternative Antimicrobial Remedy, 22*(4), 1002–1010. https://doi.org/101177/2156587217703214 [Internet]. Available from: https://journals.sagepub.com/doi/10.1177/2156587217703214 (accessed on 14 September 2023).
51. Silveira, S. R., Dornelas, M. C., & Martinelli, A. P., (2016). Perspectives for a framework to understand aril initiation and development. *Frontiers in Plant Science, 7*, 1919.
52. Zhang, C. R., Jayashree, E., Kumar, P. S., & Nair, M. G., (2015). *Antioxidant and Anti-inflammatory Compounds in Nutmeg (Myristica fragrans) Pericarp as Determined by in vitro Assays, 10*(8), 1399–1402 https://doi.org/101177/1934578X1501000822 [Internet]. Available from: https://journals.sagepub.com/doi/abs/10.1177/1934578X1501000822 (accessed on 14 September 2023).
53. Dwivedi, D. H., & Ram, R. B., (2008). Chemical composition of bitter apricot kernels from Ladakh, India. *Acta Horticulturae, 765*, 335–338.
54. Abourashed, E. A., & El-Alfy, A. T., (2016). Chemical diversity and pharmacological significance of the secondary metabolites of nutmeg (*Myristica fragrans* Houtt.). *Phytochemistry Reviews, 15*(6), 1035–1056. [Internet]. Available from: https://link.springer.com/article/10.1007/s11101-016-9469-x (accessed on 14 September 2023).
55. Rao, P. V., & Gan, S. H., (2014). Cinnamon: A multifaceted medicinal plant. *Evidence-Based Complementary and Alternative Medicine, 2014*.
56. Mateos-Martín, M. L., Fuguet, E., Quero, C., Pérez-Jiménez, J., & Torres, J. L., (2011). New identification of proanthocyanidins in cinnamon (*Cinnamomum zeylanicum* L.) using MALDI-TOF/TOF mass spectrometry. *Analytical and Bioanalytical Chemistrym., 402*(3), 1327–1336. [Internet]. Available from: https://link.springer.com/article/10.1007/s00216-011-5557-3 (accessed on 14 September 2023).
57. Wijeweera, A. A., Hewage, J. W., Jayasinghe, G. G., Wadumethrige, S. H., Hettiarachchi, S. R., & Wijesinghe, K. G. G., (2020). Maturity dependence of quality, quantity and chemical constituents of bark and leaf oil of Ceylon cinnamon (*Cinnamomum*

zeylanicum Bliume). *Ruhuna Journal of Science, 11*(1), 1. [Internet]. Available from: http://rjs.sljol.info/articles/10.4038/rjs.v11i1.82/ (accessed on 14 September 2023).
58. Ashokkumar, K., Murugan, M., Dhanya, M. K., Pandian, A., & Warkentin, T. D., (2021). Phytochemistry and therapeutic potential of black pepper [*Piper nigrum* (L.)] essential oil and piperine: A review. *Clinical Phytoscience, 7*(1), 1–11. [Internet]. Available from: https://clinphytoscience.springeropen.com/articles/10.1186/s40816-021-00292-2 (accessed on 30 September 2023).
59. Peter, K. V., & Shylaja, M. R., (2012). Introduction to herbs and spices: Definitions, trade, and applications. *Handbook of Herbs and Spices* (2nd edn., Vol. 1, pp.1–24).
60. Kaur, K., & Kaushal, S., (2019). Phytochemistry and pharmacological aspects of *Syzygium aromaticum*: A review. *Journal of Pharmacognosy and Phytochemistry, 8*(1), 398–406. [Internet]. Available from: https://www.phytojournal.com/archives/2019.v8.i1.6762/phytochemistry-and-pharmacological-aspects-of-ltemgtsyzygium-aromaticumltemgt-a-review (accessed on 14 September 2023).
61. Hussein, H. J., Hameed, I. H., & Hadi, M. Y., (2017). A Review: Anti-microbial, Anti-inflammatory effect and cardiovascular effects of garlic: *Allium sativum. Research Journal of Pharmacy and Technology, 10*(11), 4069–4078 [Internet]. Available from: https://rjptonline.org/AbstractView.aspx?PID=2017-10-11-77 (accessed on 14 September 2023).
62. Diretto, G., Rubio-Moraga, A., Argandoña, J., Castillo, P., Gómez-Gómez, L., & Ahrazem, O., (2017). Tissue-specific accumulation of sulfur compounds and saponins in different parts of garlic cloves from purple and white ecotypes. *Molecules, 22*(8), 1359. [Internet]. Available from: https://www.mdpi.com/1420-3049/22/8/1359/htm (accessed on 14 September 2023).
63. Martínez-Casas, L., Lage-Yusty, M., & López-Hernández, J., (2017). Changes in the aromatic profile, sugars, and bioactive compounds when purple garlic is transformed into black garlic. *Journal of Agricultural and Food Chemistry, 65*(49), 10804–10811 [Internet]. Available from: https://pubs.acs.org/doi/abs/10.1021/acs.jafc.7b04423 (accessed on 14 September 2023).
64. Ali, M., & Ibrahim, I. S. (2021). *Phytochemical Screening and Proximate Analysis of Garlic (Allium Sativum)* [Internet]. Available from: https://lupinepublishers.com/chemistry-journal/fulltext/phytochemical-screening-and-proximate-analysis-of-garlic-allium-sativum.ID.000180.php (accessed on 14 September 2023).
65. Baek, S. C., Nam, K. H., Yi, S. A., Jo, M. S., Lee, K. H., Lee, Y. H., et al., (2019). Anti-adipogenic effect of β-carboline alkaloids from garlic (*Allium sativum*). *Foods, 8*(12), 673. [Internet]. Available from: https://www.mdpi.com/2304-8158/8/12/673/htm (accessed on 14 September 2023).
66. Gohari, A. R., Saeidnia, S., & Mahmoodabadi, M. K., (2013). An overview on saffron, phytochemicals, and medicinal properties. *Pharmacognosy Reviews, 7*(13), 61. [Internet]. Available from: https://www.ncbi.nlm.nih.gov/pmc/articles/PMC3731881/pdf/PRev-7-61.pdf (accessed on 30 September 2023).
67. Ahrazem, O., Rubio-Moraga, A., Nebauer, S. G., Molina, R. V., & Gómez-Gómez, L., (2015). Saffron: Its phytochemistry, developmental processes, and biotechnological prospects. *Journal of Agricultural and Food Chemistry, 63*(40), 8751–8764. [Internet]. Available from: https://pubs.acs.org/doi/abs/10.1021/acs.jafc.5b03194 (accessed on 14 September 2023).

68. Shahrajabian, M. H., Sun, W., & Cheng, Q., (2020). *Chemical Components and Pharmacological Benefits of Basil (Ocimum basilicum): A Review, 23*(1), 1961–1970. https://doi.org/101080/1094291220201828456 [Internet]. Available from: https://www.tandfonline.com/doi/abs/10.1080/10942912.2020.1828456 (accessed on 14 September 2023).
69. Avetisyan, A., Markosian, A., Petrosyan, M., Sahakyan, N., Babayan, A., Aloyan, S., et al., (2017). Chemical composition and some biological activities of the essential oils from basil *Ocimum* different cultivars. *BMC Complementary and Alternative Medicine, 17*(1), 1–8. [Internet]. Available from: https://bmccomplementmedtherapies.biomedcentral.com/articles/10.1186/s12906-017-1587-5 (accessed on 14 September 2023).
70. Amalraj, A., & Gopi, S., (2017). Biological activities and medicinal properties of *Asafoetida*: A review. *Journal of Traditional and Complementary Medicine, 7*(3), 347–359.
71. Tomsone, L., Kruma, Z., Galoburda, R., & Talou, T., (2013). Composition of volatile compounds of horseradish roots (*Armoracia rusticana* L.) depending on the genotype. *Proceedings of the Latvia University of Agriculture, 29*(1), 1–10.
72. Nguyen, N. M., Gonda, S., & Vasas, G., (2013). *A Review on the Phytochemical Composition and Potential Medicinal Uses of Horseradish (Armoracia Rusticana) Root, 29*(3), 261–275. http://dx.doi.org/101080/875591292013790047 [Internet]. Available from: https://www.tandfonline.com/doi/abs/10.1080/87559129.2013.790047 (accessed on 14 September 2023).
73. Herz, C., Tran, H. T. T., Márton, M. R., Maul, R., Baldermann, S., Schreiner, M., et al., (2017). Evaluation of an aqueous extract from horseradish root (*Armoracia rusticana* Radix) against lipopolysaccharide-induced cellular inflammation reaction. *Evidence-based Complementary and Alternative Medicine, 2017.*
74. Popović, M., Maravić, A., Čulić, V. Č., Đulović, A., Burčul, F., & Blažević, I., (2020). Biological effects of glucosinolate degradation products from horseradish: A horse that wins the race. *Biomolecules, 10*(2), 343. [Internet]. Available from: https://www.mdpi.com/2218-273X/10/2/343/htm (accessed on 14 September 2023).
75. Wani, S. A., & Kumar, P., (2018). Fenugreek: A review on its nutraceutical properties and utilization in various food products. *Journal of the Saudi Society of Agricultural Sciences, 17*(2), 97–106.
76. Hoque, M., Bala, B., Hossain, M., & Uddin, M., (2013). Drying kinetics of ginger rhizome (*Zingiber officinale*). *Bangladesh Journal of Agricultural Research, 38*(2), 301–319.
77. Ghasemzadeh, A., Jaafar, H. Z. E., Rahmat, A., & Swamy, M. K., (2017). Optimization of microwave-assisted extraction of zerumbone from *Zingiber zerumbet* L. rhizome and evaluation of antiproliferative activity of optimized extracts. *Chemistry Central Journal, 11*(1).
78. Jin, W., Mujumdar, A. S., Zhang, M., & Shi, W., (2018). Novel drying techniques for spices and herbs: A review. *Food Engineering Reviews, 10*(1).
79. Bhatta, S., Stevanovic, J. T., & Ratti, C., (2020). Freeze-drying of plant-based foods. *Foods, 9*(1).
80. Prusinowska, R., & Smigielski, K., (2015). Losses of essential oils and antioxidants during the drying of herbs and spices: A review. *Nauki Inżynierskie i Technologie, 2*(17), 51–62.

81. Oluba, O. M., Mbamara, D. F. O., Akpor, O. B., Adebiyi, F. D., Alabi, O. O., Shoyombo, A., et al., (2021). *Effects of Drying Methods on Compositional Characterization and Functional Characteristics of Blighia Sapida Aril Oil* (Vol. 28). OCL.
82. Aamir, M., & Jittanit, W., (2017). Ohmic heating treatment for Gac aril oil extraction: Effects on extraction efficiency, physical properties and some bioactive compounds. *Innovative Food Science & Emerging Technologies, 41*, 224–234.
83. Srinivas, Y., Mathew, S. M., Kothakota, A., Sagarika, N., & Pandiselvam, R., (2020). Microwave assisted fluidized bed drying of nutmeg mace for essential oil enriched extracts: An assessment of drying kinetics, process optimization and quality. *Innovative Food Science & Emerging Technologies, 66*, 102541.
84. Ghafoor, K., Al Juhaimi, F., Özcan, M. M., Uslu, N., Babiker, E. E., & Mohamed, A. I. A., (2020). Total phenolics, total carotenoids, individual phenolics and antioxidant activity of ginger (*Zingiber officinale*) rhizome as affected by drying methods. *LWT, 126*, 109354.
85. Hadi, S., Artanti, A. N., Rinanto, Y., & Wahyuni, D. S. C., (2018). Curcuminoid content of *Curcuma longa* L. and *Curcuma xanthorrhiza* rhizome based on drying method with NMR and HPLC-UVD. *IOP Conference Series: Materials Science and Engineering, 349.*
86. Bernard, D., Kwabena, A., Osei, O., Daniel, G., Elom, S., & Sandra, A., (2014). The effect of different drying methods on the phytochemicals and radical scavenging activity of Ceylon cinnamon (*Cinnamomum zeylanicum*) plant parts. *European Journal of Medicinal Plants, 4*(11).
87. Baiqi, A. U., Utami, P. P., Anugrah, D., Fauzan, A. A., Ningsih, W. S., & Rusydi, M. I., (2019). Improving the quality and quantity of cinnamon drying process using art cave in Lambung Bukit West Sumatra. *IOP Conference Series: Materials Science and Engineering, 602.*
88. Kumarage, N., (2013). *Effect of Drying Temperature on the Composition of Hydro Distilled Cinnamon Bark Oil.* Sri Lanka.
89. Lakshmi, D. V. N., Muthukumar, P., & Nayak, P. K., (2021). Experimental investigations on active solar dryers integrated with thermal storage for drying of black pepper. *Renewable Energy, 167*, 728–739.
90. Paul, S., Ara, R., Rashid, A. M., Hajong, P., Paul, G., Shahriar, K. M., et al., (2021). Effect of blanching time and drying method on quality of black pepper (*Piper nigrum*). *Journal of Food Technology Research, 8*(1).
91. Roslan, N. F. M., & Yudin, A. S. M., (2020). Drying process of black pepper in a swirling fluidized bed dryer using experimental method. *IOP Conference Series: Materials Science and Engineering, 863.*
92. Zhang, X. L., Zhong, C. S., Mujumdar, A. S., Yang, X. H., Deng, L. Z., Wang, J., et al., (2019). Cold plasma pretreatment enhances drying kinetics and quality attributes of chili pepper (*Capsicum annuum* L.). *Journal of Food Engineering, 241*, 51–57.
93. Olguín, R. A., Vázquez-León, L. A., Salgado-Cervantes, M. A., Fernandez-Barbero, G., Díaz-Pacheco, A., García-Alvarado, M. A., et al., (2019). Water and phytochemicals dynamic during drying of red habanero chili pepper (*Capsicum chinense*) SLICES. *Revista Mexicana de Ingeniería Química, 18*(3).
94. Díaz-Maroto, M. C., Pérez-Coello, M. S., & Cabezudo, M. D., (2002). Effect of drying method on the volatiles in bay leaf (*Laurus nobilis* L.). *Journal of Agricultural and Food Chemistry, 50*(16).

95. Łyczko, J., Masztalerz, K., Lipan, L., Iwiński, H., Lech, K., Carbonell-Barrachina, Á. A., et al., (2021). *Coriandrum sativum* L.—Effect of multiple drying techniques on volatile and sensory profile. *Foods, 10*(2).
96. Ozdemir, N., Ozgen, Y., Kiralan, M., Bayrak, A., Arslan, N., & Ramadan, M. F., (2018). Effect of different drying methods on the essential oil yield, composition, and antioxidant activity of *Origanum vulgare* L. and *Origanum onites* L. *Journal of Food Measurement and Characterization, 12*(2).
97. Altay, K., Hayaloglu, A. A., & Dirim, S. N., (2019). Determination of the drying kinetics and energy efficiency of purple basil (*Ocimum basilicum* L.) leaves using different drying methods. *Heat and Mass Transfer, 55*(8).
98. Murni, V. W., Saepudin, E., Cahyana, A. H., Rahayu, D. U. C., Hastuti, L. T., & Haib, J. (2017). Effect of oven drying and storage on essential oil composition of clove (*Syzygium aromaticum*) from Toli-Toli. In *AIP Conference Proceedings,1862*(1). AIP Publishing.
99. Dewanti, B., Ismail, A., & Adesta, E., (2020). Alternative drying methods to improve the quality of dried cloves. *Test Engineering and Management, 83*, 6928–6939.
100. Tan, S., & Karim, M., (2018). *Effects of Three Drying Treatments on the Polyphenol Content, Antioxidant and Antimicrobial Properties of Syzygium Aromaticum Extract.* [Chiang Mai].
101. Luan, C., Zhang, M., Mujumdar, A. S., & Liu, Y., (2021). Influence of pulse-spouted infrared freeze drying on nutrition, flavor, and application of horseradish. *Drying Technology, 39*(9).
102. Gawrysiak-Witulska, M., Siger, A., Rudzińska, M., & Bartkowiak-Broda, I., (2020). The effect of drying on the native tocopherol and phytosterol content of *Sinapis alba* L. seeds. *Journal of the Science of Food and Agriculture, 100*(1).
103. Bettaieb, R. I., Bourgou, S., Ben, K. S., Aidi, W. W., Ksouri, R., Saidani, T. M., et al., (2020). On the effect of initial drying techniques on essential oil composition, phenolic compound, and antioxidant properties of anise (*Pimpinella anisum* L.) seeds. *Journal of Food Measurement and Characterization, 14*(1).
104. Wen, Y. X., Chen, L. Y., Li, B. S., Ruan, Z., & Pan, Q., (2020). Effect of infrared radiation-hot air (IR-HA) drying on kinetics and quality changes of star anise (*Illicium verum*). *Drying Technology, 39*(1).
105. Aghaei, Z., Jafari, S. M., Dehnad, D., Ghorbani, M., & Hemmati, K., (2018). Refractance-window as an innovative approach for the drying of saffron petals and stigma. *Journal of Food Process Engineering, 41*(7).
106. Edith, D. M. J., Dimitry, M. Y., Richard, N. M., Armand, A. B., Léopold, T. N., & Nicolas, N. Y., (2018). Effect of drying treatment on nutritional, functional and sensory properties of three varieties of onion powders. *Journal of Food Measurement and Characterization, 12*(4).
107. Afolabi, E. T., & Tame, V. T., (2020). Effect of drying methods and packaging on the nutritional values of onions (*Allium cepa* L.) bulbs. *Asian Plant Research Journal.*
108. Koménan, A. C. A., Bamba, B. S. B., N'gouin, J. A., Akré, M. S. H., & Soro, Y. R., (2021). Effects of drying methods and onion bulb variety on physicochemical and functional properties of onion powder. *Asian Food Science Journal.*
109. Makarichian, A., Amiri, C. R., Ahmadi, E., & Mohtasebi, S. S., (2021). Assessment the influence of different drying methods and pre-storage periods on garlic (*Allium sativum* L.) aroma using electronic nose. *Food and Bioproducts Processing, 127*, 198–211.

110. Zhou, L., Guo, X., Bi, J., Yi, J., Chen, Q., Wu, X., et al., (2017). Drying of garlic slices (*Allium sativum* L.) and its effect on thiosulfinates, total phenolic compounds and antioxidant activity during infrared drying. *Journal of Food Processing and Preservation, 41*(1).
111. Siddique, M. A. B., Tzima, K., Rai, D. K., & Brunton, N., (2020). Conventional extraction techniques for bioactive compounds from herbs and spices. In: *Herbs, Spices and Medicinal Plants.* Wiley.
112. Sánchez-Camargo, A. P., Montero, L., Mendiola, J. A., Herrero, M., & Ibáñez, E., (2020). Novel extraction techniques for bioactive compounds from herbs and spices. In: *Herbs, Spices and Medicinal Plants.* Wiley.
113. Manjare, S. D., & Dhingra, K., (2019). Supercritical fluids in separation and purification: A review. *Materials Science for Energy Technologies, 2*(3), 463–484.
114. Simoh, S., & Zainal, A., (2015). Chemical profiling of *Curcuma aeruginosa* Roxb. rhizome using different techniques of solvent extraction. *Asian Pacific Journal of Tropical Biomedicine, 5*(5), 412–417.
115. Pawar, H. A., Gavasane, J., & Choudhary, P. D., (2018). A novel and simple approach for extraction and isolation of curcuminoids from turmeric rhizomes. *Natural Products Chemistry & Research, 6*(1).
116. Sokamte, T. A., Mbougueng, P. D., Tatsadjieu, N. L., & Sachindra, N. M., (2019). Phenolic compounds characterization and antioxidant activities of selected spices from Cameroon. *South African Journal of Botany, 121*, 7–15.
117. Singh, G., Marimuthu, P., Heluani, C. S. De., & Catalan, C., (2005). Antimicrobial and antioxidant potentials of essential oil and acetone extract of *Myristica fragrans* Houtt. (Aril Part). *Journal of Food Science, 70*(2).
118. Shafeena, P., Sravan, J., Sreelatha, E., Steny, S., Theertha, M., & Prince, M., (2017). *Optimization of Microwave Assisted Process for Extraction of Nutmeg Seed Essential Oil.* Kerala, India.
119. Hassan, A., Ali, R. A., & Adel, Z., (2020). Extraction, antioxidant activity and characterization of essential oils of different spices. *International Journal of Scientific Research in Chemical Sciences, 7*(4), 21–29.
120. Ashokkumar, K., Vellaikumar, S., Murugan, M., Dhanya, M. K., Karthikeyan, A., Ariharasutharsan, G., et al., (2021). *GC/MS Analysis of Essential Oil Composition from Selected Seed Spices.* National Academy Science Letters.
121. Győri, E., Varga, A., Fábián, I., & Lázár, I., (2019). Supercritical CO_2 extraction and selective adsorption of aroma materials of selected spice plants in functionalized silica aerogels. *The Journal of Supercritical Fluids, 148*, 16–23.
122. Stanciu, G., Cristache, N., Lupsor, S., & Dobrinas, S., (2017). Evaluation of antioxidant activity and total phenols content in selected spices. *Rev Chim., 68*, 1429–1434.
123. Ganapathy, G., & Nair, A. R., (2017). Curcuminoids in *Zingiber zerumbet* rhizomes: Bioguided fractionation and chromatographic identification of antimicrobial and antioxidant metabolites. *Journal of Herbs, Spices & Medicinal Plants, 23*(2).
124. Bouchachia, C., Benkaci-Ali, F., Eppe, G., & Scholl, G., (2017). Effect of different parameters on composition of volatile components of *Myristica fragrans* seeds extracted by hydrodistillation assisted by microwave and head-space solid-phase micro-extraction. *Journal of Essential Oil Research, 29*(6).

125. Valizadeh, K. M., Najafpour-Darzi, G., Rahimnejad, M., Moghadamnia, A. A., & Valizadeh, K. M., (2016). High performance curcumin subcritical water extraction from turmeric (*Curcuma longa* L.). *Journal of Chromatography B, 1022*, 191–198.
126. Carvalho, P. I. N., Osorio-Tobón, J. F., Rostagno, M. A., Petenate, A. J., & Meireles, M. A. A., (2015). Techno-economic evaluation of the extraction of turmeric (*Curcuma longa* L.) oil and ar-turmerone using supercritical carbon dioxide. *The Journal of Supercritical Fluids, 105*, 44–54.
127. Nagavekar, N., & Singhal, R. S., (2019). Supercritical fluid extraction of *Curcuma longa* and *Curcuma amada* oleoresin: Optimization of extraction conditions, extract profiling, and comparison of bioactivities. *Industrial Crops and Products, 134*, 134–145.
128. Nagavekar, N., & Singhal, R. S., (2018). Enhanced extraction of oleoresin from *Piper nigrum* by supercritical carbon dioxide using ethanol as a co-solvent and its bioactivity profile. *Journal of Food Process Engineering, 41*(1).
129. Iqbal, M., Butt, M., & Shehzad, A., (2018). In vitro antioxidant and chromatographic quantification of supercritical fluid extracts obtained from coriander (*Coriandrum sativum* l.). *Journal of the Chemical Society of Pakistan, 40*(4), 733–741.
130. Vardakas, A. T., Shikov, V. T., Dinkova, R. H., & Mihalev, K. M., (2021). Optimization of the enzyme-assisted extraction of polyphenols from saffron (*Crocus sativus* L.) petals [pdf]. *Acta Scientiarum Polonorum Technologia Alimentaria, 20*(3).
131. Varakumar, S., Umesh, K. V., & Singhal, R. S., (2017). Enhanced extraction of oleoresin from ginger (*Zingiber officinale*) rhizome powder using enzyme-assisted three phase partitioning. *Food Chemistry, 216*, 27–36.
132. Morsy, N. F. S., & Hammad, K. S. M., (2021). Extraction of essential oil from methyl cinnamate basil (*Ocimum canum* Sims) with high yield in a short time using enzyme pretreatment. *Journal of Food Science and Technology, 58*(7).
133. Yu, H., Ren, X., Liu, Y., Xie, Y., Guo, Y., Cheng, Y., et al., (2019). Extraction of *Cinnamomum camphora* chvar. Borneol essential oil using neutral cellulase assisted-steam distillation: Optimization of extraction, and analysis of chemical constituents. *Industrial Crops and Products, 141*, 111794.
134. Gad, H. A., & Bouzabata, A., (2017). Application of chemometrics in quality control of turmeric (*Curcuma longa*) based on ultra-violet, Fourier transform-infrared and 1H NMR spectroscopy. *Food Chemistry, 237*, 857–864.
135. Liu, C., Chen, W., Wang, N., & Jin, J., (2018). Optimization of extraction of antioxidants from turmeric (*Curcuma longa* L.) using response surface methodology. *Wuhan University Journal of Natural Sciences, 23*(1), 63–69. Available from: https://springerlink.uam.elogim.com/article/10.1007/s11859-018-1295-0 (accessed on 14 September 2023).
136. Reddy, C. N., Bharate, S. B., Vishwakarma, R. A., & Bharate, S. S., (2020). Chemical analysis of saffron by HPLC based crocetin estimation. *Journal of Pharmaceutical and Biomedical Analysis, 181*, 113094.
137. Xiaobin, F., Xiaodong, Q., Shuwen, H., Chong, Y., Yumei, Y., & Guifen, Z., (2018). Extracted apocarotenoids from saffron stigmas and evaluated the quality of saffron. *Natural Product Research, 32*(2).
138. Sarfarazi, M., Jafari, S. M., Rajabzadeh, G., & Galanakis, C. M., (2020). Evaluation of microwave-assisted extraction technology for separation of bioactive components of saffron (*Crocus sativus* L.). *Industrial Crops and Products, 145*, 111978.

139. Kosar, M., Demirci, B., Goger, F., Kara, I., & Baser, K. H. C., (2017). *International Journal of Food Properties Volatile Composition, Antioxidant Activity, and Antioxidant Components in Saffron Cultivated in Turkey.* Available from: https://www.tandfonline.com/action/journalInformation?journalCode=ljfp20 (accessed on 14 September 2023).
140. Ali, Z., Saleem, M., Atta, B. M., Khan, S. S., & Hammad, G., (2019). Determination of curcuminoid content in turmeric using fluorescence spectroscopy. *Spectrochimica Acta Part A: Molecular and Biomolecular Spectroscopy, 213,* 192–198.
141. Peram, M. R., Jalalpure, S. S., Joshi, S. A., Palkar, M. B., & Diwan, P. V., (2017). Single robust RP-HPLC analytical method for quantification of curcuminoids in commercial turmeric products, Ayurvedic medicines, and nanovesicular systems. *Journal of Liquid Chromatography & Related Technologies, 40*(10).
142. Petrović, S., Drobac, M., Ušjak, L., Filipović, V., Milenković, M., & Niketić, M., (2017). Volatiles of roots of wild-growing and cultivated *Armoracia macrocarpa* and their antimicrobial activity, in comparison to horseradish, *A. rusticana. Industrial Crops and Products, 109.*
143. Bertóti, Böszörményi, Alberti, Béni, M-Hamvas, Szőke, et al., (2019). Variability of bioactive glucosinolates, isothiocyanates and enzyme patterns in horseradish hairy root cultures initiated from different organs. *Molecules, 24*(15).
144. Sarfarazi, M., Jafari, S. M., & Rajabzadeh, G., (2015). Extraction optimization of saffron nutraceuticals through response surface methodology. *Food Analytical Methods, 8*(9), 2273–2285. Available from: https://link.springer.com/article/10.1007/s12161-014-9995-3 (accessed on 14 September 2023).
145. Ahmad, R., Ahmad, N., Amir, M., Aljishi, F., Alamer, M. H., Al-Shaban, H. R., et al., (2020). Quality variation and standardization of black pepper (*Piper nigrum*): A comparative geographical evaluation based on instrumental and metabolomics analysis. *Biomedical Chromatography, 34*(3).
146. Al-Nidawi, M., Alshana, U., Caleb, J., Hassan, M., Rahman, Z. Ur., Hanoğlu, D. Y., et al., (2020). Switchable-hydrophilicity solvent liquid-liquid microextraction versus dispersive liquid-liquid microextraction prior to HPLC-UV for the determination and isolation of piperine from *Piper nigrum* L. *Journal of Separation Science, 43*(15).
147. Andriana, Y., Xuan, T. D., Quy, T. N., Tran, H. D., & Le, Q. T., (2019). Biological activities and chemical constituents of essential oils from *Piper cubeba* Bojer and *Piper nigrum* L. *Molecules, 24*(10).
148. Liang, J., Sun, J., Chen, P., Frazier, J., Benefield, V., & Zhang, M., (2021). Chemical analysis and classification of black pepper (*Piper nigrum* L.) based on their country of origin using mass spectrometric methods and chemometrics. *Food Research International, 140.*
149. Busconi, M., Lucini, L., Soffritti, G., Bernardi, J., Bernardo, L., Brunschwig, C., et al., (2017). Phenolic profiling for traceability of *Vanilla × tahitensis. Frontiers in Plant Science,* 8.
150. Santos, I. C., Smuts, J., & Schug, K. A., (2017). Rapid profiling and authentication of vanilla extracts using gas chromatography-vacuum ultraviolet spectroscopy. *Food Analytical Methods, 10*(12).
151. Liang, Y., Liu, J., Zhong, Q., Shen, L., Yao, J., Huang, T., et al., (2018). Determination of major aromatic constituents in vanilla using an on-line supercritical fluid extraction coupled with supercritical fluid chromatography. *Journal of Separation Science, 41*(7).

152. Yeh, C. H., Chen, K. Y., Chou, C. Y., Liao, H. Y., & Chen, H. C., (2021). New insights on volatile components of *Vanilla planifolia* cultivated in Taiwan. *Molecules, 26*(12).
153. Ahmed, H. M., Ramadhani, A. M., Erwa, I. Y., Ishag, O. A. O., & Saeed, M. B., (2020). Phytochemical screening, chemical composition, and antimicrobial activity of cinnamon verum bark. *International Research Journal of Pure and Applied Chemistry.*
154. Liyanage, N. M. N., Bandusekara, B. S., Kanchanamala, R. W. M. K., Hathurusinghe, H. A. B. M., Rathnayaka, A. M. R. W. S. D., Pushpakumara, D. K. N. G., et al., (2021). Identification of superior *Cinnamomum zeylanicum Blume germplasm* for future true cinnamon breeding in the world. *Journal of Food Composition and Analysis, 96.*
155. Sedighi, M., Nazari, A., Faghihi, M., Rafieian-Kopaei, M., Karimi, A., Moghimian, M., et al., (2018). Protective effects of cinnamon bark extract against ischemia-reperfusion injury and arrhythmias in rat. *Phytotherapy Research, 32*(10).
156. Qabaha, K., Abu-Lafi, S., & Al-Rimawi, F., (2017). Anti-inflammatory activities of ethanolic extracts of *Curcuma longa* (turmeric) and cinnamon (*Cinnamomum verum*). *Journal of Food and Nutrition Research, 5*(9), 668–673.
157. Kaavya, R., Pandiselvam, R., Mohammed, M., Dakshayani, R., Kothakota, A., Ramesh, S. V., et al., (2020). Application of infrared spectroscopy techniques for the assessment of quality and safety in spices: A review. *Applied Spectroscopy Reviews* (Vol. 55, pp. 593–611). Taylor and Francis Inc.
158. Danciu, V., Hosu, A., & Cimpoiu, C., (2018). Thin-layer chromatography in spices analysis. *Journal of Liquid Chromatography and Related Technologies, 41*(6), 282–300.
159. Oliveira, M. M., Cruz-Tirado, J. P., & Barbin, D. F., (2019). Nontargeted analytical methods as a powerful tool for the authentication of spices and herbs: A review. *Comprehensive Reviews in Food Science and Food Safety* (Vol. 18, pp. 670–689). Blackwell Publishing Inc.
160. Kuballa, T., Brunner, T. S., Thongpanchang, T., Walch, S. G., & Lachenmeier, D. W., (2018). Application of NMR for authentication of honey, beer, and spices. *Current Opinion in Food Science* (Vol. 19, pp. 57–62). Elsevier Ltd.
161. Kucharska-Ambrożej, K., & Karpinska, J., (2020). The application of spectroscopic techniques in combination with chemometrics for detection adulteration of some herbs and spices. *Microchemical Journal* (Vol. 153). Elsevier Inc.
162. Pacholczyk-Sienicka, B., Ciepielowski, G., Albrecht, Ł., Zanardi, E., & Husáková, L., (2021). The application of NMR spectroscopy and chemometrics in authentication of spices. *Molecules*. Available from: https://doi.org/10.3390/molecules26020382.
163. Dubey, N. K. *Chapter Metabolomics* [Internet]. Available from: www.intechopen.com (accessed on 14 September 2023).
164. Corell, L., Armenta, S., Esteve-Turrillas, F. A., & De La Guardia, M., (2018). Flavonoid determination in onion, chili and leek by hard cap espresso extraction and liquid chromatography with diode array detection. *Microchemical Journal, 140*, 74–79.
165. Martinez-Sena, M. T., De La Guardia, M., Esteve-Turrillas, F. A., & Armenta, S., (2017). Hard cap espresso extraction and liquid chromatography determination of bioactive compounds in vegetables and spices. *Food Chemistry, 237*, 75–82.
166. Khazaei, K. M., Jafari, S. M., Ghorbani, M., Kakhki, A. H., & Sarfarazi, M., (2015). Optimization of anthocyanin extraction from saffron petals with response surface methodology. *Food Analytical Methods, 9*(7), 1993–2001. Available from: https://link.springer.com/article/10.1007/s12161-015-0375-4 (accessed on 14 September 2023).

CHAPTER 8

Technological Aspects of Bioactive Compounds and Nutraceuticals from Microbial Sources

ROUF AHMAD DAR,[1,2] MAWISH MAHROOF,[1] MANPREET KAUR,[3] RUQEYA NAZIR,[1] KHALID GUL,[4] and ARAKKAVEETTIL KABEER FARHA[5]

[1]*PG Program of Microbiology, Center of Research for Development, University of Kashmir, Srinagar, Jammu and Kashmir, India*

[2]*Department of Industrial Microbiology, Sam Higginbottom University of Agriculture, Technology, and Sciences, Prayagraj, Uttar Pradesh, India*

[3]*Department of Biochemistry, Punjab Agricultural University, Ludhiana, Punjab, India*

[4]*Department of Food Process Engineering, National Institute of Technology, Rourkela, Odisha, India*

[5]*Department of Food Science and Engineering, School of Agriculture and Biology, Shanghai Jiaotong University, Shanghai, China*

ABSTRACT

Natural bioactive compounds (NBCs) and nutraceuticals confer health-promoting and medical benefits to humans. There is a growing demand for such compounds, and their search is currently an interest of scientific and food industrial people. Microorganisms are known to produce such compounds. Extraction and purification of these compounds is labor time consuming. The conventional extraction processes require specific temperatures, organic solvents, more volume consumption, and other related conditions. Also, the

Bioactive Compounds and Nutraceuticals from Dairy, Marine, and Nonconventional Sources: Extraction Technology, Analytical Techniques, and Potential Health Prospects.
Charanjit Singh Riar, Parmjit S. Panesar (Eds.)

toxic solvents or chemicals used in these conventional methods cause harm to the people involved in it. However, advanced techniques like ultrasound-assisted extraction (UAE), supercritical fluid extraction (SFE), pressurized liquid extraction (PLE), microwave-assisted extraction (MAE), and pulsed electric field-assisted extraction (PEF) alleviate these issues. The advances in their extraction procedures had made their wide utilization possible. These nonconventional methods of the extraction are rapid and less time-consuming extraction procedures. Further, the incorporation of bioactive compounds in food has rendered success to this sector as they are used as preservatives, texturizers, flavor enhancers, emulsifiers, acidulants, surfactants or thickners. Also, these compounds are nontoxic, noncarcinogenic, and biodegradable as compared to synthetic compounds. However, this area still demands exploration and characterization of such compounds from the vast diversity of microbes present and the innovation in extraction and purification procedures to enhance their wider applicability and utility.

8.1 INTRODUCTION

The industrialization has changed the lifestyle of the people. It has increased the sedentary mode of lifestyle and fast-food consumption pattern. This has resulted in various diseases like obesity, cardiovascular, and hypertension, etc. Therefore, people in recent years have started noticing such things, and awareness is growing among the masses to look for a healthy diet that is nutritive with functional properties [1]. Thus, the search for natural bioactive compounds (NBCs) and nutraceuticals with the potential to meet food needs and the prevention of human diseases is currently an interest of scientific and food industrial people. These compounds efficiently interact with proteins, DNA, and other biological molecules to produce the desired outcome, which could be exploited for designing natural products-derived therapeutic agents [2]. Development and production of functional foods with the incorporation of bioactive compounds of the microbial source is the priority of the present time. These types of food products have wide acceptance due to the growing interest of people in healthy and nutritive food. Therefore, the characterization of bioactive compounds from microbial sources is of common interest for the food and pharmaceutical sector as these can be used as functional food ingredients and drugs.

Microbes synthesize bioactive in small quantities, and their extraction and purification processes are time-consuming and labor-intensive. Also, the chemical synthesis of these bioactives is unprofitable due to their structural complexity and diversity [3]. To address this issue, the development of advanced technologies and methods have revolutionized the field of bioactive compound screening, extraction, separation, and characterization, which could be potentially utilized [4, 5]. These technologies mainly consist of metabolic engineering and optimizing the isolation methods of bioactive compounds. These technologies could provide an innovative approach to increase the production of the desired compounds. Keeping in mind the substantial potential of bioactives from microbes as functional ingredients, this chapter aims in summarizing and presenting an overview of the potential microbial sources of bioactive compounds and technological advances in the extraction technologies. It also encompasses importance in food industries and incorporation in food products.

8.1.1 IMPORTANCE OF BIOACTIVE COMPOUNDS AND NUTRACEUTICALS IN INDUSTRIES

A marked trend has been observed in food industries nowadays towards the development and manufacture of functional food products. Bioactive compounds and nutraceuticals have seen great importance in the market due to the growing interest of consumers in 'healthy' food. Therefore, pharmaceuticals and food domains have similarities in obtaining and characterizing new NBCs that are used as functional food ingredients, drugs, or nutraceuticals. But their limited supply from natural resources affects the large-scale use of these compounds. Though microorganisms have been of great interest in the production of such useful compounds [6]. Almost 23000 NBCs are found to be obtained from microorganisms and very few groups of microorganisms are involved in their isolation [7].

8.1.1.1 BIOACTIVE COMPOUNDS IN FOOD INDUSTRIES

Incorporation of bioactive compounds in food as nutritional supplements which have been produced from microorganisms are nowadays of great success in the food industry as they are used as preservatives, texturizers, flavor enhancers, emulsifiers, acidulants, surfactants or thickeners [8]. A wide range of microorganisms is used to produce bioactive compounds like fungi, bacteria, and microalgae. But microalgae have been seen to produce a variety of bioactive compounds that are used as functional ingredients including carotenoids, polyphenols, and other antioxidant pigments, flavonoids, acid derivatives and dipeptides [3].

Various bioactive compounds derived from microbial sources that play an important role in the food industry are:

1. **Carotenoids:** Many bacteria (*Flavobacterium* spp., *Agrobacterium* spp., *Pseudomonas aeruginosa, Serrantia marcescens*, etc.) and microalgae (*Chlorella, Dunaliella, Scenedesmus*, and *Haematococcus*) produce important metabolites that have been used in food industries like carotenoids (β carotene and lycopene) [2, 9–11].

   Carotenoids are natural antioxidants in food and confer red, orange, or yellow color; it is a precursor for aromatic compounds. It also helps in extending shelf life of food. β-carotene is converted into vitamin A (retinol) in the body that is an essential nutrient for vision and plays an important role in cell growth, maintaining healthy organs such as kidneys, heart, and lungs. Thus, it can be incorporated in food to perform such vital activities of life.

2. **Phenyl Propanoids:** and their derivatives that are produced by genetically modified microorganisms including *Pseudomonas putida, Streptomyces lividans, Escherichia coli* and *Saccharomyces cerevisiae* are found to possess versatile activities such as antimicrobial, antioxidant, and photoprotective and therefore extensively used in the food industry as packaging films, preservatives, and edible coating. Phenyl propanoids are also available in commercially dietary supplements [12]. Foodborne illness that is caused by eating microbial contaminated food, these phenyl propanoids act antagonistic towards those microbes which cause foodborne illness, Also, the emergence of resistance to conventional antibiotics in bacteria is considered as one of the grave threats to human health [13]. There is a need for proper antibiotic therapy [14]. Phenyl propanoids and their derivatives have been found to have a wide range of antimicrobial activity [12].

   FA and Cinnamic acid (CiA), along with their derivatives, have been recommended to be used as flavoring agents or food additives in foods [15]. The U.S. Food and Drug Administration (USFDA) has recognized CiA and some of its derivatives as safe (GRAS) in food [16]. Furthermore, phenyl propanoids are promising in decreasing food's allergenicity in ovalbumin from the egg while caffeic acid

(CA) helps in lowering the egg white protein in mouse [17]. Oxidation of lipids during the processing and storage of food products is of major concern because it gives rise to volatile compounds that result in rancidity and off-flavor of food products [12]. Hydroxycinnamic acid inhibits lipid oxidation of *Trachurus trachurus*, commonly Atlantic horse mackerel [18]. Lipid oxidation of fish emulsions during storage is reduced by using CA-integrated films [19]. FA has also been of great importance in inhibiting the growth of pathogenic bacteria that cause foodborne illness, like it inhibits the growth *of Listeria monocytogenes* in ready-to-eat food (smoked salmon and cheese) [20].

3. **Organic Acids:** A wide variety of fungi and bacteria, namely *Penicillium, Aspergillus*, and *Streptomyces*, are exploited to produce bioactive molecules like organic acids (acetic acid, citric acid, formic acid, lactic acid, propionic acid, sorbic acid and benzoic acid). These acids are being used in food preservation to reduce the spoilage of food products and thus assure safe and quality foods [8, 21].

8.1.1.2 NUTRACEUTICALS IN FOOD INDUSTRIES

A variety of microorganisms like *Escherichia coli, Saccharomyces cerevisiae* has been genetically modified for the production of value-added chemicals such as phenolic compounds, polysaccharides, and poly-amino acids which have been incorporated in food supplements for preservation, packaging, and enhancing flavor and texture of food [22]. The various nutraceuticals which are used in the food industry are:

1. **Phenolic Compounds:** These have been widely used as bio-preservatives in food industries. These compounds are studied because of their wide application in the food industry for improving the shelf life of perishable food products. Food industries face many challenges in improving the shelf life of perishable foods. The biological activity of phenolic compounds protects against the deterioration of foods. Thus, phenolic compounds in the food industry can be used to inhibit or delay the oxidation and growth of microorganisms that cause the deterioration of foods. Phenolic compounds can also be used in food packing and edible films. Food products can be deteriorated biologically, physically as well as chemically [23]. Traditional food packaging concepts are reaching their limits in prolonging

shelf life. Therefore, novel packaging concepts, as active packaging, enable a further prolongation of shelf life and improve safety or sensory properties while maintaining the quality of food products [24].

2. **Polysaccharides (Xanthan, Pullulan, Gellan, Dextran, and Alginates):** These are the polymers of sugar molecules of multifaceted structures that are widely synthesized by different microorganisms like fungi, bacteria, and yeast. The polysaccharides derived from microbial sources are good sources of nutraceuticals with many health benefits. The genetically engineered microorganisms like *Streptococcus* and *Lactococcus* are preferentially used in the production of exopolysaccharides (EPS), these EPS have wide applications in dairy products [22]. Microbial polysaccharides play a vital role in the food industry as they are used as texturizers, thickeners, gelling agents, and stabilizers, thereby enhancing the flavor, quality, and texture of food [25–27] and with all these qualities, the microbial polysaccharides find a vital role in the food industry [28–31]. Various microbial polysaccharides that are used in the food industry for various purposes are as under:

   i. Xanthan is produced from *Xanthomonas campestries* and is used in the food industry as gelling agents for cheese spreads, ice creams, puddings, and other desserts.

   ii. Pullulan is produced from *Aureobasidium pullans* and is used in low-calorie drinks and foods as food additives. Pullulan also has a good moisture retention capacity, hence inhibits fungal growth, and therefore, can be used in food preservation.

   iii. Gellan is produced from *Pseudomonas elodea* and is used in jams, jellies, frostings as gelling agents.

   iv. Dextran is produced from *Leuconostoc mesenteroids* and is used in frozen foods, particularly in ice creams.

   v. Alginates are produced from *Azotobacter* and *Pseudomonas* and are used in jams, sauces, soups, and in dairy products as stabilizers, thickeners, and gelling agents [32].

3. **Polyaminoacids:** Microorganisms synthesize polyaminoacids either from single or different kinds of amino acids through ribosome independent enzymatic processes that differentiate them from

protein synthesis. In nature, three types of polyamino acids viz poly-E-lysine (E-PL), multi-L-arginyl-poly(L-aspartic acid) and poly-ϒ glutamic acid (ϒ-PGA) are present. ϒ-PGA can be largely produced by *Bacillus* species when feeding L-glutamic acid. It can also be produced from glucose by optimization of co-cultural conditions of *Corynebacterium glutamicum* and *Bacillus subtilus* [33]. *Streptomyces* spices such as *Streptomyces albulus* produces EP-L and in the United States and Japan, E-PL has been approved for food preservation because it possesses antibacterial activities or as a dietary supplement [34]. Amino acids are used either or in combination to enhance the flavor. In the food industry, monosodium glutamate is most often used in enhancing the flavor and to increase the taste and savory aroma of processed food like stews, meats, soups, and many more food products. Alanine and glycine are also used as taste enhancers. Aspartic acid, in combination with phenyl aniline, produces a dipeptide known as aspartame, which is used as a low-calorie artificial sweetener in the soft drink industry because it is about 200 times sweeter than sucrose. Table 8.1 summarizes the various bioactive compounds used in food industries.

8.2 MICROBES AS POTENTIAL SOURCES OF BIOACTIVE COMPOUNDS

Consumer's demand for food has changed considerably in the last decades. Food is now not only needed for hunger satisfaction and to provide required nutrients, but it should also contribute to the health of an individual. In this concern, functional foods have been described as ingredients with extra nutritional value on human health and hence play an important role. Thus, there is an increase in the demand for such important foods [60]. However, their limited supply and difficulties in extraction from natural sources like animals or plants have restricted their extensive use. Therefore, the production of such compounds from microbial origin has proved as an eco-friendly, advanced, and alternative method [22]. In this aspect, bacteria, archaea, fungi, protozoa, microalgae, and viruses are the diverse group of microorganisms that produce unique chemicals in massive quantity, which accords an important source for innovational biotechnology [61]. Microorganisms are known to produce about 23,000 secondary metabolites (SMs), out of

TABLE 8.1 Applications of Various Bioactive Compounds and Nutraceuticals in the Food Industry from Microbial Sources

Bioactive Compound/ Nutraceuticals	Name of Microorganisms	Applications in the Food Industry	References
Bacteriocins	*Shigella, Lactococcus, Leuconostoc, Escherichia coli, Lactobacillus, Pseudomonas, Serrentia*	Generally used for food preservation in canned foods, salads, milk products, meat, fish, cheese, etc.	[35]
Flavonoids	*Saccharomyces cerevisiae, Corynebacterium glutamicum, Escherichia coli, Streptomyces venezuelae.*	It acts as a natural antioxidant in fats and oils and thus, helps in prevention of oxidation of lipids. It also helps in prolonging the shelf life of food as they protect foods and beverages from exposure to light.	[36–39]
Lycopene	*Escherichia coli*	It is widely used for human nutrition as dietary supplements and in products with high lipid content, and it is also used as an antioxidant and colorant.	[40]
Cellulose	*Agrobacterium, Aerobacter, Gluconacetobacter, Rhizobium, Salmonella, Acetobacter, Salmonella, Rhizobium, Sarcina*	Cellulose, in combination with water, acts as a thickener and stabilizer in food. It helps also in the formation of thick and creamy food products without the addition of fat. Powdered cellulose is used as an anticaking agent.	[41]
Collagen	*Streptococcus pyogenes*	It is used for improving the rheological properties of food like sausages; it also ensures an adequate amount of nutrients in food. Also, it is used as an edible film for a prolonged shelf life of food, and collagen coating acts as an oxidizing agent in food.	[42, 43]
Vanillin	*Bacillus subtilis*	It is used for the preservation of food because of its antimicrobial and antioxidant properties and is also used in sweet foods like cakes, muffins, cookies, ice creams as a flavoring agent.	[44, 45]
Cinnamaldehyde	*Escherichia coli*	It is used in making fruit essences like cherry essence and apple essence, which can be used as a flavoring agent in candies, cakes, beverages, etc.	[46]

TABLE 8.1 *(Continued)*

Bioactive Compound/ Nutraceuticals	Name of Microorganisms	Applications in the Food Industry	References
Astaxanthin	*Xanthophyllomyces dendrorchous, Haematococcus pluvialis*	It is used as a colorant agent in foods and used for the production of dietary supplements as raw material.	[47, 48]
Folic acid	*Lactococcus lactis* and *Streptococcus thermophilus*	It is used for enhancing nutrient content in food.	[49]
β-Glucan	*Lastodiplodia theobromae, Botryosphaeria rhodina, Ganoderma lucidum.*	It is used as a thickener in skimmed milked yogurt.	[50, 51]
Ascorbic acid	*Saccharomyces cerevisiae*	It is used to fortify or to add citrus flavor in candies, fruit juices, dried fruit, cereals, and frozen fruits and it is also used in the preservation of jams, meats, bread, and jellies.	[52]
α-Tocopherols	*Euglena gracilis*	It is an antioxidant and is used as antioxidant packaging in foods and hence improves the stability and quality of food items for a long duration of time.	[53, 54]
Casein	*Lactococcus lactis, Streptococcus thermophiles* and *lactobacillus* sp.	It is used in nutritional fortification and to give the required texture to high protein food items. It is also used as a coating material or to decrease the absorption of oil while cooking/frying.	[55, 56]
Crocetin	*Saccharomyces cerevisiae*	In many food items like sweets, it is used as a colorant.	[57]
Zeaxanthin	*Chlorella pyrenoidosa*	It is used as a nutritional supplement and as a coloring agent in many food items like chewing gum, dairy product, frozen dairy desserts, hard candies and in breakfast cereals.	[58, 59]

these fungi and actinomycetes exclusively produce approximately 42% each and 16% which is remaining has been produced by eubacteria [62].

Also, microalgae have significant potential to fulfill the need of the population towards sustainable food supply, especially in fulfilling the protein demand. Microalgae with the potentiality to have benefits for the health of an individual [63]. The protein-rich microalgae *Arthrospira* (usually known as *Spirulina*), *Nostoc* and *Aphanizonmenon* since thousands of years are used as a part of the human diet [64] Proteins based on the origin of microalgae have a significant role to fulfill the need of the population for protein, with a wide range of benefits over the protein sources currently used. Apart from their protein benefits, these are also sources for many beneficial compounds which provides benefits to human health like essential minerals, vitamins, polyunsaturated fatty acids (PUFAs), and carbohydrates [65–67] that increases the nutrients in the food upon incorporation. Bioactive compounds that have been produced from marine fungi have potential and are promising in the novel antibacterial, anti-plasmodial, anti-inflammatory, and antiviral activities [68, 69]. Also, some microorganisms like *Saccharomyces cerevisiae* and *Escherichia coli* have been genetically modified to produce a complex and wide variety of value-added chemicals like probiotics, phytochemicals, polyamino acids and polysaccharides [22] which have health benefits. The various microorganisms that are used to produce bioactive compounds and nutraceuticals are given in Table 8.2.

8.3 TECHNOLOGICAL ACHIEVEMENTS FOR EXTRACTION AND PROCESSING OF BIOACTIVES AND NUTRACEUTICALS

In recent years, consumption of microalgal biomass, microbes, or its usage in the form of bioactives has fascinated a great deal of attention. Bioactives have been conventionally extracted in several ways [86]. However, each method has various disadvantages such as less recovery, more laborious, time-consuming, less efficient, essential components degradation, concerns regarding health and environmental factors due to the usage of certain solvents or chemicals. It has been reported that the yield of bioactives, composition, stability, and their functions vary significantly with the type of extraction method used [87]. This has urged a persistent search to find the right extraction method with ideal persuading factors which may ultimately lead to the maximum yield of bioactives.

TABLE 8.2 Various Microorganisms That are Used to Produce Bioactive Compounds and Nutraceuticals

Bioactive Compound/ Nutraceuticals	Microbial Source	Applications	References
Dolastatin	*Dolabella auricularia*	It is used in cancer chemotherapy.	[70]
Curacin A	*Cyanobacterium Lyngbya majuscula*	It possesses potentially antimitotic and antiproliferative activities against renal, colon, and breast cancer.	[71]
Toyocamycin	*Stretomyces diastatochromogenes*	Antifungal activity in the prevention of plant diseases.	[72]
Daryamides	*Streptomyces strain CNQ085*	It possesses antifungal activity against *Candida albicans* and anticancer activity against colon cancer.	[73]
Marinomycin	Marine actinomycetes	Antibiotic and antitumor activity.	[74]
Eicosapentanoic acid (EPA)	*Mododus subteraneus UTEX 151, Chorella minutissima UTEX 2341 and Phacodactylum tricornutum UTEX 642.*	It shows anti-inflammatory activity and also helps in the treatment of heart disease.	[75, 76]
Calothrixin-A	*Nostoc, Calothrix, Lyngbya*, and *Sympora*	Anti-cancerous and antimicrobial activity.	[77, 78]
7-O-malonyl macrolactin A	*Bacillus subtilis*	It is a bacteriostatic antibiotic that inhibits the growth of gram-positive pathogenic bacteria.	[79]
2,4-diacetylphorogluc-inol (DAPG)	*Pseudomonas flurescens*	It possesses activity against soil-borne bacteria and fungi.	[80]
Arachidonic acid	*Mortierella alpina*	Aggregation and vasoconstriction of platelets.	[81, 82]
Gamma-linolenic acid	*Spirulina maxima, Spirulina platensis, Mucor javanicus, Mortierella isabellina*	delays aging and maintain the integrity of the tissue.	[81, 83]
Okadaic acid	*Dinoflagellates*	It possesses antifungal activity and promotes the secretion of nerve growth factor.	[81, 84]
Decosahexanoic acid	*Schizochytrium and Aurantiochyrtrium*	It prevents chronic diseases like certain cancers, heart disease, Alzheimer's disease, depression, etc.	[85]
Vitamins	*Arthrospira, Isochrysis galbana, Porphyridium crventum and Tetraselmis*	Vitamins possess strong antioxidant properties.	[81]

The microbes derived bioactives can be extracted by both conventional and non-conventional methods [88]. The conventional methods require specific temperatures, organic solvents, more volume consumption, and other related conditions. Nevertheless, harm caused by the usage of toxic solvents or chemicals, there is an uprising demand for sustainable, efficient, and green methods. Currently, innovative alternative techniques like ultrasound-assisted extraction (UAE), supercritical fluid extraction (SFE), pressurized liquid extraction (PLE), microwave-assisted extraction (MAE), and pulsed electric field-assisted extraction (PEF) are being considered to extract desired bioactives [89]. They require low temperatures, less sample consumption, minimal or no use of organic solvents, stable extracted compounds, fewer processing steps, efficient use of energy, and highly purified end products ultimately depicting their superiority in comparison to traditional methods.

8.3.1 CONVENTIONAL METHODS

Various types of conventional extraction methods have been reported in the literature for the extraction of bioactives which includes hydro-distillation, solvent/Soxhlet extraction, infusion, percolation, decoction, and hot continuous extraction. The extraction of various organic compounds, such as polycyclic aromatic hydrocarbons, phenols, and other lipophilic compounds, is widely carried out by using solvent/Soxhlet extraction [90]. The solvent extraction (SE) method involves the mixing of solvent and biomass to extract the solute on reaching the equilibrium state and the solution attained at the last step is known as an extract which is the mixture of solute and solvent [91]. In this method, the solvent is recirculated through the sample throughout the whole process till the extraction is completed. The role of solvent is to solubilize the desired bioactives in the biomass sample, which may have restricted solubility and while selecting a solvent, bioactives solubility, harmfulness, and biological influence are amongst the imperative points to be considered. Chemical affinities are involved in the transference from one stage to the other. Chlorophyll was extracted efficiently from green microalgae *Dunaliella primolecta* through SE method by using methanol as solvent [92]. Through solvent-based extraction, cyclopentene was purified and identified from *Streptomyces olivaceus* LEP7, a tree bark inhabiting lichen *Leptogium* sp. [93]. MC21A bioactive compound was extracted from *Pseudoalteromonas phenolica* through SE method by using methanol [94]. Lutein and sulfated polysaccharide was extracted through SE technique from

Scenedesmus obliquus and *Gyrodinium impudium*, respectively [95, 96]. Likewise, laminarin was obtained from *Laminaria* and *Saccharina* species by this method [97]. Later, anticonvulsant inosine was obtained through the SE method from *Skeletonema marinoi* [98]. Two steroids, ergosterol and cerevisterol were isolated from the ethyl acetate extract of fungi obtained from leaves of *Ocimum bascilicum* [99]. The leading benefits of this method are cost-effectiveness, easy infrastructure, and operating procedures but its efficiency is measured by yield, quality, and composition of bioactives obtained. The SE method has remained the most widely for bioactives extraction from biomass of microalgal species which cannot tolerate harsh cell disruption methods. Furthermore, this method has several advantages over advanced methods such as basic procedure, no advanced training required and can be altered by changing solvent, size, and time of extraction [100]. On the other hand, this method demands a high volume of organic solvents, a long time of processing in case of carotenoids and phycobiliproteins extraction, incapable to separate the solvent from bioactives, and degradation of temperature-sensitive bioactives.

8.3.2 NON-CONVENTIONAL OR ADVANCED GREEN METHODS

Non-conventional methods of bioactives extraction, also known as green methods, are favored over conventional techniques in food processing and pharmacological drug industries.

8.3.2.1 ULTRASOUND-ASSISTED EXTRACTION (UAE)

Ultrasound-assisted extraction (UAE) method is a more resilient, high reproducibility, low investment and low-maintenance technique than conventional methods and is based on the usage of sound waves (20 kHz to 100 MHz) and used to extract the high molecular weight (MW) bioactives such as pigments, carotenoids, steroids, and SMs [101]. UAE is demarcated by two key processes, i.e., cavitation and sonication and later one is involved in the ultrasonic waves regulation. Parniakov et al. [102] extracted chlorophyll and phenolics efficiently from *Nannochloropsis* spp. by UAE. The bioactive compounds and myricitrin were isolated from leaves of lichen *Elaeocarpus serratus* L. by UAE and shaking-assisted extraction (SAE) for total phenols, total flavonoids, myricitrin, and antioxidant capacity and revealed the content of extracted bioactive compounds were 92.35 mg-gallic-acid-equivalent/g-DL of phenols, 53.14 mg-rutin-equivalent/g-DL of flavanoids, and 4.41 mg/g-DL of myricitrin, which were 1 to 1.5 folds higher than SAE indicating UAE a better method of extraction [103]. Likewise, Adam et al. [104] described efficient lipids extraction from *Nannochloropsis oculata* under this method. Then again, Vernès et al. [105] and Hildebrand et al. [106] suggested UAE as a proficient tool for rapid proteins extraction from *Spirulina* and *Chlorella vulgaris*, respectively.

8.3.2.2 SUPERCRITICAL FLUID EXTRACTION (SFE)

SFE has been progressively employed in the food, nutraceutical, and pharmaceutical industries, as a green, sustainable, and contamination-free extraction technique. SFE is based on a vapor-liquid critical point mechanism and combines separation and extraction through severely regulated factors like pressure, rate of flow, time of processing and temperature which allows end products or bioactives to be obtained with superior yield and purity [107]. The supercritical solvents (CO_2, n-propane, ethanol) above the critical point exhibit liquid-like properties such as negligible surface tension, less viscosity, harmless, easily spreadable, improved transfer activity and certainly penetrant into matrix thus more advanced and efficient than conventional solvents [108]. Amongst all, CO_2 is non-toxic, inflammable, cost-effective, and environmentally friendly, has less risk of contamination, low critical temperature, and pressure (31.2°C, 7.29 MPa) due to which it is highly preferable SFE solvent [150]. Mendiola et al. [109] used CO_2 as SFE solvent to extract various bioactives such as pigments, phenolics, and PUFAs from *Dunaliella salina*. The SMs were extracted from the lichen of *Cladonia* genus by SFE and the highest content was obtained by using CO_2 as solvent than acetone, ethanol, and petroleum ether on the Soxhlet apparatus, the content of isolated target component—usnic acid (UA)—in the extract (91%, yield—2.5% of absolutely dry raw material) were obtained 35 MPa, 40C, in 40 min [110]. Guedes et al. [111] extracted carotenoids and chlorophyll from *Scenedesmus obliquus* through CO_2 SFE to be use used in food processing but the maximum yield was obtained using ethanol (7.7%, v/v) as co-solvent. Vitamin E and pigments were extracted from *Spirulina platensis* with a yield of 8.08% and 7.94%, respectively by SFE using CO_2 and ethanol as solvent [112, 113] and lipids (12.08%) by using CO_2 SFE [114]. Likewise, Nobre et al. [115] extracted astaxanthin from *H. pluvialis* and reported that maximum yield was obtained (92%) using ethanol as a co-solvent (10%) obtained at

300 bar and 60°C whereas Di Sanzo et al. [116] showed the highest recovery at 50°C and 550 bar without the use of co-solvent.

8.3.2.3 PRESSURIZED LIQUID EXTRACTION (PLE)

In recent years, the pressurized liquid extraction (PLE) method is implemented as a new green technology technique to augment the bioactives production in the food industry. In this method, extraction is based on the solvent separation process in which suspension (solvent + biomass) is employed at high temperature and pressure (100–200°C and 140 bar) [117]. At such conditions, the stability of the cell wall was decreased, solvent surface tension reduced, and the flux of solvent increased into the cell, ultimately enhancing the analyte transfer. The technique has been successfully employed by Cha et al. [118] to isolate pigments like carotenoids and chlorophylls from *Chlorella vulgaris*. Carotenoids were extracted from *Dunaliella salina* and *Neochloris oleoabundans* by PLE [119, 120]. Similarly, fucoxanthin isolated from *Phaeodactylum tricornutum* by PLE [96]. But the main limitations of this method are the use of high temperature and high-cost operation.

8.3.2.4 MICROWAVE-ASSISTED EXTRACTION (MAE)

Microwave-assisted extraction (MAE) extraction method is based on the use of microwaves combined with the conventional extraction methods to enhance the solvent penetration and transfer into the microalgal biomass to promote the phytoconstituents dissolution [121]. Likewise, in UAE, ultrasonic waves break the cell walls and progress the bioactives release into the solvent. The foremost advantages of MAE are the less time of extraction and consumption of solvent, making it a green extraction method. There are various variants of MAE, such as ultrasonic, high-pressure, vacuum, nitrogen-protected, dynamic, and solvent-free for obtaining higher yields and concentration in less time as compared to conventional extraction methods [122, 123]. Ionic liquids (ILs) based MAE methods were used to extract norstictic acid, a common depsidone from *Pertusaria pseudocorallina*, a crustose lichen and the extraction amount of norstictic acid was determined after recovery on high performance thin layer chromatography (HPTLC) with a spectrophotodensitometer [124]. Phycobiliproteins and phycocyanins were extracted from *Porphyridium purpureum* using MAE at 40°C and 100°C, respectively in microwave implementation of 10 seconds by Juin et al. [125]. Chlorophylls and carotenoids were extracted from *Cylindrotheca closterium*, at 56°C through MAE [126].

8.4 ANALYTICAL TECHNIQUES FOR ANALYSIS OF BIOACTIVE COMPOUNDS DERIVED FROM MICROBIAL SOURCES

The separation of bioactives is a process of isolating the constituent's one after the other and then purifying them into monomer compounds by different physical and chemical methods. The extraction process should be evaluated in terms of both the quality, quantity, and activity of the bioactives obtained. Classical isolation methods, including solvent extraction (SE), precipitation, crystallization, and fractional distillation are used at present. In contrast, modern separation technologies such as column chromatography, affinity chromatography, ion-exchange chromatography (IEC), size exclusion chromatography (SEC), gas chromatography (GC), high-performance liquid chromatography (HPLC), ultrafiltration (UF), and liquid column chromatography coupled with mass spectrometry (LCMS) also play an important role in the separation of bioactive.

Phytochemical analysis of the acetone extract of the lichen (*Pleurosticta acetabulum*) was carried out by HPLC, salazinic, norstictic, protocetraric, evernic acid and atranorin were identified from the extract and exhibited free radical scavenging activity (half-maximum inhibitory concentration, IC_{50} of 151.7301 μg/mL) [127]. Several bioactive compounds were isolated from three rhizobacterial species (*B. amyloliquefaciens*, *B. thuringiensis* and *Bacillus* sp.) obtained from the rhizosphere of *Bambara* groundnut and gas chromatography-mass spectrometry (GC-MS) was used to carry out the analysis using seven extraction solvents and through this method, 68 compounds were identified based on peak area percentage, retention time and structure [128]. UA was isolated from the extract of lichen *Romanian Cetraria islandica*, and was characterized using FTIR spectroscopy and HPTLC techniques. The extracts and pure UA showed high antioxidant activity and better antimicrobial activity against certain Gram-positive and Gram-negative bacteria and fungi such as *Candida albicans* [129]. A total of 58 endolichenic fungi (ELF) were isolated from 29 lichens collected from mangrove and mangrove-associated plants of Negombo lagoon, Sri Lanka and the SMs were characterized from the crude extract by liquid chromatography-mass spectrometry (LC-MS) and the extracts of *Daldinia eschscholtzii* and *Hypoxylon lividipigmentum* showed highest antioxidant activity with SC_{50} values 14.27±0.24 μg/mL and 18.34±1.37 μg/mL, respectively [130].

8.5 INCORPORATION OF MICROBE-DERIVED BIOACTIVE COMPOUNDS AND NUTRACEUTICALS INTO FOOD

Food is used as a medium for incorporating bioactive compounds and nutraceuticals with benefits for human health [131]. Incorporation of bioactive compounds in food as nutritional supplements which have been produced from microorganisms are nowadays of great success in the food industry as they are used as preservatives, texturizers, flavor enhancers, emulsifiers, acidulants, surfactants or thickeners [8]. Variety of microorganisms like *Escherichia coli, Saccharomyces cerevisiae* has been genetically modified for the production of value-added chemicals such as phenolic compounds, polysaccharides, and polyamino acids which have been incorporated in food supplements for preservation, packaging, and enhancing flavor and texture of food [22].

Also, microalgae can be incorporated into dairy products for delivering bioactive active compounds [132]. Incorporation of *Arthrospira* spp. into yogurts and fermented milk, helps in stimulating the growth of desired probiotic bacteria and hence increases the viability of the probiotics [133]. Cookies are being incorporated with *Chlorella vulgaris* as a potential antioxidant supplement [134]. *Dunaliella* has been incorporated into white wheat bread as a protein supplement [135]. *Chlorella vulgaris* and *Arthropira maxima* are incorporated into pasta for enhancing its nutritional content into it [136]. By incorporating *Isochrysis albana, Diacronema vlkianum* and *Nannochloropsis* sp. into pasta will deliver ω-3 PUFAs and antioxidants which has potential benefits for the health of an individual [137, 138].

Bacteriocins, another type of bioactive compound, have a wide range of applications in food bio-preservation. Recently, it has been found bacteriocins are used to prevent foodborne pathogens in dairy products by inoculating the lactic acid bacteria that produce bacteriocin into food or adding semi-purified or purified bacteriocins directly into food. But in most cases, bacteriocins get absorbed into the food and can be easily degraded, thus can lose their antibacterial property. Therefore, the incorporation of bacteriocins in food packaging is an alternative method that improves their activity and stability in food products [139]. The use of potential immobilized bacteriocins to develop an antimicrobial coating film has proved helpful in preventing foodborne diseases caused by *Listeria monocytogenes* [140, 141]. Table 8.3 summarizes the bacteriocins, their source of production and their functions.

TABLE 8.3 Various Microorganisms that Produce Different Types of Bacteriocins

Type of Bacteriocin	Microorganism	Function	References
Nisin	*Lactococcus lactis, Streptococcus* strain	It possesses antimicrobial activity against many gram-positive bacteria including lactic acid bacteria, pathogenic bacteria like *Staphylococcus* and *listeria,* and *Bacillus* and *Clostridium,* which are spore-forming bacteria.	[142]
Pediocins	*Pediococcus spp*	Pediocin exhibit more effectiveness against foodborne pathogenic bacteria like *Staphylococcus aureus, Listeria monocytogenes* and gram-negative bacteria *Escherichia coli* and *Pseudomonas.*	[143–145]
Lacticin	*Lactococcus lactis*	Lacticin possesses antimicrobial properties against numerous foodborne pathogens and food spoiling bacteria. A lacticin 3147 powder has been shown effective in preventing *Listeria* and *Bacillus* in natural yogurt, infant milk formulation and cottage cheese.	[146]
Enterocins	*Enteriococcus faecalis*	It possesses antimicrobial activity against certain microorganisms like *Clostridium* spp. and *Bacillus*. It is the most potent bacteriocin, which is highly stable to heat and pH, making it a significant candidate in food application.	[147, 148]
Lactococcin Bz	*Lactococcus lactis*	It is active against gram-negative and gram-positive bacteria, possess great antilisterial activity in milk.	[149]

8.6 CHALLENGES AND FUTURE PERSPECTIVES

There are various challenges associated with the production of bioactives and nutraceuticals from microbes like low yield, difficulty in their isolation and identification. Also, the assimilation of end products inside the microbe can result in stress development and inhibition of cell growth. However, different strategies like genome editing, mutagenesis, overexpression of structural genes along with conventional recombinant technology can be used for efficient production of bioactives and nutraceuticals in microbes. The determination and improving the quality of these compounds from the microbes is and should remain an active area of research. A wide variety of engineering strategies can be used with the conventional recombinant DNA technologies, including genome editing, ribosome engineering, precursor engineering, mutagenesis, and overexpression of structural genes, making it possible to facilitate the efficient production of natural products and pharmaceuticals in microbial systems. Understanding the role of the bioactive compounds and the discovery of new metabolites is a ray of hope to combat the growing antimicrobial resistance and lifestyle diseases. There is a need for extensive quantitative study and optimization of these compounds to determine their potential roles and efficacy. Also, the extraction and purification procedures should be designed to provide optimum product quality of new bioactives and nutraceuticals. Prediction of the behavior of these compounds through computational models could be rewarding in this research. Also, there is a dire need to educate people about the harmful impact of polluting the environment, which is directly affecting microbes producing such novel bioactive compounds.

KEYWORDS

- **biodegradable**
- **conventional extraction processes**
- **extraction nonconventional method**
- **natural bioactive compounds**
- **noncarcinogenic**
- **nutraceuticals**
- **preservatives**

REFERENCES

1. Granato, D., Barba, F. J., Bursać, K. D., Lorenzo, J. M., Cruz, A. G., & Putnik, P., (2020). Functional foods: Product development, technological trends, efficacy testing, and safety. *Annu. Rev. Food Sci. Technol., 11*, 93–118.
2. Ajikumar, P. K., Tyo, K., Carlsen, S., Mucha, O., Phon, T. H., & Stephanopoulos, G., (2008). Terpenoids: Opportunities for biosynthesis of natural product drugs using engineered microorganisms. *Mol. Pharm., 5*, 167–190.
3. Lam, K. S., (2007). New aspects of natural products in drug discovery. *Trends Microbiol., 15*, 279–289.

4. Van, L. S. G., & Shen, B., (2006). Microbial genomics for the improvement of natural product discovery. *Curr. Opin. Microbiol., 9*, 252–260.
5. Wang, L., & Weller, C. L., (2006). Recent advances in extraction of nutraceuticals from plants. *Trends Food Sci. Technol., 17*, 300–312.
6. Demain, A. L., & Adrio, J. L., (2008). Contributions of microorganisms to industrial biology. *Mol. Biotechnol., 38*, 41–55.
7. Olano, C., Lombó, F., Méndez, C., & Salas, J. A., (2008). Improving production of bioactive secondary metabolites in actinomycetes by metabolic engineering. *Metab. Eng., 10*, 281–292.
8. Gil-Chávez, G. J., Villa, J. A., Ayala-Zavala, J. F., Basilio, H. J., Sepulveda, D., Yahia, E. M., et al., (2013). Technologies for extraction and production of bioactive compounds to be used as nutraceuticals and food ingredients: An overview. *Compr. Rev. Food Sci. Food Saf., 12*, 5–23.
9. Chang, M. C. Y., & Keasling, J. D., (2006). Production of isoprenoid pharmaceuticals by engineered microbes. *Nat. Chem. Biol., 212*(2), 674–681.
10. Klein-Marcuschamer, D., Ajikumar, P. K., & Stephanopoulos, G., (2007). Engineering microbial cell factories for biosynthesis of isoprenoid molecules: Beyond lycopene. *Trends Biotechnol., 25*, 417–424.
11. Donnez, D., Jeandet, P., Clément, C., & Courot, E., (2009). Bioproduction of resveratrol and stilbene derivatives by plant cells and microorganisms. *Trends Biotechnol., 27*, 706–713.
12. Neelam, Khatkar, A., & Sharma, K. K., (2020). Phenylpropanoids and its derivatives: Biological activities and its role in food, pharmaceutical and cosmetic industries. *Crit. Rev. Food Sci. Nutr., 60*, 2655–2675.
13. Nićiforović, N., & Abramovič, H., (2014). Sinapic acid and its derivatives: Natural sources and bioactivity. *Compr. Rev. Food Sci. Food Saf., 13*, 34–51.
14. Kon, K. V., & Rai, M. K., (2012). Plant essential oils and their constituents in coping with multidrug-resistant bacteria. *Expert. Rev. Anti. Infect. Ther., 10*, 775–790.
15. The Japan Food Chemical Research Foundation, (n.d.). https://www.ffcr.or.jp/en/tenka/list-of-existing-food-additives/list-of-existing-food-additives.html (accessed on 14 September 2023).
16. Zabalza, A., Orcaray, L., Fernández-Escalada, M., Zulet-González, A., & Royuela, M., (2017). The pattern of shikimate pathway and phenylpropanoids after inhibition by glyphosate or quinate feeding in pea roots. *Pestic. Biochem. Physiol., 141*, 96–102.
17. Tong, P., Chen, S., Gao, J., Li, X., Wu, Z., Yang, A., et al., (2018). Caffeic acid-assisted cross-linking catalyzed by polyphenol oxidase decreases the allergenicity of ovalbumin in a Balb/c mouse model. *Food Chem. Toxicol., 111*, 275–283.
18. Medina, I., González, M. J., Iglesias, J., & Hedges, N. D., (2009). Effect of hydroxycinnamic acids on lipid oxidation and protein changes as well as water holding capacity in frozen minced horse mackerel white muscle. *Food Chem., 114*, 881–888.
19. Yu, S. H., Hsieh, H. Y., Pang, J. C., Tang, D. W., Shih, C. M., Tsai, M. L., et al., (2013). Active films from water-soluble chitosan/cellulose composites incorporating releasable caffeic acid for inhibition of lipid oxidation in fish oil emulsions. *Food Hydrocoll., 32*, 9–19.
20. Takahashi, H., Kashimura, M., Koiso, H., Kuda, T., & Kimura, B., (2013). Use of ferulic acid as a novel candidate of growth-inhibiting agent against *Listeria monocytogenes* in ready-to-eat food. *Food Control, 33*, 244–248.

21. Anyasi, T. A., Jideani, A. I. O., Edokpayi, J. N., & Anokwuru, C., (2017). Application of organic acids in food preservation. *Org. Acids Charact. Prop. Synth.* (pp. 1–45). Nova Science Publishing.
22. Wang, J., Guleria, S., Koffas, M. A. G., & Yan, Y., (2016). Microbial production of value-added nutraceuticals. *Curr. Opin. Biotechnol., 37*, 97–104.
23. Martillanes, S., Rocha-Pimienta, J., Cabrera-Bañegil, M., Martín-Vertedor, D., & Delgado-Adámez, J., (2017). Application of phenolic compounds for food preservation: Food additive and active packaging. *Phenolic Compd. – Biol. Act.* doi: 10.5772/66885.
24. Brody, A. L., Bugusu, B., Han, J. H., Sand, C. K., & McHugh, T. H., (2008). Scientific status summary. *J. Food Sci., 73*, R107–116.
25. Kimmel, S. A., Roberts, R. F., & Ziegler, G. R., (1998). Optimization of exopolysaccharide production by *Lactobacillus delbrueckii* subsp. bulgaricus RR grown in a semidefined medium. *Appl. Environ. Microbiol., 64*, 659–664.
26. Nwodo, U. U., Green, E., & Okoh, A. I., (2012). Bacterial exopolysaccharides: Functionality and prospects. *Int. J. Mol. Sci., 13*, 14002–14015.
27. Öner ET. (2013). Microbial Production of Extracellular Polysaccharides from Biomass. *Green Energy Tech., 115*, 35–56.
28. Khan, T., Park, J. K., & Kwon, J. H., (2007). Functional biopolymers produced by biochemical technology considering applications in food engineering. *Korean J. Chem. Eng., 245*(24), 816–826.
29. Ahmad, N. H., Mustafa, S., & Man, Y. B. C., (2014). Microbial polysaccharides and their modification approaches: A review. *Int. J. Food Prop., 18*, 332–347.
30. Patel, S., Majumder, A., & Goyal, A., (2012). Potentials of exopolysaccharides from lactic acid bacteria. *Indian J. Microbiol., 52*, 3.
31. Chidambaram, R., Priya, J., Shraddha. (2014). Applications of microbial polysaccharides in food industry. *Int. J. Pharm. Sci. Rev. Res., 27*, 322–324.
32. Jindal, N., & Singh, K. J., (2018). Microbial polysaccharides in food industry. *Biopolym Food Des.*, 95–123.
33. Shi, F., Xu, Z., & Cen, P., (2007). Microbial production of natural poly amino acid. Sci. *China Ser. B Chem., 503*(50), 291–303.
34. El-Sersy, N. A., Abdelwahab, A. E., Abouelkhiir, S. S., Abou-Zeid, D. M., & Sabry, S. A., (2012). Antibacterial and anticancer activity of ε-poly-L-lysine (ε-PL) produced by a marine *Bacillus subtilis* sp. *J. Basic Microbiol., 52*, 513–522.
35. Karpiński, T. M., & Szkaradkiewicz, A. K., (2013). Characteristic of bacteriocines and their application. *Polish J. Microbiol., 62*, 223–235.
36. Yilmaz, Y., (2006). Novel uses of catechins in foods. *Trends Food Sci. Technol., 17*, 64–71.
37. Huvaere, K., & Skibsted, L. H., (2015). Flavonoids protecting food and beverages against light. *J. Sci. Food Agric., 95*, 20–35.
38. Kallscheuer, N., Vogt, M., Bott, M., & Marienhagen, J., (2017). Functional expression of plant-derived O-methyltransferase, flavanone 3-hydroxylase, and flavonol synthase in *Corynebacterium glutamicum* for production of pterostilbene, kaempferol, and quercetin. *J. Biotechnol., 258*, 190–196.
39. Pandey, R. P., Parajuli, P., Koffas, M. A. G., & Sohng, J. K., (2016). Microbial production of natural and non-natural flavonoids: Pathway engineering, directed evolution and systems/synthetic biology. *Biotechnol. Adv., 34*, 634–662.

40. Dobrzyńska, M. M., Gajowik, A., & Radzikowska, J., (2019). The effect of lycopene supplementation on radiation-induced micronuclei in mice reticulocytes in vivo. *Radiat. Environ. Biophys., 58*, 425–432.
41. Rangaswamy, B. E., Vanitha, K. P., & Hungund, B. S., (2015). Microbial cellulose production from bacteria isolated from rotten fruit. *Int. J. Polym. Sci., 2015*. doi: 10.1155/2015/280784.
42. Peng, Y. Y., Howell, L., Stoichevska, V., Werkmeister, J. A., Dumsday, G. J., & Ramshaw, J. A. M., (2012). Towards scalable production of a collagen-like protein from *Streptococcus pyogenes* for biomedical applications. *Microb. Cell Fact., 11*, 1–8.
43. Hashim, P., Mohd, R. M. S., Bakar, J., & Mat, H. D., (2015). Collagen in food and beverage industries. *Int. Food Res. J., 22*, 1–8.
44. Burri, J., Graf, M., Lambelet, P., & Löliger, J., (1989). Vanillin: More than a flavouring agent—A potent antioxidant. *J. Sci. Food Agric., 48*, 49–56.
45. Naidu, A., (2019). *Natural Food Antimicrobial Systems* (1st edn.). CRC Press.
46. Xiao, H., (2019). *Application Research of Cinnamic Aldehyde in Fragrance & Flavor, Daily Chemicals and Food Additives* | Foreverest Resources.
47. Zhou, T., Wang, X., Ju, Y., Shi, C., & Kan, G., (2018). Stability application and research of astaxanthin integrated into food. *IOP Conf. Ser. Mater. Sci. Eng.*, 394.
48. Choubert, G., Cravedi, J. P., & Laurentie, M., (2009). Effect of alternate distribution of astaxanthin on rainbow trout (*Oncorhynchus mykiss*) muscle pigmentation. *Aquaculture, 286*, 100–104.
49. Rossi, M., Amaretti, A., & Raimondi, S., (2011). Folate production by probiotic bacteria. *Nutrients, 3*, 118–134.
50. Raikos, V., Grant, S. B., Hayes, H., & Ranawana, V., (2018). Use of β-glucan from spent brewer's yeast as a thickener in skimmed yogurt: Physicochemical, textural, and structural properties related to sensory perception. *J. Dairy Sci., 101*, 5821–5831.
51. Philippini, R. R., Martiniano, S. E., Santos, J. C. D., Silva, S. S. Da., & Chandel, A. K., (2019). Fermentative production of beta-glucan: Properties and potential applications. *Bioprocess Biomol. Prod.*, 303–320.
52. Bremus, C., Herrmann, U., Bringer-Meyer, S., & Sahm, H., (2006). The use of microorganisms in L-ascorbic acid production. *J. Biotechnol., 124*, 196–205.
53. Ogbonna, J. C., (2009). Microbiological production of tocopherols: Current state and prospects. *Appl. Microbiol. Biotechnol., 84*, 217–225.
54. Scarfato, P., Avallone, E., Galdi, M. R., Di Maio, L., & Incarnato, L., (2017). Preparation, characterization, and oxygen scavenging capacity of biodegradable α-tocopherol/PLA microparticles for active food packaging applications. *Polym. Compos., 38*, 981–986.
55. Southward, C. R., (2003). Casein and caseinates | use in the food industry. *Encycl. Food Sci. Nutr.*, 948–958.
56. Kumar, A., & Chordia, N., (2017). Role of microbes in dairy industry. *Nutr. Food Sci. Int. J., 3*, 1–3.
57. Song, T., Wu, N., Wang, C., Wang, Y., Chai, F., Ding, M., et al., (2020). Crocetin overproduction in engineered *Saccharomyces cerevisiae* via tuning key enzymes coupled with precursor engineering. *Front Bioeng. Biotechnol., 8*, 1061.
58. Stankovic, I., (2003). *Curcumin Chemical and Technical Assessment (CTA).* Nfao, Rome.

59. Inbaraj, B. S., Lu, H., Hung, C. F., Wu, W. B., Lin, C. L., & Chen, B. H., (2008). Determination of carotenoids and their esters in fruits of *Lycium barbarum* Linnaeus by HPLC–DAD–APCI–MS. *J. Pharm. Biomed. Anal., 47*, 812–818.
60. Roberfroid, M. B., (2000). Prebiotics and probiotics: Are they functional foods? *Am. J. Clin. Nutr., 71.*
61. Rani, A., Saini, K. C., Bast, F., Mehariya, S., Bhatia, S. K., Lavecchia, R., et al., (2021). Microorganisms: A potential source of bioactive molecules for antioxidant applications. *Mol., 26*, 1142.
62. De Castro, I., Mendo, S., & Caetano, T., (2020). Antibiotics from haloarchaea: What can we learn from comparative genomics? *Mar. Biotechnol., 22*, 308–316.
63. Caporgno, M. P., & Mathys, A., (2018). Trends in microalgae incorporation into innovative food products with potential health benefits. *Front Nutr., 5*, 58.
64. Spolaore, P., Joannis-Cassan, C., Duran, E., & Isambert, A., (2006). Commercial applications of microalgae. *J. Biosci. Bioeng., 101*, 87–96.
65. Becker, E. W., (2007). Micro-algae as a source of protein. *Biotechnol. Adv., 25*, 207–210.
66. Chronakis, I. S., & Madsen, M. (2011). Algal proteins. *Handbook of Food Proteins,* pp. 53–94.
67. Wells, M. L., Potin, P., Craigie, J. S., Raven, J. A., Merchant, S. S., Helliwell, K. E., et al., (2017). Algae as nutritional and functional food sources: Revisiting our understanding. *J. Appl. Phycol., 29*, 949–982.
68. Bhadury, P., Mohammad, B. T., & Wright, P. C., (2006). The current status of natural products from marine fungi and their potential as anti-infective agents. *J. Ind. Microbiol. Biotechnol., 33*, 325–337.
69. Newman, D. J., & Hill, R. T., (2006). New drugs from marine microbes: The tide is turning. *J. Ind. Microbiol. Biotechnol., 33*, 539–544.
70. Segovia-Mendoza, M., Lemini, C., García-Becerra, R., & Morales-Montor, J., (2021). Immunoconjugates as immune canoes to kill breast cancer cells. *Immunother. Resist. Cancer from Lab Bench Work to Its Clin Perspect,* 11–31.
71. Townsend, C. A., & Ebizuka, Y., (2010). *Natural Products Structural Diversity-I, Secondary Metabolites: Organization and Biosynthesis* (eBook, 2010). WorldCat.org.
72. Shentu, X. P., Cao, Z. Y., Xiao, Y., Tang, G., Ochi, K., & Yu, X. P., (2018). Substantial improvement of toyocamycin production in *Streptomyces diastatochromogenes* by cumulative drug-resistance mutations. *PLoS One*, 13.
73. Asolkar, R. N., Jensen, P. R., Kauffman, C. A., & Fenical, W., (2006). Daryamides A-C, weakly cytotoxic polyketides from a marine-derived actinomycete of the genus *Streptomyces* strain CNQ-085. *J. Nat. Prod., 69*, 1756–1759.
74. Kwon, H. C., Kauffman, C. A., Jensen, P. R., & Fenical, W., (2006). Marinomycins A-D, antitumor-antibiotics of a new structure class from a marine actinomycete of the recently discovered genus "marinispora." *J. Am. Chem. Soc., 128*, 1622–1632.
75. Vazhappilly, R., & Chen, F., (1998). Eicosapentaenoic acid and docosahexaenoic acid production potential of microalgae and their heterotrophic growth. *JAOCS; J. Am, Oil Chem, Soc., 75,* 393–397.
76. Singh, S., Kate, B. N., & Banecjee, U. C., (2005). Bioactive compounds from cyanobacteria and microalgae: An overview. *Crit. Rev. Biotechnol., 25*, 73–95.
77. Dixit, R. B., & Suseela, M. R., (2013). Cyanobacteria: Potential candidates for drug discovery. *Antonie Van Leeuwenhoek, 103*, 947–961.

78. Rickards, R. W., Rothschild, J. M., Willis, A. C., De Chazal, N. M., Kirk, J., Kirk, K., et al., (1999). Calothrixins A and B, novel pentacyclic metabolites from *Calothrix* cyanobacteria with potent activity against malaria parasites and human cancer cells. *Tetrahedron, 55*, 13513–13520.
79. Romero-Tabarez, M., Jansen, R., Sylla, M., Lünsdorf, H., Häußler, S., Santosa, D. A., et al., (2006). 7-O-malonyl macrolactin A, a new macrolactin antibiotic from *Bacillus* subtilis active against methicillin-resistant *Staphylococcus aureus*, vancomycin-resistant enterococci, and a small-colony variant of *Burkholderia cepacia*. *Antimicrob Agents Chemother., 50*, 1701–1709.
80. Zhao, M. M., Lyu, N., Wang, D., Wu, X. G., Zhao, Y. Z., Zhang, L. Q., et al., (2020). PhlG mediates the conversion of DAPG to MAPG in *Pseudomonas fluorescens* 2P24. *Sci. Reports, 101*(10), 1–7.
81. Gupta, C., & Prakash, D., (2019). Nutraceuticals from microbes of marine sources. *Nutraceuticals – Past, Present Future.*
82. Asadi, S. Z., Khosravi-Darani, K., Nikoopour, H., & Bakhoda, H., (2018). Production of arachidonic acid and eicosapentaenoic acid by *Mortierella alpina* CBS 528.72 on date waste. *Food Technol. Biotechnol., 56*, 197.
83. Mishra, D., & Singh, S., (2011). Microbial production of γ-linolenic acid: An overview. *J. Pharm Res., 4*, 199–203.
84. Reguera, B., Riobó, P., Rodríguez, F., Díaz, P. A., Pizarro, G., Paz, B., et al., (2014). Dinophysis toxins: Causative organisms, distribution and fate in shellfish. *Mar. Drugs, 12*, 394.
85. Jovanovic, S., Dietrich, D., Becker, J., Kohlstedt, M., & Wittmann, C., (2021). Microbial production of polyunsaturated fatty acids — High-value ingredients for aquafeed, superfoods, and pharmaceuticals. *Curr. Opin. Biotechnol., 69*, 199–211.
86. Azmir, J., Zaidul, I. S. M., Rahman, M. M., Sharif, K. M., Mohamed, A., Sahena, F., et al., (2013). Techniques for extraction of bioactive compounds from plant materials: A review. *J. Food Eng., 117*, 426–436.
87. Zhang, Q. W., Lin, L. G., & Ye, W. C., (2018). Techniques for extraction and isolation of natural products: A comprehensive review. *Chinese Med. (United Kingdom), 13*, 1–26.
88. Ummat, V., Sivagnanam, S. P., Rajauria, G., O'Donnell, C., & Tiwari, B. K., (2021). Advances in pre-treatment techniques and green extraction technologies for bioactives from seaweeds. *Trends Food Sci. Technol., 110*, 90–106.
89. Pagano, I., Campone, L., Celano, R., Piccinelli, A. L., & Rastrelli, L., (2021). Green non-conventional techniques for the extraction of polyphenols from agricultural food by-products: A review. *J. Chromatogr. A., 1651*, 462295.
90. Abdel-Shafy, H. I., & Mansour, M. S. M., (2016). A review on polycyclic aromatic hydrocarbons: Source, environmental impact, effect on human health and remediation. *Egypt. J. Pet., 25*, 107–123.
91. Cañadas, R., González-Miquel, M., González, E. J., Díaz, I., & Rodríguez, M., (2021). Evaluation of bio-based solvents for phenolic acids extraction from aqueous matrices. *J. Mol. Liq., 338*, 116930.
92. Pourkarimi, S., Hallajisani, A., Nouralishahi, A., Alizadehdakhel, A., & Golzary, A., (2020). Factors affecting production of beta-carotene from *Dunaliella salina* microalgae. *Biocatal. Agric. Biotechnol., 29*, 101771.
93. Rajaram, S. K., Ahmad, P., Sujani, S. K. S., Jeya, C. P., Ganesh, M. I., & Suresh, R. S. S., (2020). Extraction and purification of an antimicrobial bioactive element from

lichen-associated *Streptomyces olivaceus* LEP7 against wound-inhabiting microbial pathogens. *J. King Saud. Univ. – Sci., 32*, 2009–2015.

94. Isnansetyo, A., & Kamei, Y., (2003). MC21-A, a bactericidal antibiotic produced by a new marine bacterium, *Pseudoalteromonas phenolica* sp. Nov. O-BC30T, against methicillin-resistant *Staphylococcus aureus. Antimicrob Agents Chemother., 47*, 480.
95. Chan, M. C., Ho, S. H., Lee, D. J., Chen, C. Y., Huang, C. C., & Chang, J. S., (2013). Characterization, extraction and purification of lutein produced by an indigenous microalga *Scenedesmus obliqus* CNW-N. *Biochem. Eng. J., 78*, 24–31.
96. Kim, S. M., Jung, Y. J., Kwon, O. N., Cha, K. H., Um, B. H., Chung, D., et al., (2012). A potential commercial source of fucoxanthin extracted from the microalga *Phaeodactylum tricornutum. Appl. Biochem. Biotechnol., 166*, 1843–1855.
97. Kadam, S. U., Tiwari, B. K., & O'Donnell, C. P., (2015). Extraction, structure and biofunctional activities of laminarin from brown algae. *Int. J. Food Sci. Technol., 50*, 24–31.
98. Brillatz, T., Lauritano, C., Jacmin, M., Khamma, S., Marcourt, L., Righi, D., et al., (2018). Zebrafish-based identification of the antiseizure nucleoside inosine from the marine diatom *Skeletonema marinoi. PLoS One, 13*.
99. Haque, M. A., Hossain, M. S., Rahman, M., Rahman, M. R., Hossain, M. S., Mosihuzzaman, M., et al., (2005). Isolation of bioactive secondary metabolites from the endophytic fungus of *Ocimum basilicum. Dhaka Univ. J. Pharm. Sci.*, 4.
100. López-Bascón-Bascon, M. A., & Luque De, C. M. D., (2019). Soxhlet extraction. *Liq. Extr.*, 327–354.
101. Kumar, K., Srivastav, S., & Sharanagat, V. S., (2021). Ultrasound-assisted extraction (UAE) of bioactive compounds from fruit and vegetable processing by-products: A review. *Ultrason. Sonochem., 70*, 105325.
102. Parniakov, O., Apicella, E., Koubaa, M., Barba, F. J., Grimi, N., Lebovka, N., et al., (2015). Ultrasound-assisted green solvent extraction of high-added value compounds from microalgae *Nannochloropsis* spp. *Bioresour. Technol., 198*, 262–267.
103. Chen, Y. H., & Yang, C. Y., (2020). Ultrasound-assisted extraction of bioactive compounds and antioxidant capacity for the valorization of *Elaeocarpus serratus* L. Leaves. *Process, 8*, 1218.
104. Adam, F., Abert-Vian, M., Peltier, G., & Chemat, F., (2012). "Solvent-free" ultrasound-assisted extraction of lipids from fresh microalgae cells: A green, clean, and scalable process. *Bioresour. Technol., 114*, 457–465.
105. Vernès, L., Abert-Vian, M., El Maâtaoui, M., Tao, Y., Bornard, I., & Chemat, F., (2019). Application of ultrasound for green extraction of proteins from spirulina. Mechanism, optimization, modeling, and industrial prospects. *Ultrason. Sonochem., 54*, 48–60.
106. Hildebrand, G., Poojary, M. M., O'Donnell, C., Lund, M. N., Garcia-Vaquero, M., & Tiwari, B. K., (2007). Result. *J. Appl. Phycol., 32*, 1709–1718.
107. Villacís-Chiriboga, J., Vera, E., Van, C. J., Ruales, J., & Elst, K., (2021). Valorization of byproducts from tropical fruits: A review, Part 2: Applications, economic, and environmental aspects of biorefinery via supercritical fluid extraction. *Compr. Rev. Food Sci. Food Saf., 20*, 2305–2331.
108. Manjare, S. D., & Dhingra, K., (2019). Supercritical fluids in separation and purification: A review. *Mater. Sci. Energy Technol., 2*, 463–484.

109. Mendiola, J. A., Santoyo, S., Cifuentes, A., Reglero, G., Ibáñez, E., & Javier, S. F., (2008). Antimicrobial activity of sub- and supercritical CO_2 extracts of the green alga *Dunaliella salina*. *J. Food Prot.*, *71*, 2138–2143.
110. Brovko, O. S., Ivakhnov, A. D., Palamarchuk, I. A., & Boitsova, T. A., (2017). Supercritical fluid extraction of usnic acid from lichen of *Cladonia* genus. *Russ. J. Phys. Chem B.*, *118*(11), 1306–1311.
111. Guedes, A. C., Gião, M. S., Matias, A. A., Nunes, A. V. M., Pintado, M. E., Duarte, C. M. M., et al., (2013). Supercritical fluid extraction of carotenoids and chlorophylls a, b and c, from a wild strain of *Scenedesmus obliquus* for use in food processing. *J. Food Eng.*, *116*, 478–482.
112. Mendiola, J. A., Marín, F. R., Hernández, S. F., Arredondo, B. O., Señoráns, F. J., Ibañez, E., et al., (2005). Characterization via liquid chromatography coupled to diode array detector and tandem mass spectrometry of supercritical fluid antioxidant extracts of *Spirulina platensis* microalga. *J. Sep. Sci.*, *28*, 1031–1108.
113. Mendiola, J. A., García-Martínez, D., Rupérez, F. J., Martín-Álvarez, P. J., Reglero, G., Cifuentes, A., et al., (2008). Enrichment of vitamin E from *Spirulina platensis* microalga by SFE. *J. Supercrit. Fluids*, *43*, 484–489.
114. Leone, G. P., Balducchi, R., Mehariya, S., Martino, M., Larocca, V., Sanzo, G. D., et al., (2019). Selective extraction of ω-3 fatty acids from *Nannochloropsis* sp. using supercritical CO_2 extraction. *Molecules*, *24*.
115. Nobre, B., Marcelo, F., Passos, R., Beirão, L., Palavra, A., Gouveia, L., et al., (2006). Supercritical carbon dioxide extraction of astaxanthin and other carotenoids from the microalga *Haematococcus pluvialis*. *Eur. Food Res. Technol.*, *223*, 787–790.
116. Di Sanzo, G., Mehariya, S., Martino, M., Larocca, V., Casella, P., Chianese, S., et al., (2018). Supercritical carbon dioxide extraction of astaxanthin, lutein, and fatty acids from *Haematococcus pluvialis* microalgae. *Mar. Drugs;16*.
117. Ruiz-Domínguez, M. C., Fuentes, J. L., Mendiola, J. A., Cerezal-Mezquita, P., Morales, J., Vílchez, C., et al., (2021). Bioprospecting of cyanobacterium in Chilean coastal desert, *Geitlerinema* sp. molecular identification and pressurized liquid extraction of bioactive compounds. *Food Bioprod. Process*, *128*, 227–239.
118. Kwang, H. C., Lee, H. J., Koo, S. Y., Song, D. G., Lee, D. U., & Pan, C. H., (2010). Optimization of pressurized liquid extraction of carotenoids and chlorophylls *from Chlorella vulgaris*. *J. Agric. Food Chem.*, *58*, 793–797.
119. Herrero, M., Jaime, L., Martín-Álvarez, P. J., Cifuentes, A., & Ibáñez, E., (2006). Optimization of the extraction of antioxidants from *Dunaliella salina* microalga by pressurized liquids. *J. Agric. Food Chem.*, *54*, 5597–5603.
120. Castro-Puyana, M., Herrero, M., Urreta, I., Mendiola, J. A., Cifuentes, A., Ibáñez, E., et al., (2013). Optimization of clean extraction methods to isolate carotenoids from the microalga *Neochloris oleoabundans* and subsequent chemical characterization using liquid chromatography tandem mass spectrometry. *Anal. Bioanal. Chem.*, *405*, 4607–4616.
121. Hamid, N. A., Ruth, O. A., Hamid, N. A., Suliman, O. M., & Ahmed, N., (2021). Microwave-assisted extraction of bioactive compounds (review). *Microw Heat – Electromagn Fields Causing Therm. Non-thermal Eff.*
122. Chan, C. H., Yusoff, R., Ngoh, G. C., & Kung, F. W. L., (2011). Microwave-assisted extractions of active ingredients from plants. *J. Chromatogr. A.*, *1218*, 6213–6225.

123. Fierascu, R. C., Fierascu, I., Ortan, A., Georgiev, M. I., & Sieniawska, E., (2020). Innovative approaches for recovery of phytoconstituents from medicinal/aromatic plants and biotechnological production. *Mol., 25*, 309.
124. Bonny, S., Paquin, L., Carrié, D., Boustie, J., & Tomasi, S., (2011). Ionic liquids-based microwave-assisted extraction of lichen compounds with quantitative spectrophoto densitometry analysis. *Anal. Chim. Acta, 707*, 69–75.
125. Juin, C., Chérouvrier, J. R., Thiéry, V., Gagez, A. L., Bérard, J. B., Joguet, N., et al., (2015). Microwave-assisted extraction of phycobiliproteins from *Porphyridium purpureum. Appl Biochem. Biotechnol., 175*, 1–15.
126. Pasquet, V., Chérouvrier, J. R., Farhat, F., Thiéry, V., Piot, J. M , Bérard, J. B., et al., (2011). Study on the microalgal pigments extraction process: Performance of microwave assisted extraction. *Process Biochem., 46*, 59–67.
127. Tomović, J., Kosanić, M., Ristić, S., Ranković, B., Stanojković, T., & Manojlović, N., (2018). Chemical composition and bioactive properties of the lichen, *Pleurosticta acetabulum. Trop J. Pharm. Res., 16*, 2977–2984.
128. Ajilogba, C. F., & Babalola, O. O., (2019). GC–MS analysis of volatile organic compounds from Bambara groundnut rhizobacteria and their antibacterial properties. *World J. Microbiol. Biotechnol., 35*, 1–19.
129. Patriche, S., Ghinea, I. O., Adam, G., Gurau, G., Furdui, B., Dinica, R. M., et al., (2019). Characterization of bioactive compounds from Romanian *Cetraria islandica* (L) *Ach. Rev Chim., 70*, 2186–2191.
130. Weerasinghe, R. H., Maduranga, K., Attanayake, R. N., Shevkar, C., Kate, A. S., Weerakoon, G., et al., (2021). Bioactive properties and metabolite profiles of endolithic fungi in mangrove ecosystem of Negombo Lagoon, Sri Lanka. *Nat Prod Commun., 16*.
131. Carrasco, E., Morales-Rueda, A., & García-Gimeno, R. M., (2012). Cross-contamination and recontamination by *Salmonella* in foods: A review. *Food Res. Int., 45*, 545–556.
132. Beheshtipour, H., Mortazavian, A. M., Mohammadi, R., Sohrabvandi, S., & Khosravi-Darani, K., (2013). Supplementation of *Spirulina platensis* and *Chlorella vulgaris* algae into probiotic fermented milks. *Compr. Rev. Food Sci. Food Saf., 12*, 144–154.
133. Varga, L., Szigeti, J., Kovács, R., Földes, T., & Buti, S., (2002). Influence of a *Spirulina platensis* biomass on the microflora of fermented ABT milks during storage (R1). *J. Dairy Sci., 85*, 1031–1038. doi: 10.3168/JDS.S0022-0302(02)74163-5.
134. Gouveia, L., Batista, A. P., Raymundo, A., & Bandarra, N., (2008). *Spirulina maxima* and *Diacronema vlkianum* microalgae in vegetable gelled desserts. *Nutr. Food Sci., 38*, 492–501. doi: 10.1108/00346650810907010/FULL/XML.
135. Finney, K., Pomeranz, Y., & Bruisnma, B., (1984). Use of algae Dunaliella as a protein supplement in bread. *Cereal Chem.*, 61, 402–406.
136. Fradique, M., Batista, A. P., Nunes, M. C., Gouveia, L., Bandarra, N. M., & Raymundo, A., (2010). Incorporation of *Chlorella vulgaris* and *Spirulina maxima* biomass in pasta products. Part 1: Preparation and evaluation. *J. Sci. Food Agric., 90*, 1656–1664.
137. Fradique, M., Batista, A. P., Nunes, M. C., Gouveia, L., Bandarra, N. M., & Raymundo, A., (2013). *Isochrysis galbana* and *Diacronema vlkianum* biomass incorporation in pasta products as PUFA's source. *LWT – Food Sci Technol., 50*, 312–319.
138. Rodríguez De, M., E., Steffolani, M. E., Martínez, M., & León, A. E., (2018). The use of *Nannochloropsis* sp. as a source of omega-3 fatty acids in dry pasta: chemical, technological and sensory evaluation. *Int. J. Food Sci. Technol., 53*, 499–507.

139. Salgado, P. R., Ortiz, C. M., Musso, Y. S., Di Giorgio, L., & Mauri, A. N., (2015). Edible films and coatings containing bioactives. *Curr. Opin. Food Sci., 5*, 86–92.
140. Ibarguren, C., Céliz, G., Díaz, A. S., Bertuzzi, M. A., Daz, M., & Audisio, M. C., (2015). Gelatine based films added with bacteriocins and a flavonoid ester active against food-borne pathogens. *Innov. Food Sci. Emerg. Technol., 28*, 66–72.
141. Narsaiah, K., Wilson, R. A., Gokul, K., Mandge, H. M., Jha, S. N., Bhadwal, S., et al., (2015). Effect of bacteriocin-incorporated alginate coating on shelf-life of minimally processed papaya (*Carica papaya* L.). *Postharvest Biol. Technol., 100*, 212–218. doi: 10.1016/J.POSTHARVBIO.2014.10.003.
142. and, H. C., & Hoover, D. G., (2003). Bacteriocins and their food applications. *Compr. Rev. Food Sci. Food Saf., 2*, 82–100.
143. Cintas, L. M., Casaus, P., Fernández, M. F., & Hernández, P. E., (1998). Comparative antimicrobial activity of enterocin L50, pediocin PA-1, nisin A and lactocin S against spoilage and foodborne pathogenic bacteria. *Food Microbiol., 15*, 289–298.
144. Eijsink, V. G. H., Axelsson, L., Diep, D. B., Håvarstein, L. S., Holo, H., & Nes, I. F., (2002). Production of class II bacteriocins by lactic acid bacteria; an example of biological warfare and communication. *Antonie Van Leeuwenhoek, 81*, 639–654.
145. Jamuna, M., & Jeevaratnam, K., (2004). Isolation and characterization of *lactobacilli* from some traditional fermented foods and evaluation of the bacteriocins. *J. Gen. Appl. Microbiol., 50*, 79–90.
146. Morgan, S. M., Galvin, M., Ross, R. P., & Hill C., (2001). Evaluation of a spray-dried lacticin 3147 powder for the control of Listeria monocytogenes and *Bacillus cereus* in a range of food systems. *Lett. Appl. Microbiol., 33*, 387–391.
147. Egan, K., Field, D., Rea, M. C., Ross, R. P., Hill, C., & Cotter, P. D., (2016). Bacteriocins: Novel solutions to age-old spore-related problems? *Front Microbiol.*, 7. doi: 10.3389/FMICB.2016.00461.
148. Muñoz, A., Maqueda, M., Gálvez, A., Martínez-Bueno, M., Rodríguez, A., & Valdivia, E., (2004). Biocontrol of psychrotrophic enterotoxigenic *Bacillus cereus* in a nonfat hard cheese by an enterococcal strain-producing enterocin AS-48. *J. Food Prot., 67*, 1517–1521.
149. Şahingil, D., Işleroğlu, H., Yildirim, Z., Akçelik, M., & Yildirim, M., (2011). Characterization of lactococcin BZ produced by *Lactococcus lactis* subsp. lactis BZ isolated from boza. *Turkish J. Biol., 35*, 21–33.
150. Sun, J., Wang, Y., Liu, J., Xu, Q., & Yin, J. (2021). Highly selective separation of CO_2/N_2 using [Emim][TF2N] supported ionic liquid membranes prepared by supercritical fluid deposition. *J Supercrit Fluids, 2021,* 170.

Index

D

G

H

P

T

U

V

W

X

Y

Z